HUMAN DEVELOPMENT 89/90

Seventeenth Edition

Editor
Hiram E. Fitzgerald
Michigan State University

Hiram E. Fitzgerald is a professor and associate chairperson in the Department of Psychology at Michigan State University. He received a B.A. in 1962 from Lebanon Valley College, and an M.A. in 1964 and a Ph.D. in 1967 from the University of Denver. In addition to holding memberships in a variety of scientific associations, he is the Executive Director of the International Association for Infant Mental Health, and editor of the *Infant Mental Health Journal*. He has authored and edited over one hundred and twenty publications.

Editor
Michael G. Walraven
Jackson Community College

Michael G. Walraven is Dean of Instruction and Professor of Psychology at Jackson Community College. He received a B.A. from the University of Maryland in 1966, an M.A. from Western Michigan University in 1968, and a Ph.D. from Michigan State University in 1974. He is affiliated with the American Psychological Association, the Association for Behavior Analysis, and the Biofeedback Society of Michigan.

A Library of Information from the Public Press

Cover illustration by Mike Eagle

The Dushkin Publishing Group, Inc.
Sluice Dock, Guilford, Connecticut 06437

The Annual Editions Series

Annual Editions is a series of over forty-five volumes designed to provide the reader with convenient, low-cost access to a wide range of current, carefully selected articles from some of the most important magazines, newspapers, and journals published today. Annual Editions are updated on an annual basis through a continuous monitoring of over 200 periodical sources. All Annual Editions have a number of features designed to make them particularly useful, including topic guides, annotated tables of contents, unit overviews, and indexes. For the teacher using Annual Editions in the classroom, an Instructor's Resource Guide with test questions is available for each volume.

VOLUMES AVAILABLE

Africa
Aging
American Government
American History, Pre-Civil War
American History, Post-Civil War
Anthropology
Biology
Business and Management
Business Ethics
China
Comparative Politics
Computers in Education
Computers in Business
Computers in Society
Criminal Justice
Drugs, Society, and Behavior
Early Childhood Education
Economics
Educating Exceptional Children
Education
Educational Psychology
Environment
Geography
Global Issues
Health
Human Development
Human Resources
Human Sexuality
Latin America
Macroeconomics
Marketing
Marriage and Family
Middle East and the Islamic World
Nutrition
Personal Growth and Behavior
Psychology
Social Problems
Sociology
Soviet Union and Eastern Europe
State and Local Government
Third World
Urban Society
Western Civilization, Pre-Reformation
Western Civilization, Post-Reformation
Western Europe
World History, Pre-Modern
World History, Modern
World Politics

Library of Congress Cataloging in Publication Data
Main entry under title: Annual Editions: Human development.
    1. Child study—Addresses, essays, lectures. 2. Socialization—Addresses, essays, lectures.
3. Old age—Addresses, essays, lectures. I. Title: Human development.
HQ768.A55   155'.05   72-91973
ISBN 0-87967-785-6

Seventeenth Edition

Manufactured by The Banta Company, Harrisonburg, Virginia 22801

To The Reader

In publishing ANNUAL EDITIONS we recognize the enormous role played by the magazines, newspapers, and journals of the *public press* in providing current, first-rate educational information in a broad spectrum of interest areas. Within the articles, the best scientists, practitioners, researchers, and commentators draw issues into new perspective as accepted theories and viewpoints are called into account by new events, recent discoveries change old facts, and fresh debate breaks out over important controversies. Many of the articles resulting from this enormous editorial effort are appropriate for students, researchers, and professionals seeking accurate, current material to help bridge the gap between principles and theories and the real world. These articles, however, become more useful for study when those of lasting value are carefully *collected, organized, indexed,* and *reproduced* in a *low-cost format,* which provides easy and permanent access when the material is needed. That is the role played by *Annual Editions.* Under the direction of each volume's *Editor,* who is an expert in the subject area, and with the guidance of an *Advisory Board,* we seek each year to provide in each *ANNUAL EDITION* a current, well-balanced, carefully selected collection of the best of the public press for your study and enjoyment. We think you'll find this volume useful, and we hope you'll take a moment to let us know what you think.

Any history of the field of human development will reflect the contributions of the many individuals who helped craft the topical content of the discipline. For example, Binet launched the intelligence test movement, Freud focused attention on personality development, and Watson and Thorndike paved the way for the emergence of social learning theory. However, the philosophical principles that give definition to the field of human development have their direct ancestral roots in the evolutionary biology of Darwin, Wallace, and Spencer, and in the embryology of Preyer. Each of the two most influential developmental psychologists of the early twentieth century, James Mark Baldwin and G. Stanley Hall, was markedly influenced by questions about phylogeny (species' adaptation) and ontogeny (individual adaptation or fittingness). Baldwin's persuasive arguments challenged the assertion that changes in species precede changes in individual organisms. Instead, Baldwin argued, ontogeny not only precedes phylogeny but is the process that shapes phylogeny. Thus, as Robert Cairns points out, developmental psychology has always been concerned with the study of the forces that guide and direct development. Early theories stressed that development was the unfolding of already formed or predetermined characteristics. Many contemporary students of human development embrace the epigenetic principle which asserts that development is an emergent process of active, dynamic, reciprocal, and systemic change. This systems perspective forces one to think about the historical, social, cultural, interpersonal, and intrapersonal forces that shape the developmental process.

The study of human development involves all fields of inquiry comprising the social, natural, and life sciences and professions. The need for depth and breadth of knowledge creates a paradox: While students are being advised to acquire a broad-based education, each discipline is becoming more highly specialized. One way to combat specialization is to integrate the theories and findings from a variety of disciplines with those of the parent discipline. This, in effect, is the approach of *Annual Editions: Human Development 89/90.* This anthology includes articles that discuss the problems, issues, theories, and research findings from many fields of study; the common element is that they all address issues relevant to understanding human development. In most instances, the articles were not prepared for technical professional journals but were written specifically to communicate information about recent scientific findings or controversial issues to the general public. As a result, the articles tend to blend the history of a topic with the latest available information. In many instances, the reader is challenged to consider the personal and social implications of the topic. The articles included in this anthology were selected by the editors, with valued advice and recommendations from an advisory board consisting of faculty from community colleges, small liberal arts colleges, and large universities. Evaluations obtained from students, instructors, and advisory board members influenced the decision to retain or replace specific articles. Throughout the year we screen many articles for accuracy, interest value, writing style, and recency of information.

Human Development 89/90 is organized into six major units. Unit 1 focuses on the origins of life, including prenatal development, and unit 2 focuses on development during infancy and early childhood. Unit 3 is divided into subsections addressing social, emotional, and cognitive development. Unit 4 addresses issues related to family, school, and cultural influences on development. Units 5 and 6 cover human development from adolescence to old age. In our experience, this organization provides great flexibility for those using the anthology with any standard textbook. The units can be assigned sequentially, or instructors can devise any number of arrangements of individual articles to fit their specific needs. In large lecture classes this anthology seems to work best as assigned reading to supplement the basic text. In smaller sections, articles can stimulate instructor-student discussions. Regardless of the instructional style used, we hope that our excitement for the study and teaching of human development is evident and catching as you read the articles in this seventeenth edition of *Human Development.*

Hiram E. Fitzgerald

Michael G. Walraven

Editors

Contents

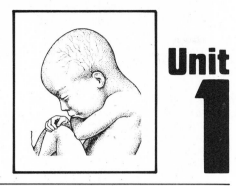

Unit 1

Psychobiology of Human Development

Four selections discuss genetic influences on development, focusing on reproductive technology, prenatal development, and twins.

Unit 2

Development During Infancy and Early Childhood

Six selections discuss development of the brain and development of communication, emotions, and cognitive systems during the first years of life.

To the Reader iv
Topic Guide 1
Overview 4

1. **The Mitochondrial Eve,** R. L. Cann, *The World & I,* September 1987. 6
   Molecular biologists study transmission of **mitochondrial DNA** in females and conclude that **human evolution** originated in Africa, perhaps 140,000 to 290,000 years ago. The author of this article calls for an interdisciplinary coalition of molecular biologists, archaeologists, and paleontologists to further specify the course of human evolution.

2. **Perfect People?** Amy Virshup, *New York,* July 27, 1987. 12
   Genetic screening via techniques such as **chorionic villus sampling**, **alpha-fetoprotein analysis**, and **amniocentesis** permits identification of a wide range of inherited and congenitally acquired **chromosomal dysfunctions**. This article describes how advances in **reproductive technology** create moral dilemmas for scientists, practitioners, and parents with no simple solutions in sight.

3. **The New Origins of Life,** *Time,* September 10, 1984. 19
   Infertility is a serious threat to **marital stability** for many couples, and for unknown reasons it seems to be increasing, according to this article. Today, **reproductive technology** allows couples to select from a variety of "artificial" techniques for creating babies, including **artificial insemination** and **in vitro fertilization**.

4. **All About Twins,** *Newsweek,* November 23, 1987. 24
   Similarities and differences in **twins** continue to intrigue and mystify human development researchers. This article reviews what is known about the psychobiology of **identical, fraternal, half-identical**, and **different father twins**. Closeness between twins is often beneficial, but sometimes it can be pathological, interfering with other developmental processes.

Overview 30

5. **The First 9 Months of School,** William Poole, *Hippocrates,* July/August 1987. 32
   Research with newborns, documented in this article, confirms that the **fetus** is responsive to a variety of **sensory stimuli** and suggests the possibility that some rudimentary **learning** occurs prenatally. **Fetal educators**, however, may be promising more than graduates of "Prenatal University" can deliver.

6. **Making of a Mind,** Kathleen McAuliffe, *Omni,* October 1985. 36
   Changes in **brain organization** during **infancy** are linked to the quality, quantity, and timing of **environmental stimulation**. Such changes are evident behaviorally in the areas of **language acquisition**, **learning**, and **socialization**. This article maintains that specification of sensitive periods for brain growth and development may have implications for **education**.

7. **What Do Babies Know?** *Time,* August 15, 1983. 40
   Historical descriptions of the infant vastly underestimated both the infant's abilities and its influence on its caregivers. A baby's abilities develop rapidly as a result of complicated interactions among genetic codes, reflexive behaviors, caregiving practices, and other forms of **environmental stimulation**. This article indicates that research may have important implications for diagnosis of atypical **growth patterns** in **infant development**.

The concepts in bold italics are developed in the article. For further expansion please refer to the Topic Guide and the Index.

8. **The Art of Talking to a Baby,** Alice Sterling Honig, *Baby,* Vol. III, No. 3.   46

Language empowers humans, granting them the creative ability to speak with others about any topic of experience, no matter how abstract. As the author points out, long before mastering the intricacies of speaking, babies play **communication games** with parents.

9. **The Child-Care Dilemma,** *Time,* June 22, 1987.   50

The demand for **day care** has reached unprecedented levels, but experts disagree sharply about its long-term effects on child development. Over 100 countries have national policies regulating child care; the United States is not one of them. According to this article, research on **infant day care**, **separation effects**, and **staff training** suffers from political debate fueled by moral questions.

10. **"What Is Beautiful Is Good": The Importance of Physical Attractivenes in Infancy and Childhood,** Linda A. Jackson and Hiram E. Fitzgerald, *Child International,* 1988.   54

The authors contend that children, parents, and strangers behave differently toward infants they perceive as attractive or unattractive. Children's **negative attributions** are directed toward attractive peers, whereas teachers' negative attributions are directed toward children they perceive as unattractive. **Physical attractiveness** is not a stable characteristic, either within or between cultures.

Unit 3

Development During Childhood

Twelve selections examine human development during childhood, paying specific attention to social and emotional development and cognitive and language development.

Overview   58

A. *SOCIAL AND EMOTIONAL DEVELOPMENT*

11. **Born to Be Shy?** Jules Asher, *Psychology Today,* April 1987.   60

Physiological and behavioral measures of **temperament** suggest that **shyness**, or **"behavioral inhibition,"** has genetic origins. **Inhibited children** may be susceptible to **stress-related illnesses**. The author claims that favorable environments, especially those provided by good parenting, can help overcome biological predispositions to shyness.

12. **Compliance, Control, and Discipline, Part 1,** Alice Sterling Honig, *Young Children,* January 1985.   64

Compliance with adult requests, the development of **self-control and self-regulation**, and **affiliative relations** with strangers have their origins in the secure attachments of infancy. According to this article, **inductive reasoning** is a more effective disciplinary technique than **love withdrawl** or **power assertion**, whether used at home or in the **day-care** setting.

13. **Compliance, Control, and Discipline, Part 2,** Alice Sterling Honig, *Young Children,* March 1985.   71

Achieving inner control is a long and often uneven developmental process in very young children. They need all the help they can get from perceptive adults in order to achieve self-regulation. The author recommends adult techniques for encouraging cooperation and compliance.

14. **Beyond Selfishness,** Alfie Kohn, *Psychology Today,* October 1988.   76

We start **helping others** early in life, but we are not always consistent in this behavior. In this article the author examines what makes us helpful sometimes and not helpful other times, and traces the development of caringness.

15. **Aggression: The Violence Within,** Maya Pines, *Science Digest,* July 1985.   **79**

Investigators from many disciplines search for the causes of physical **aggression**. The author argues that physical aggression is related to **child-rearing practices** and cultural levels of aggression, as well as to **brain chemistry** and **hormone levels** in the individual. The fact that extreme aggression runs in families lends credence to the importance of genetic origins. Intervention during childhood, maintains the author, is imperative if aggression is to be suppressed in adulthood.

16. **Depression at an Early Age,** Joseph Alper, *Science 86,* May 1986.   **83**

Depression is evident as early as infancy and early childhood; diagnosis, however, is not easy. Although the **affective disorders** (depression, **suicide**, **bulimia**) have a genetic and hormonal basis, they are precipitated by various environmental **stressors**. This article presents evidence to support the argument that affective disorders may be the major public health problem of the next decade.

17. **Resilient Children,** Emmy E. Werner, *Young Children,* November 1984.   **87**

Resilient children are not victimized by severe or chronic psychological **stress**, reports Emmy E. Werner in this article. The ability to elicit positive responses from others, the presence of a close **attachment** to a caregiver, and strong **self-confidence** are characteristic of the developmental histories of invulnerable children.

18. **The Importance of Play,** Bruno Bettelheim, *The Atlantic,* March 1987.   **92**

The author contrasts **rule-bound games** with **rule-free play** and argues that both are maximally beneficial when children are allowed to be spontaneous and to engage in their own **problem solving**. Parents should not **sex type** toys or overreact to their children's **aggressive** play. **Psychoanalytic theory** links children's play to the development of self-esteem.

B. *COGNITIVE AND LANGUAGE DEVELOPMENT*

19. **Practical Piaget: Helping Children Understand,** Sharon L. Pontious, *American Journal of Nursing,* January 1982.   **101**

The author provides an overview of Piaget's **stage theory** of **cognitive development**, focusing on perception, thought, reasoning, and language aspects of the preoperational and concrete operational periods. Specific examples illustrating how nurses can enhance **communication** with children may have broader implications as well.

20. **Three Heads Are Better Than One,** Robert J. Trotter, *Psychology Today,* August 1986.   **105**

The author reviews Robert Sternberg's **triarchic theory of intelligence** and contrasts it with the traditional IQ approach. The triarchic theory de-emphasizes speed and accuracy, focusing instead on **executive processes** related to planning and evaluating. The emphasis on **practical intelligence** and **multi-dimensional intelligence** has revitalized the study of individual differences in mental abilities.

21. **How the Brain Really Works Its Wonders,** *U.S. News & World Report,* June 27, 1988.   **111**

Neutral networks, computer simulations of the brain's interconnected nerve cells, are providing neuroscientists with new insights into brain structure and function. This article describes how neural network simulations are being applied to the analysis of **smell, perception, information processing**, and recovery from **brain injury**.

22. **Dyslexia,** Frank R. Vellutino, *Scientific American,* March 1987.   **115**

The author challenges traditional views that **dyslexia** is a **visual-spatial dysfunction**, presenting research results to support his theory that dyslexia reflects problems with storage and retrieval of **linguistic information**. He notes various "causes" of dyslexia, linking them to their associated treatments; to date, however, there is no substitute for direct remedial instruction in reading.

The concepts in bold italics are developed in the article. For further expansion please refer to the Topic Guide and the Index.

Unit 4

Family, School and Cultural Influences on Development

Seven selections discuss the impact of home, school, and culture on child rearing and child development. The topics include parenting styles and family structure, as well as the role that the educational system plays in the social and cognitive development of the child.

Overview   124

23. **Family Ties: The Real Reason People Are Living Longer,** Leonard A. Sagan, *The Sciences,* March/April 1988.   126

The author contends that it is impossible to trace the hardiness of modern people directly to improvements in medicine, sanitation, or diet. Personal relationships and the dynamics of a strong family, on the other hand, do have a direct impact. *Good health* is as much a *social and psychological achievement* as a physical one.

24. **The Child Yesterday, Today, and Tomorrow,** David Elkind, *Young Children,* May 1987.   132

The author reviews historical images of the child as a background for his consideration of three contemporary views of the child, labeled the *sensual child*, the *malleable child*, and the *competent child*. Although embedded in the current socio-historical context, such views lead to distortions of developmental processes; however, future generations will certainly produce similar distortions.

25. **The Importance of Fathering,** Alice Sterling Honig, *Dimensions,* October 1980.   138

Father absence and *father presence* have important implications for children's *sex-role development*, *cognitive achievement*, and *personal-social adjustment*. Alice Sterling Honig advocates programs for teaching paternal behavior that are analogous to those currently available for teaching maternal behavior.

26. **Helping Children Cope with Divorce,** Ralph F. Ranieri, *Marriage and Family Living,* April 1980.   144

One million children a year are affected by the divorces of their parents; *depression, anger, acting-out behavior, and guilt* are common reactions of these children. This article offers suggestions for helping children cope during and after a divorce.

27. **Project Day-Care,** Robert J. Trotter, *Psychology Today,* December 1987.   148

In this article Edward Zigler, one of the architects of Project Head Start, calls for a national commitment to quality *day care* for infants, preschool and school-age children. Staffing public school-based child-care centers with *child development associates*, combined with *parent effectiveness training* programs, will help to alleviate the child-care crisis and to reduce the variety of *social problems* affecting our children.

28. **Rumors of Inferiority,** Jeff Howard and Ray Hammond, *The New Republic,* September 9, 1985.   152

Studies cited in this article link deficiencies in the process of *cognitive development* to continued *social-economic underdevelopment* in America's black population. The authors argue that differences in performance between blacks and whites are related to self-doubt and fear of intellectual abilities on the part of blacks. Internalized negative expectations affect competitive behaviors and cause blacks to avoid situations that might reinforce their feelings of inferiority.

29. **Alienation and the Four Worlds of Childhood,** Urie Bronfenbrenner, *Phi Delta Kappan,* February 1986.   157

Disorganized *families* and disorganized *environments* contribute to the adolescent's sense of *alienation* from family, friends, school, and work. This article links dramatic changes in the structure of the family to social upheaval in the United States, which ranks first among the industrialized countries of the world in teen pregnancy, poverty, and teen drug use.

The concepts in bold italics are developed in the article. For further expansion please refer to the Topic Guide and the Index.

Unit 5

Development During Adolescence and Early Adulthood

Six selections examine some of the effects of social environment, sibling relationships, sex differences, jealousy, and suicide on human development during adolescence and early adulthood.

Overview   **162**

30. The Magic of Childhood, Paul Chance and Joshua Fischman, *Psychology Today,* May 1987.   **164**

Memory strategies, problem solving, creativity, and social interaction skills develop at different rates, contrary to predictions based on stage theories of cognitive development. *Adolescence* often marks a new burst of creativity coincident with changes in *abstract reasoning* and *language skills*. This article features psychologist Mihaly Csikszentmihalyi's studies of teenage social-emotional life.

31. Puberty and Parents: Understanding Your Early Adolescent, Bruce A. Baldwin, *Piedmont Airlines,* October 1986.   **172**

The striving of adolescents for *independence* is linked to parental apprehensions focused on *risk taking, societal dangers,* and *lack of control*. The author proposes 3 *stages* of adolescence spanning the years from 10 to 30, lists 15 adolescent *attitudes,* links them to 5 *aroused emotions* in parents, and tops off the article by listing 3 teen needs.

32. Jealousy and Envy: The Demons Within Us, Jon Queijo, *Bostonia,* May/June 1988.   **177**

Social psychologists view the interrelated *emotions* of jealousy and envy in terms of *motivation and self-esteem*. Hostility and privacy distinguish jealousy and envy. According to this article, three strategies—*self-reliance, selective ignoring,* and *self-bolstering*—help individuals to cope with negative emotions.

33. The Prime of Our Lives, Anne Rosenfeld and Elizabeth Stark, *Psychology Today,* May 1987.   **182**

The authors examine the essence of change, referring to *Sheehy's midlife crisis, Erikson's eight stages of human development,* and *Levinson's ladder*. They suggest that *cohort effects* may provide a better explanation for changes during adulthood. The ages-and-stages approach to adult development, popular in the 1970s, clearly is under attack.

34. Suicide, George Howe Colt, *Harvard Magazine,* September/October 1983.   **190**

This article reveals that *adolescent suicide* has increased threefold during the past 25 years. Overall, 30,000 to 100,000 people commit suicide annually. Despite a lack of government support for research on suicide, suicidologists continue to search for causes and to evaluate prevention programs. Causes of suicide range from *depression, hopelessness,* and *despair* to *alienation* and the erosion of faith.

35. The Emotional Brain, Laurence Miller, *Psychology Today,* February 1988.   **200**

Studies of *cerebral lateralization* of brain function cited in this article suggest that injury to the brain's left hemisphere is related to *depression* and *aphasia,* whereas injury to the right hemisphere is related to *euphoria* and *aprosodia*. Such injuries seem to interfere with the synchronization of interhemispheric control of *emotional expression* and *communication*.

The concepts in bold italics are developed in the article. For further expansion please refer to the Topic Guide and the Index.

Unit 6

Development During Adulthood and Aging

Eight selections explore how family lifestyles, loneliness, and depression relate to development during adulthood and consider how physical, cognitive, and social changes affect the aged.

Overview   204

36. **The Measure of Love,** Robert J. Sternberg, *Science Digest,* April 1985.   206
    Despite its pervasive importance for human behavior, psychologists know remarkably little about love. Although love for lovers, siblings, and parents differs, the basic nature of *love* is constant.

37. **The Fear of Forgetting,** Jeanne Toal, *American Health,* October 1986.   210
    Although *memory* weakens with age, the author notes that techniques are available to fight forgetfulness. Even though brain chemistry changes, arteries harden, and information overload occurs, today's older generation has benefited from better education and nutrition and also benefits from memory research, which has demonstrated important links between *emotions* and memory.

38. **Toward an Understanding of Loneliness,** Richard Booth, *Social Work,* March/April 1983.   214
    Loneliness is an emotional state that nearly everyone, regardless of age, experiences at one time or another. However, little is known about this frequently occurring condition. According to the author, loneliness seems to be related to unrealistic expectations, dwelling on the past, and *depression*.

39. **Biology, Destiny, and All That,** Paul Chance, *Across the Board,* July/August 1988.   217
    The author reviews evidence for sex differences in *aggression, self-confidence, rational thinking*, and *emotionality*, and concludes that, although some differences seem to exist, they are small and in many instances are differences in quality, not quantity. Some behavioral differences may be linked to differences in *hormones*, but biology, in this case, is not necessarily destiny.

40. **The Vintage Years,** Jack C. Horn and Jeff Meer, *Psychology Today,* May 1987.   222
    As a group, America's 28 million citizens over 65 years of age are healthy, vigorous, and economically independent—contrary to many *stereotyped* views of the aged. Most elderly people live in their own homes or apartments, usually with a partner. Nevertheless, many of the elderly do suffer from disorders such as *Alzheimer's disease*, and providing care for them can be stressful.

41. **Erikson, In His Own Old Age, Expands His View of Life,** Daniel Goleman, *The New York Times,* June 14, 1988.   230
    Almost 40 years ago Erik M. Erikson expanded the psychological model of the life cycle and human development. He and his wife, his lifelong collaborator, explain in an interview how *the lessons of each major stage of life* can mature into the many facets of *wisdom in old age*.

42. **A Vital Long Life: New Treatments for Common Aging Ailments,** Evelyn B. Kelly, *The World & I,* April 1988.   233
    The number of persons 65 and older has increased from 4 percent of the population in 1900 to 12 percent in 1985. In this article the author reviews some of the new treatments and attitudes that have contributed to this trend.

43. **Aging: What Happens to the Body as We Grow Older?** *Bostonia,* February/March 1986.   238
    This article challenges the stereotypical view that all systems deteriorate during *aging*. Although there is a strong *genetic* component to aging, abuse of the body speeds deterioration, whereas diet and exercise slows it down. The article gives special attention to *Alzheimer's disease*, osteoporosis, presbyopia, and presbyacusis.

Index   242
Article Rating Form   245

The concepts in bold italics are developed in the article. For further expansion please refer to the Topic Guide and the Index.

Topic Guide

This topic guide suggests how the selections in this book relate to topics of traditional concern to human development students and professionals. It is very useful in locating articles which relate to each other for reading and research. The guide is arranged alphabetically according to topic. Articles may, of course, treat topics that do not appear in the topic guide. In turn, entries in the topic guide do not necessarily constitute a comprehensive listing of all the contents of each selection.

TOPIC AREA	TREATED AS AN ISSUE IN:	TOPIC AREA	TREATED AS AN ISSUE IN:
Adolescence/ Adolescent Development	29. Alienation and the Four Worlds of Childhood 30. The Magic of Childhood 32. Puberty and Parents 34. Suicide	Competence	9. The Child-Care Dilemma 12. Compliance, Control, and Discipline, Part 1 13. Compliance, Control, and Discipline, Part 2 17. Resilient Children 24. The Child 27. Project Day-Care
Adulthood	4. All About Twins 32. Jealousy and Envy 33. The Prime of Our Lives 34. Suicide 35. The Emotional Brain 36. The Measure of Love 37. The Fear of Forgetting 38. Toward an Understanding of Loneliness 39. Biology, Destiny, and All That	Coping Skills	12. Compliance, Control, and Discipline, Part 1 13. Compliance, Control, and Discipline, Part 2 17. Resilient Children 26. Helping Children Cope 32. Jealousy and Envy
Aggression/Violence	12. Compliance, Control, and Discipline, Part 1 13. Complaince, Control, and Discipline, Part 2 15. Aggression 18. The Importance of Play 25. The Importance of Fathering 39. Biology, Destiny, and All That	Creativity	20. Three Heads Are Better Than One 30. The Magic of Childhood
		Cultural Influences	1. The Mitochondrial Eve 10. "What is Beautiful is Good" 27. Project Day-Care
Aging	37. The Fear of Forgetting 38. Toward an Understanding of Loneliness 40. The Vintage Years 41. Erikson, In His Own Old Age 42. A Vital Long Life 43. Aging: What Happens to the Body	Day Care	9. Child-Care Dilemma 10."What is Beautiful is Good" 12. Compliance, Control, and Discipline, Part 1 13. Compliance, Control, and Discipline, Part 2 27. Project Day-Care
Alienation	29. Alienation and the Four Worlds of Childhood 32. Jealousy and Envy 34. Suicide 38. Toward an Understanding of Loneliness	Depression/Despair	16. Depression 26. Helping Children Cope 34. Suicide 35. The Emotional Brain 38. Toward an Understanding of Loneliness
Alzheimer's Disease	37. The Fear of Forgetting	Developmental Disabilities	2. Perfect People 7. What Do Babies Know 16. Depression 17. Resilient Children 35. The Emotional Brain 43. Aging: What Happens to the Body
Attachment	12. Compliance, Control, and Discipline, Part 1 13. Compliance, Control, and Discipline, Part 1 17. Resilient Children 23. Family Ties 25. The Importance of Fathering	Divorce	25. The Importance of Fathering 26. Helping Children Cope
Birth Defects	2. Perfect People	Drugs	29. Alienation and the Four Worlds of Childhood
Brain Organization/ Function/Chemistry	5. The First 9 Months of School 6. Making of a Mind 15. Aggression 16. Depression 21. How the Brain Really Works 22. Dyslexia 35. The Emotional Brain	Education/Educators	6. Making of a Mind 10. "What is Beautiful is Good" 12. Compliance, Control, and Discipline, Part 1 13. Compliance, Control, and Discipline, Part 2 19. Practical Piaget 20. Three Heads Are Better Than One 22. Dyslexia 25. The Importance of Fathering 27. Project Day-Care 28. Rumors of Inferiority 29. Alienation and the Four Worlds of Childhood
Cognitive Development	4. All About Twins 6. Making of a Mind 19. Practical Piaget 20. Three Heads Are Better Than One 25. The Importance of Fathering 27. Project Day-Care 28. Rumors of Inferiority 30. The Magic of Childhood 37. The Fear of Forgetting 39. Biology, Destiny, and All That 40. The Vintage Years	Emotional Development	4. All About Twins 9. Child Care Dilemma 11. Born to Be Shy? 14. Beyond Selfishness 16. Depression 21. How the Brain Really Works

TOPIC AREA	TREATED AS AN ISSUE IN:	TOPIC AREA	TREATED AS AN ISSUE IN:
Emotional Development (cont.)	23. Family Ties 26. Helping Children Cope with Divorce 29. Alienation and the Four Worlds of Childhood 32. Jealousy and Envy 34. Suicide 35. The Emotional Brain 36. The Measure of Love 39. Biology, Destiny, and All That	Intelligence	6. Making of a Mind 19. Practical Piaget 20. Three Heads Are Better Than One 24. The Child 37. The Fear of Forgetting
Environmental Factors/Stimulation	4. All About Twins 5. The First 9 Months of School 6. Making of a Mind 25. The Importance of Fathering 27. Project Day-Care 43. Aging: What Happens to the Body	Language Development	4. All About Twins 6. Making of a Mind 8. The Art of Talking to a Baby 19. Practical Piaget 22. Dyslexia 35. The Emotional Brain
Erikson's Theory	12. Compliance, Control, and Discipline, Part 1 13. Compliance, Control, and Discipline, Part 2 33. The Prime of Our Lives 41. Erikson, In His Own Old Age	Learning	5. The First 9 Months of School 6. Making of a Mind 12. Compliance, Control, and Discipline, Part 1 13. Compliance, Control, and Discipline, Part 2 21. How the Brain Really Works 22. Dyslexia 24. The Child 27. Project Day-Care 28. Rumors of Inferiority 37. The Fear of Forgetting 40. The Vintage Years
Evolutionary Theory	1. The Mitochondrial Eve 11. Born to Be Shy?	Loneliness	32. Jealousy and Envy 38. Toward an Understanding of Loneliness
Family Development	2. Perfect People 4. All About Twins 5. The First 9 Months of School 9. Child-Care Dilemma 10. "What Is Beautiful Is Good" 12. Compliance, Control, and Discipline, Part 1 13. Compliance, Control, and Discipline, Part 2 23. Family Ties 26. Helping Children Cope With Divorce 27. Project Day-Care 29. Alienation and the Four Worlds of Childhood	Love/Marriage	26. Helping Children Cope With Divorce 32. Jealousy and Envy 36. The Measure of Love
		Maternal Employment	9. The Child-Care Dilemma 27. Project Day-Care 29. Alienation and the Four Worlds of Childhood 39. Biology, Destiny, and All That
Fertilization/Infertility	2. Perfect People 3. The New Origins of Life 4. All About Twins	Memory	6. Making of a Mind 7. What Do Babies Know? 21. How the Brain Really Works 22. Dyslexia 30. The Magic of Childhood 37. The Fear of Forgetting
Genetics	1. The Mitochondrial Eve 2. Perfect People 4. All About Twins 11. Born to Be Shy? 16. Depression 22. Dyslexia 43. Aging: What Happens to the Body	Midlife Crisis	33. The Prime of Our Lives
Giftedness	20. Three Heads Are Better Than One	Moral Development	2. Perfect People 12. Compliance, Control, and Discipline, Part 1 13. Compliance, Control, and Discipline, Part 2 25. The Importance of Fathering
High-Risk Infants	2. Perfect People 4. All About Twins 5. The First 9 Months of School 12. Compliance, Control, and Discipline, Part 1 13. Compliance, Control, and Discipline, Part 2	Nervous System Organization	5. The First 9 Months of School 6. Making of a Mind 16. Depression 21. How the Brain Really Works 35. The Emotional Brain 37. The Fear of Forgetting
Hormones	15. Aggression 16. Depression 39. Biology, Destiny, and All That	Parenting	4. All About Twins 5. The First 9 Months of School 7. What Do Babies Know? 8. The Art of Talking to a Baby 9. The Child-Care Dilemma
Infant Development	6. Making of a Mind 7. What Do Babies Know? 8. The Art of Talking to a Baby 9. The Child-Care Dilemma 10. "What Is Beautiful Is Good" 12. Compliance, Control, and Discipline, Part 1 13. Compliance, Control, and Discipline, Part 2 24. The Child 25. The Importance of Fathering		
Information Processing	5. The First 9 Months of School 6. Making of a Mind		

TOPIC AREA	TREATED AS AN ISSUE IN:	TOPIC AREA	TREATED AS AN ISSUE IN:
Parenting (cont.)	10. "What Is Beautiful Is Good" 11. Born to Be Shy? 12. Compliance, Control, and Discipline, Part 1 13. Compliance, Control, and Discipline, Part 2 17. Resilient Children 18. The Importance of Play 25. The Importance of Fathering 27. Project Day-Care 29. Alienation and the Four Worlds of Childhood 31. Puberty and Parents 36. The Measure of Love	Reproductive Technology	2. Perfect People 3. The New Origins of Life
		Retirement	40. The Vintage Years
		Self-Esteem/ Self-Control	12. Compliance, Control, and Discipline, Part 1 13. Compliance, Control, and Discipline, Part 2 17. Resilient Children 23. Family Ties 32. Jealousy and Envy 34. Suicide 39. Biology, Destiny, and All That
Peers	10. "What Is Beautiful Is Good" 12. Compliance, Control, and Discipline, Part 1 13. Compliance, Control, and Discipline, Part 2 29. Alienation and the Four Worlds of Childhood 31. Puberty and Parents	Selfishness	14. Beyond Selfishness
		Separation	9. The Child-Care Dilemma 11. Born to Be Shy? 38. Toward an Understanding of Loneliness
Perception	5. The First 9 Months of School 6. Making of a Mind 21. How the Brain Really Works 22. Dyslexia	Sex Differences/ Roles/Behavior/ Characteristics	15. Aggression 16. Depression 18. The Importance of Play 25. The Importance of Fathering 39. Biology, Destiny, and All That
Personality Development	4. All About Twins 7. How to Understand Your Baby Better 10. "What Is Beautiful Is Good" 11. Born to Be Shy? 12. Compliance, Control, and Discipline, Part 1 13. Compliance, Control, and Discipline, Part 2 17. Resilient Children 18. The Importance of Play 25. The Importance of Fathering 28. Rumors of Inferiority 32. Jealousy and Envy 33. The Prime of Our Lives 35. The Emotional Brain 36. The Measure of Love 39. Biology, Destiny, and All That	Shyness	11. Born to Be Shy? 12. Compliance, Control, and Discipline, Part 1 13. Compliance, Control, and Discipline, Part 2
		Siblings	4. All About Twins 26. Helping Children Cope With Divorce
		Socialization	29. Alienation and the Four Worlds of Childhood
		Social Skills	12. Compliance, Control, and Discipline, Part 1 13. Compliance, Control, and Discipline, Part 2 14. Beyond Selfishness 25. The Importance of Fathering 27. Project Day-Care 28. Rumors of Inferiority 30. The Magic of Childhood 32. Jealousy and Envy 33. The Prime of Our Lives
Piagetian Theory	19. Practical Piaget 30. The Magic of Childhood		
Play	18. The Importance of Play 27. Project Day-Care	Stress	7. What Do Babies Know? 11. Born to Be Shy? 16. Depression 17. Resilient Children 26. Helping Children Cope With Divorce 29. Alienation and the Four Worlds of Childhood 34. Suicide 38. Toward an Understanding of Loneliness
Pregnancy	2. Perfect People 3. The New Origins of Life 4. All About Twins		
Prenatal Development	2. Perfect People 4. All About Twins 5. The First 9 Months of School		
Preschoolers	9. The Child-Care Dilemma 11. Born to Be Shy? 12. Compliance, Control, and Discipline, Part 1 13. Compliance, Control, and Discipline, Part 2 24. The Child 26. Helping Children Cope With Divorce 27. Project Day-Care	Suicide	16. Depression 34. Suicide
		Values	2. Perfect People 17. Resilient Children
Psychoanalytic Theory	18. The Importance of Play 24. The Child 29. Alienation and the Four Worlds of Childhood		
Puberty	32. Puberty and Parents		
Racism/Prejudice	1. The Mitochondrial Eve 10. "What Is Beautiful Is Good" 27. Project Day-Care 28. Rumors of Inferiority		

Psychobiology of Human Development

In human reproduction, sexual intercourse brings together two living cells: a sperm, contributed by the male, and an ovum, contributed by the female. The unique union of these two cells marks the beginning of development, as well as the beginning of organism-environment interaction. Knowledge of the existence of sperm and ova was made possible by the discovery and refinement of the microscope, about 300 years ago, but no era can rival the contemporary one for drama with respect to advances in our knowledge of reproductive biology. Molecular biologists are at the forefront of such research, one line of which provides new information about the development of the species. In "The Mitochondrial Eve," the first article in this unit, a geneticist draws on evidence from molecular biology, archaeology, and paleontology to advance the thesis that the evolution of *Homo sapiens* began in Africa perhaps 140,000 to 290,000 years ago.

From the moment of conception, the newly formed organism exists in relationship to its environment. The quality of the prenatal environment and subsequent postnatal care received by the developing individual establish the foundation for development. In addition to discussing species evolution, the articles in this unit cover genetic screening, reproductive technology and related ethical issues, and prenatal development. During each of the periods of prenatal development, the fetus is vulnerable to a variety of biological and environmental stresses. Biological stresses include a variety of chromosomal anomalies and malformations. Environmental stresses include infectious disease, malnutrition, blood incompatibility, drugs, radiation, parental age, maternal emotional state, and chemical toxins. The article "Perfect People" describes genetic screening and gene therapy—technology that would have seemed like science fiction to the prospective parent of the 1960s. For the prospective parent of the 1990s, however, reproductive technology is real, and so are the related ethical issues.

Although couples may elect to try "artificial" methods for achieving fertilization, some of which are discussed in "The New Origins of Life," the sperm and ovum retain their dynamic characteristics; only the environment in which fertilization occurs or the technique for inserting the sperm into the womb changes. Nevertheless, reproductive technology has created a variety of ethical, legal, and social issues that will not easily be resolved. One such issue involves selection of sperm based on known characteristics of the donor. Although sperm banks provide hope for many infertile couples and single women, selecting sperm based on the donor's athletic, intellectual, or artistic accomplishments may lead to unreasonable expectations for the child's subsequent achievements.

Often, attempts to enhance fertility result in multiple births. Regardless of why multiple pregnancies occur, monozygotic twins hold as great a fascination for developmentalists as they do for the general public. Accounts of the biological, behavioral, and life experience similarities of monozygotic twins are not only fascinating, but they provide the data for intensive study of the relationship between genetic and environmental influences on development. The article "All About Twins" reviews much of this research and points out that, in addition to identical and fraternal twins, there are also half-identical and different father twins.

Technological advances have produced various techniques for assessing the developmental status of the fetus. Whereas each of these techniques contributes to a more complete evaluation of the structural and biological viability of the fetus, each technique also raises ethical and moral questions such as whether to retain or abort the fetus. Ethical questions are also raised with respect to preterm birth. Thirty years ago, prematurity generally referred to infants born no more than two or three months prior to the expected date. Today infants born at less than seven months gestational age are brought to term with the assistance of biomedical technology. However, the quality of caregiving intervention received by prematurely born infants may not be equivalent to the quality of life-sustaining technology. But it is clear that, without modern technology, most premature infants weighing less than 1500 grams would die, regardless of the quality of caregiving they receive.

The articles in this section discuss many of the discoveries that have been made about the early development of the fetus, and challenge the reader to consider the questions, "Where did life begin?" and "When does life begin?"

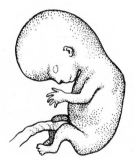
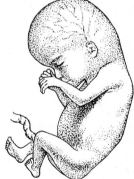
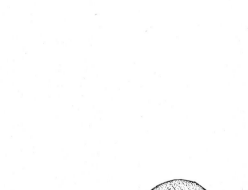
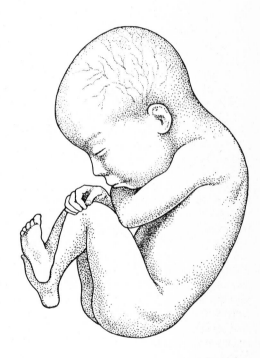

Looking Ahead: Challenge Questions

If scientists are able to confirm that the evolution of the human species began in Africa as a result of the survival of a common mutation, what implications might this have for modern humankind? Would irrefutable evidence for common evolution have any effect on current race relations?

Does artificial insemination or in vitro fertilization alter the dynamic relationship between the ovum and sperm during fertilization? At what point in development would you feel secure in concluding that such alternatives to normal sexual intercourse have no long-term consequences?

Consider your current beliefs about abortion, genetic engineering, and socialized medicine. How do you think these views would be challenged if you learned that your baby-to-be was expected to be profoundly retarded?

If manipulation of genetic material can prevent the appearance of physical dysfunctions, might not similar manipulations be used to engineer intellectual abilities, personality traits, or socially desirable behaviors? What factors would constrain a society that attempted to actively and explicitly practice eugenics?

Do you think that parents should support the commonalities of their identical twins, or should they strive to help each twin establish his or her individuality? What factors may counter an attempt by parents to ensure the uniqueness of their identical twins?

THE MITOCHONDRIAL EVE

A mapping of genes and ancestral linkages indicates that all people alive today carry some genes from the same African woman.

R.L. CANN

R.L. Cann is assistant professor of genetics at the University of Hawaii.

Most urban North Americans are a bit like West African Pygmies. They have neither the interest nor the ability to trace their kinship back further than 3 generations. In contrast, some Samoans can produce pedigrees 20 generations long. History may count for so little to so many, but what about the future? Demographers estimate that an American male probably has less than a 20 percent chance that a direct descendant with the same last name will be alive 250 years from now. Human families just don't persist very long.

If most lineages that we can actually identify ultimately die out, how is it possible that our species has survived for thousands of years? Recent research suggests that we are all directly linked through maternal genes to a nameless, faceless great-grandmother 200,000 years old, and scientists must now grapple with the implications of this deep penetration into human kinship. Astonishing as it may seem, molecular anthropology shows that all people alive today trace some of their genes to a single woman who was probably from Africa. This female may be the only common link in our species. She contributed a set of genes that has been passed on only by women, and through those genes, we are uncovering areas of the past that have been hidden from us by history, culture, and our own eyes.

Scientific advances rarely proceed in straight lines, and the study of human evolution has been no exception. Yet the traditional portrayal of the fossil evidence of our ancestry gives the impression that scientists have reliably mapped the evolution of humans. It seems that once some messiness was sorted out 5 million years ago, there was a smooth steady development of the line. Although new molecular refinements provide counterevidence, human evolution is still diagrammed in a series of simple, ancestor-descendant relationships. Dioramas in a natural history museum can be counted on to show some primitive ape-human (*Australopithecus*) merging into some early species of our own genus, *Homo*. Then *Homo erectus* evolves into *Homo sapiens*, along with mortgages, star wars, and ice cream. Such schemes imply a genetic continuity in space and time that contains more fantasy than Spiderman's best escapades. Accurate reconstructions of human phylogeny depend on the successful mating of two individuals, a process we can only infer for the past. A *Homo erectus* couple never gave birth to a Homo sapiens baby, so how did modern humans evolve? Contrary to the smooth development shown in the dioramas, we cannot yet explain these transitions.

All scientists make assumptions about their systems. Paleontologists ask us to believe that they can accurately piece human remains into neat categories that denote direct ancestry and illustrate the processes by which evolution proceeded on this planet. They have been slow to admit, however, the assumptions that enter into their evolutionary reconstructions from fossil evidence. Many paleontologists fear that if they expose the legitimate scientific limits of the certainty of their theories, fundamentalists and creation "scientists" may misrepresent these data to dispute the fact that evolution occurred.

Molecular biologists in turn are often highly critical of paleontologists and draw attention to the lack of explicitly stated hypotheses about how evolution works at the level of DNA and fundamental life processes. Nevertheless, paleontologists and molecular biologists are both attempting to explain the continuity of life, and collaboration between them should prove to be fruitful.

GENES AND CHROMOSOMES

Hereditary information in each cell is contained in the chemical DNA, deoxyribonucleic acid. Genes, discrete linear sequences of DNA, determine traits such as eye color and whether or not we can curl our tongues. In the past, biologists have concentrated on those genes found in the nucleus of the cell. For many years scientists assumed that these "nuclear" genes, which are contributed by both parents, contained all of the genetic material.

Chromosomes are long, orderly assemblies of genes

attached end to end. Chromosomes contain DNA, so when we get one chromosome from each parent, we also get a copy of whatever DNA sequences are present on that chromosome. One law in biology is that with

few exceptions each parent makes an equal contribution to the genetic makeup of an offspring. Mendel worked out the mathematical relationships for the expected outcome of crossing different-colored peas in 1865, but his rules hold for people as well as for plants.

Starting in the 1950s, however, scientists began to accumulate information that suggested there might be some exceptions to Mendel's rules. Some of the most significant and most puzzling phenomena were factors, such as antibiotic resistance, which showed strictly maternal inheritance—something not predicted or allowed in the Mendelian genetic schema. Maternal inheritance was demonstrated if the offspring showed only a particular characteristic of the female parent and none of that characteristic matching from the male parent. This was not simply due to the dominance or recessiveness of traits, which Mendel had described for pea color.

In the 1970s, thanks to advances in biology, scientists began to isolate DNA originating from different parts of the cell. They found that most DNA was in the cell nucleus as expected. This corresponded to the DNA in the chromosomes that behaved in the expected Mendelian fashion. More puzzling was the discovery that some DNA could be isolated from cellular cytoplasm, the sap outside the cell nucleus.

MITOCHONDRIAL DNA

Among the several different components of the cytoplasm are mitochondria—small energy-producing bodies inside animal and plant cells. Scientists now believe that the mitochondria were once free-living organisms that invaded the primitive cells of some early life forms and became permanent residents. The mitochondria provide energy for the cell, and the cell provides nutrients for the mitochondria. Most of the genes within the mitochondria have been lost in the last billion years, but a few essential ones remain. Mitochondrial DNA (mtDNA) became the prime suspect for the source of maternally inherited characteristics.

Above: Contrary to the conclusions of paleontology, recent research in molecular biology suggests that we are all directly linked through maternal genes to an African great-grandmother who lived 200,000 years ago.

Below: Although museum dioramas developed by paleontologists suggest a smooth development of the human line, molecular biology reminds us that we cannot yet explain the transitions from one humanlike form to another.

Suspicions were confirmed when developmental biologists examined what happened during the fertilization of mammalian embryos. The egg contains thousands of mitochondria ready to deliver bursts of energy needed in the rapidly growing embryo. Sperm, on the other hand, contain fewer than 10 mitochondria. If these male mitochondria escape deactivation or outright destruction upon fertilizing the egg, they are soon overwhelmed and eliminated by the sheer numbers of female mitochondria. It is possible to use genetic tags to identify mitochondria contributed by males versus females and to show that only female mitochondria are present in new offspring.

Both males and females have mtDNA in every cell, but only females pass it on. For mtDNA there appears to be none of the genetic recombination that is responsible for mixing up the nuclear genes of father and mother, and for complicating paternity suits. Instead, mitochondrial descent is simple and unilineal. Any trait related to the DNA contained in those mitochondria shows a pattern of female transmission. MtDNA carries the codes for about 37 genes, and we understand the role of these genes in the cell.

Pure DNA exists as a long thread. Each thread has a partner strand running alongside it and winding slightly. Hydrogen bonds weakly connect the two strands and give the structure stability. In mitochondria this arrangement of DNA has an added twist, for the DNA is one big circle instead of just a long, folded double helix as it is in nuclear DNA. Chemical isolation of pure mtDNA is relatively simple due to this circular structure.

The length of the mtDNA circle in humans is 16,569 base pairs, or 37 genes, long. The genes contained in mtDNAs are so essential to the cells of mammals that the order in the circle has been conserved for 80 million

years of evolutionary history. Yet in spite of this extreme conservation of the order of the genes, humans express great variability within every one of these 37 genes. Such diversity means that if we could read all of the base pairs we might discover that for a given sequence of DNA, every human female lineage (i.e., matriline) is unique. Nonetheless, females descended from the same mother carry DNA sequences that identify their common genealogy, and all people, male and female, can be related to each other by connecting female lineages based on the individual mutations that are shared between lineages. Mutations in mtDNA are not repaired like those in nuclear genes, so those changes that occur in ova are automatically passed on. A child will inherit whatever new mutations may have accumulated in the ova of its mother.

Although the word *mutation* is often thought to have a bad connotation, in the context of mtDNA the word can also have positive connotations. To a geneticist, mutation implies variation in the gene pool, and heritable variation is the raw material that natural selection acts on. Species without genetic variation in the form of mutations may be more vulnerable to environ-

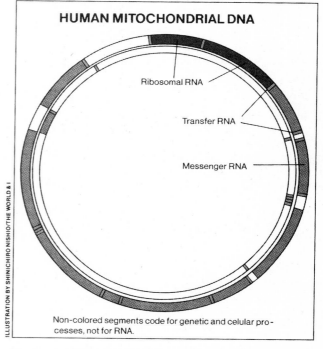

HUMAN MITOCHONDRIAL DNA

Ribosomal RNA

Transfer RNA

Messenger RNA

Non-colored segments code for genetic and celular processes, not for RNA.

The human mitochondrial DNA, mtDNA, consists of one double-stranded chromosome that forms a closed loop or ring. MtDNA is particularly valuable for mapping genealogies because of its transmission through females and its rapid rate of mutation. RNA units of three different types are transcribed from the DNA. The RNA carries the codes for making the diverse kinds of protein units needed for that particular cell.

mental stresses, risking extinction. Before the development of methods of molecular analysis, this genetic variation was so hard to detect that many evolutionists doubted its presence.

Several recent advances in recombinant DNA technology help us find mutations in human mtDNA. Individual variants that are present in a population at a frequency greater than 5 percent are called poly-

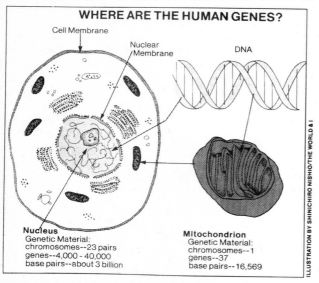

WHERE ARE THE HUMAN GENES?

Cell Membrane

Nuclear Membrane

DNA

Nucleus
Genetic Material:
chromosomes--23 pairs
genes--4,000 - 40,000
base pairs--about 3 billion

Mitochondrion
Genetic Material:
chromosomes--1
genes--37
base pairs--16,569

Most of the genetic material, DNA, for a particular species resides in the nucleus of each cell. However, DNA has also been found to reside in a second site—the mitochondria of each cell. In humans, one cell may have up to 10,000 mitochondria, and each mitochondrion can have four to five copies of the mitochondrial DNA. While our nuclear genes may have come from many women and men, the mitochondrial genes have come from only one woman.

morphisms. If molecular polymorphisms in humans could be grouped, they might reveal the evolutionary history of our species more accurately than could be inferred from the indirect evidence of fossil reconstructions. As molecular biologists, we at least knew that the genes available from present-day specimens came from some ancestor. In contrast, paleontologists can never be certain that a given fossil has left descendants.

COMPARING THE SEQUENCES

Our study (Cann, Stoneking, and Wilson, *Nature* 325 (1987): 31–36) focused on the mtDNA of 147 people from all major geographic regions of the world. Most of these samples were prepared from placentas donated by the mothers of newborn babies. The placenta is a large organ that is rich in DNA, and it is usually discarded by the hospital. During fetal development, the placenta acts as the life-support system between mother and baby, and the cells making up the placenta have the same genetic make-up as the infant. Yet taking a placenta after a normal delivery does not put either the mother or child at risk, and the mothers often gave us information about their maternal ancestors.

Bringing these placentas from labor and delivery back to the lab, or running to the airport to intercept an arrival frozen on dry ice from the Papuan New Guinea Highlands, we proceeded to isolate mtDNA from each donor.

MtDNA can serve as a tag to identify groups of females descended from the same mother, and so is a useful genealogical tool. Genealogies traced through maternal lines may be more accurate than those traced through paternal lines, because mothers and babies are rarely mixed up while fathers and their children are often separated or mixed.

We used a new class of molecules, called restriction enzymes, to find differences in mtDNA sequences between human lineages. These enzymes recognize a particular unique sequence, called a recognition sequence, and cut both strands of DNA at precise points in relation to that sequence. Many such enzymes are known. By using a battery of them with each specific for different short sequences, we looked for mutations in all 37 of the mitochondrial genes. Altogether we studied about 10 percent of the entire mtDNA circle, randomly cut with these enzymes, in great detail.

To describe a piece of DNA, it is now fashionable to use a genetic map as a shorthand way of referring to all of the possible, precise locations for which a recognition sequence might be either present or absent. One human mtDNA has been completely mapped using restriction enzymes and is now used as the standard reference for further studies of mtDNA. By comparing the maps of two individuals, we might find that they share a new sequence that no one else has. Or, alternatively, they could share a mutation that has abolished the recognition sequence for a site on the map. Both types of mutations are commonly seen in humans.

By using the restriction sequence mapping proce-

dure, we gathered data that was coded, entered into a computer, and ordered in reference to an arbitrary starting point on the DNA circle. Using DNA polymorphisms, basic information about human mutation rates in different cultural groups is now being accumulated.

We used data from 398 mutations to do something further. We compared each individual with every other individual in order to ask whether groupings based on shared mutations correspond with groupings based on geography, ethnicity, or morphology.

RECONSTRUCTING THE GIANT TREE

Like paleontologists, molecular biologists make assumptions to guide their work. The first assumption is that evolution proceeds in as few steps as possible; it goes in the simplest mode scientists can imagine. This is called parsimony, and as a first approximation, is often correct. We all know it is possible to go from New York to Boston via San Francisco, but this is not the usual route. Parsimony dictates that if a direct route is shorter, that is the one travelers most often choose. In the same way, if fewer mutations are required to connect Africans with New Guinea Highlanders than Africans with all other Africans or Papua New Guineans with all other Papuans, we chose the direct, simplest route. We assumed that the giant tree which connects all human mtDNA mutations by the fewest number of events is most likely the correct one for sorting humans into groups related through common female ancestry. What does the structure of this tree look like?

To find a common root, we followed the usual practice of placing the root at a point half-way between the two most divergent lineages. This separates the human species in sub-Saharan Africa because in our sample the two individuals with the most divergent ancestries both had an African ancestry. By evolutionary processes, genetic diversity accumulates through time from a common ancestor. The great divergence of lineages in Africa suggests that Africa is the cradle of human polymorphism.

Our tree of the human species contains two major branches. On the first branch are only Africans. The second branch, which connects to the first, contains Africans in addition to individuals from all other parts of the world. No other arrangement is more parsimonious, or requires fewer mutations. This pattern strongly supports the view that modern humans stem from African lineages. Considering climatic variation in the last 600,000 years, it is now well-accepted that the ebb and flow of glaciers encroaching on Europe and Asia has periodically driven our ancestors to take refuge in Africa, the Middle East, and perhaps India. With the genetic evidence supported by the climatic record, Africa emerges as our homeland, and the most ancient reservoir of human life. Africa has been a source of continual renewal for our species over the thousands of years of human existence and development.

The second branch contains a mix of people from all parts of the world, including several "bushes" in which

1. PSYCHOBIOLOGY OF HUMAN DEVELOPMENT

many people with diverse traits all originate from a common lineage. Yet if one expects a simple grouping of all Europeans or all Chinese, one is in for a surprise. It takes a lot of branching to get to the point where a single limb in the tree connects only to people from one remote area, such as highland Papua New Guinea (PNG). Even then, it is possible to find some people from PNG who do not connect to that single limb. They

THE HUMAN ANCESTRAL TREE

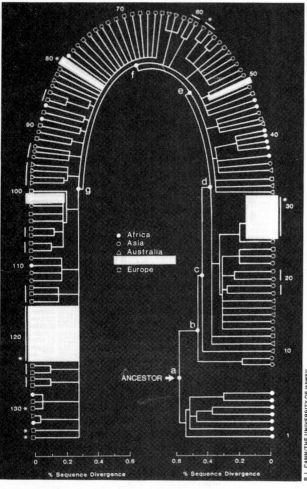

Genealogies traced through maternal lines may be more accurate than those traced through paternal lines. This tree of human ancestry was produced by examining the mitochondrial DNA from placentas from 134 different lineages of mothers. The sections colored in blue indicate mothers native to Papua New Guinea who revealed remarkable diversity of ancestry connecting back to an African root.

connect at many places on the second branch. The complexity of the human tree of lineages suggests that females were highly mobile in our history. Multiple female lineages can be found in all of the major races recognized today. This underscores the variation that geneticists suspected but had a hard time resolving when they just concentrated on nuclear genes.

A SINGLE FEMALE ANCESTOR?

A single female as our mitochondrial ancestor is not the same thing as saying all humans stem from a single female ancestor. As contemporaries of our common mitochondrial ancestor, there were many women as well as men alive, perhaps as many as 10,000–50,000 humans altogether. Many of the people, men and women, contributed nuclear genes to us today. However, many of the women had only sons, and their mitochondrial DNA reached a dead end. As the generations passed and the lineages spread, the interplay of populations eventually resulted in the ending of all mitochondrial genetic lines except one. Our research does not suggest why this may have happened or if it was at all necessary that it happen. Our research merely suggests that, in fact, the survival of only one line of mitochondrial DNA has occurred.

In order to estimate when people began to leave Africa, we relied on careful estimates of the rate at which mtDNA mutations occur in humans as well as in our closest living relatives, the great apes (chimps, gorillas, orangutans). This rate was based on comparing homologous DNA sequences in humans and apes and assuming that the human-great ape split occurred five to seven million years ago. This gave us a preliminary calibration of 2 to 4 percent change in DNA sequence for every million years of human history. Such a rate suggests that the primary fork in the human mtDNA tree may be 140,000–290,000 years old. Humans from different parts of the world freely exchange genes, so it is not completely accurate to equate presumed barriers to gene flow between races with barriers to gene flow between species. For this reason, to calibrate internally the human mtDNA tree, Mark Stoneking, Kuldeep Bhatia, and Allan Wilson at the University of California at Berkeley have considered colonization events in well-known parts of the world where archaeologists can estimate the first appearance of humans. Their comparisons include the populations that settled the New World from Siberia, and Old World colonizers in New Guinea and Australia. From these estimates, it appears that our ancestors may have left Africa anywhere from 180,000 to 230,000 years ago.

MtDNA data simply cannot be used to say precisely when such migrations actually took place, or the routes that were taken. Archaeologists will have to supply this data.

THE RESPONSE OF SPECIALISTS

What has been the response of specialists from various fields to this new interpretation of human evolution? Objections to our conclusions have come from two fronts. Previous work on the part of human geneticists had suggested that mtDNA might support an Asian origin for modern humans. This was due to a study that included a greater number of subjects but examined them in much less detail. Scientists working at Stanford University (M.J. Johnson, D.C. Wallace, S.D. Ferris, M.C. Rattazzi, and L.L. Cavilli-Sforza, *Journal of Molecular Evolution* 19 (1983): 255) saw a level of genetic polymorphism among Africans similar to that derived from our study. To explain this high level of genetic polymorphism, however, they chose to postulate that the African mutation rate for mtDNA was extreme, at least 10 times higher than mutation rates elsewhere. On close inspection of actual DNA sequences, there is no indication that this is indeed true. One might reasonably expect that both nuclear DNA and mtDNA would be affected by mutagenic agents such as strong ultraviolet light, aflatoxins on peanut shells, and extreme heat, which are widespread in the African environment. However, these extreme effects have not been seen in the nuclear DNA of Africans relative to Asians or Europeans. Thus, the hypothesis of an Asian origin for modern humans rests on the sharing among Asians of a most-frequent class of mutations. This may be due to a simple effect of sampling distortion.

Second, their grouping of people into racially specific mitochondrial types, or morphs, has not stood the test of time. It is possible to break these morphs into subsets that contain individuals of many races. Our study, therefore, is consistent with earlier data but allows us to see how wrong guesses can be made with insufficient evidence. This is simply a demonstration of the scientific tradition.

Recent data on the molecular basis of genetic polymorphisms for the nuclear gene B-hemoglobin, which codes for the protein that carries oxygen in red blood cells, further questions the Asian origin hypothesis. Hemoglobins are the best-studied nuclear genes in humans, and such close scrutiny should have revealed evidence of an accelerated rate of mutation among Africans. Hematologists at Oxford University (J.S. Wainscoat, A.V.S. Hill, A.L. Boyce, J. Flint, M. Hernandez, S.L. Thien, J.M. Old, J.R. Lynch, A.G. Falusi, D.J. Weatherall, and J.B. Clegg, *Nature* 319 (1986): 491) found no significant difference in mutation rates between different populations of the world. This data therefore supports an African origin for human populations also.

Finally, a few paleontologists have objected to the dismissal of some of their best-loved fossils as immediate ancestors of modern humans. They point to the long history of human occupation in Asia, and similarities in the teeth and skulls of modern Chinese compared to ancient Asian fossils. It is difficult for them to accept a conclusion that such traits may have evolved twice from various genetic backgrounds. Yet such examples of convergence due to environmental factors are common. Specialists in the African Middle Pleistocene period, which started some 600,000 years ago, are now assembling evidence of the transition between *Homo erectus* and *Homo sapiens* in Africa. However, from *Homo erectus* fossils, there is little evidence to support the hypothesis of a separate transition to modern Chinese that cannot be countered by the alternative hypothesis of migration from Africa.

Furthermore, one might argue that in the past we failed to sample an adequate representation of modern Asians, so that we may have missed some important mitochondrial lineages from this area of the world. As additional labs report their data, however, it is clear that the roots of Asian mtDNA diversity are simply not as ancient as those in Africa. Significant genetic contributions of people from this ancient Asian line of descent should have produced modern Asian maternal lineages which at least match the age of the African ones. Since they did not, we conclude that even though fossils of considerable antiquity are known from Asia, they were replaced by lines of African origin.

Cultural innovations allow groups to dominate and replace even well-adapted indigenous populations. Archaeologists again could provide important clues through discovering large changes in either tools or morphology associated with dietary shifts of Asian people. Even so archaeologists may miss a rapid turnover of an invading population.

Consider European colonial expansion in both Asia and Africa. Since the 1500s, this great movement of people and ideas has completely changed the geographic distribution of Caucasian genes from what a spaceship might have encountered if aliens had taken blood samples from Earth's inhabitants in the year 1491. Yet this is but an instantaneous moment in the history of a species, and so pervasive are the effects of fast food, video games, and plastic that an archaeologist from space might be severely challenged to document the direction of this cultural transfer or invasion if modern humans were wiped out by epidemics or warfare in the next 20 years.

Research on modern human mtDNA diversity has challenged some entrenched ideas from paleoanthropology about the history of anatomically modern humans but is consistent with new interpretations of both stones and bones from Africa. People who look like us first appear in Africa, and their tools do too. All it takes is some stretching of the imagination to grant them the benefit of the brainpower they obviously must have had to survive and expand. Not everyone can leave a string of descendants to carry on a family name. An example of this can be seen in the extraordinary Neanderthals of western Europe, who for 12,000 years fought off the cold, cave bears, and the depression of active minds trapped in arthritic bodies. They too were the products of a long line of fossil humans in Europe, but modern Europeans show no evidence in their mtDNA of ancient maternal lineages contributed by them.

The power of paleontology lies not so much in its ability to accurately reflect the evolutionary history of extant groups. Instead, it has the unique focus that captures for us the world as it might have been, and indeed for a moment, once was. Understanding how and when modern people left Africa remains a major scientific issue. Archaeologists and paleontologists will contribute the information that settles this issue. It is crucial to remember, however, that the question has only been raised because of insights from molecular genetics. Biology puts real constraints on hypotheses that deal with living populations. Thus a new, interdisciplinary coalition of molecular biology, archaeology, and paleontology holds greater promise for deciphering the real course of human evolution than any individual discipline. This could be the start of a wonderful working relationship.

THE PROMISE AND THE PERIL OF GENETIC TESTING

PERFECT PEOPLE ?

Amy Virshup

SUSAN AND TOM MURPHY* HAVE GLIMPSED THE future. In the spring of 1984, when Susan was four months pregnant with their second child, they learned that their firstborn, sixteen-month-old Sarah*, had cystic fibrosis, an inherited, incurable disease that can mean a short, difficult life for its victims. Sarah would likely need constant medication and daily sessions of strenuous physical therapy. To make matters worse, no one could say whether the child Susan was carrying also had the disease.

Their doctor advised the Murphys to consider having an abortion. But they decided to have the baby, and after five tense months, their son was born—without the disease. Though both Tom and Susan wanted a large family, neither was willing to risk conceiving another child who might get CF.

Within two years, the Murphys changed their minds. By then, researchers abroad had found a way to home in on the gene that causes CF, and a prenatal test had become available in the United States for families who already had a child with the disease. A door had been opened for the Murphys, and they decided they could chance another pregnancy. Tom, Susan, and Sarah had preliminary analyses of their DNA—the body's basic genetic material—and Susan became pregnant in September 1986.

Just nine weeks into her pregnancy, she had chorionic villus sampling (CVS), a recently developed twenty-minute procedure similar to amniocentesis. The CVS material was sent to the lab, and the Murphys could only wait—and worry—for ten days. They'd never decided just what they would do if the news was bad, but abortion was a real option. The day before Thanksgiving, they got the news: Their unborn child did not have CF.

SUDDENLY, WITHOUT MUCH TIME TO PONDER THE MORAL and ethical implications, Americans are being thrust into the age of the tentative pregnancy. For many, the decision to have a child is made not at conception but when the lab sends back the test results. "The difference between having a baby twenty years ago and having a baby today," says ethics specialist Dr. John Fletcher of the National Institutes of Health, "is that twenty years ago, people were brought up to accept what random fate sent them. And if you were religious, you were trained to accept your child as a gift of God and make sacrifices. That's all changing."

Over the last decade, genetics has been revolutionized: Using remarkable new DNA technology, molecular biologists can now diagnose in the womb inherited diseases like CF, hemophilia, Huntington's chorea, and Duchenne muscular dystrophy. Genetics is advancing at unprecedented speed, and important breakthroughs are announced almost monthly.

Since each person's DNA is distinct, the potential uses of genetic testing seem limitless: It could settle paternity suits, take the place of dental records or fingerprints in forensic medicine, identify missing children, warn about a predisposition to diseases caused by workplace health hazards. By the mid-1990s, geneticists should be able to screen the general population for harmful genes and test—at birth—a person's likelihood of developing certain types of cancer, high blood pressure, and heart disease.

Besides the prenatal tests now commonly in use, doctors may have a blood test that screens fetal cells in a mother's bloodstream and determines the fetus's sex and whether it has any chromosomal disorders or inherited genetic diseases. Work toward the ultimate goal—gene therapy—has already begun.

But this technological wizardry has some vehement opponents. A French government committee made up of doctors and lawyers recently called for a three-year moratorium on prenatal genetic tests because of the fear that they would lead to "ethically reprehensible attempts to standardize human reproduction for reasons of health and convenience." This is similar to the objection that anti-abortionists have always raised and now extend to genetic testing. (The Vatican, in its recent "instruction" on birth technology, condemned any prenatal test that might lead to abortion.) These days, though, right-to-lifers are finding themselves with some odd allies— feminists, ethics specialists, and advocates for the handicapped who are unsettled by the implications of the new genetics. The tests, they argue, will winnow out fetuses so that only "acceptable products" will be born, thus devaluing the lives of the handicapped.

* Names marked with an asterisk have been changed.

WITHOUT MUCH DISCUSSION, AMERICANS ARE BEING THRUST INTO THE AGE OF THE TENTATIVE PREGNANCY.

Other critics fear that services for the handicapped will be cut back and that people will be saddled with new responsibilities rather than new opportunities—after all, never before have parents had such an ability to choose whether to accept a child with an inherited condition. In addition, they claim that the sophisticated tests will give employers and insurers the ability to discriminate on genetic grounds. (A 1982 study found that eighteen major American companies—Dow Chemical and du Pont among them—had done some sort of genetic testing.)

Strong as the opposition is, it's not likely that the genetic revolution will be stopped: Few people, given the chance to avoid the emotional, physical, and economic burdens of raising a handicapped child, are likely to refuse it. The widespread use of prenatal tests parallels the advance of feminism. As more and more women made careers, the tests took on added importance—especially since many working women are putting off childbirth until their mid-thirties, when chromosomal abnormalities are more common.

"I didn't feel I had a choice," says advertising executive Judith Liebman, who became pregnant for the second time at 40. "It wasn't 'Is CVS sophisticated enough that I want to take the risk?' The choice was 'Look, they're doing this test, they're recommending I take it because of my age, so I have to go with the medical profession and say that it's okay to take it.'"

As sociologist Barbara Katz Rothman writes in *The Tentative Pregnancy*, her study of genetic testing, "In gaining the choice to control the quality of our children, we may rapidly lose the choice not to control the quality, the choice of simply accepting them as they are."

"The Luddites had a point," says Rothman. "Not all technology is good technology, and not everything needs to be done faster and better."

Mount Sinai geneticist Fred Gilbert is tracking the CF gene. Like Gilbert, his lab is casual, low-key; the most spectacular thing about it is its slightly begrimed view. Notes are taped to shelves, a stack of papers and manila folders sits atop a file cabinet, and several odd-looking blue plastic boxes lined with paper towels are arrayed on the counter (they are used to separate pieces of DNA). Flipping through a looseleaf binder of test results, Gilbert, a burly man with a dark, full beard, talks about his patients like an old-fashioned family doctor—this father drinks, that mother is overburdened, another family's religious views have made it impossible for them to abort—but he knows most of them only from the reports of their genetic counselors and from their DNA, which is shipped to him from cities all over the country.

Two years ago, Gilbert began offering a biochemical enzyme test developed in Europe for cystic fibrosis; then, when DNA probes became available, in January 1986, he started running DNA diagnoses

---

THE DNA TEST

Scientists rarely know the location of a harmful gene among the 50,000 to 100,000 in human DNA, so they look instead for a genetic marker—a seemingly harmless variation in DNA that is inherited along with a disease gene. DNA looks like a twisted rope ladder, and its "rungs" are pairs of nucleic acids, which always join in the same way. In order to find a marker, a scientist must first extract DNA from an individual's cells. Then the DNA ladder is broken into two separate strands and cut into fragments by enzymes that are able to "read" the order of the nucleic acids.

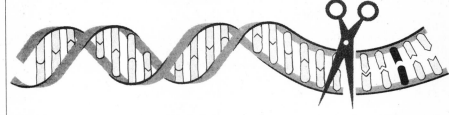

The cut-up DNA is separated and then exposed to a gene probe—a cloned piece of DNA that includes the variation being looked for. The probe is tagged with radioactivity.

If the probe bonds to a piece of DNA, then the genetic marker—and therefore a disease gene—is present.

---

as well (both tests are about 95 percent accurate and are used to back each other up). Now Gilbert and his team extract DNA from the cells of families at risk for CF—his patients include a tenant farmer, an executive of a large firm, and several welfare families—expose it to gene probes that have been tagged with radioactivity, and then study photographs of the DNA, looking for the pattern that means the disease is present. Since CF is inherited recessively, there is a one-in-four chance in each pregnancy that the child might have the disease (AE p. 16).

This test can be used only by families who already have a child with the disease, but many people are betting on a CF screening test for the general population. After all, CF is the most common inherited disease in whites, striking about one child in 2,000. And now that scientists have moved in on the gene—in April, a British research team announced that it had found a marker that is even closer to it and may turn out to be the gene itself—development may be imminent. Since the test will be highly lucrative for its developer, there's a great incentive to come up with one; by one estimate, half the genetics labs in the country are looking for the CF gene. Gilbert, however, is not making money. Since the test he is using is still experimental, he is offering it at a minimal fee; for those who cannot afford it, the test is done free.

Even so, response to the test has been slow. The eight labs in the country offering DNA diagnosis have done about 400 tests, and Gilbert has completed just fifteen prenatal diagnoses using DNA. One problem is that some cystic-fibrosis specialists have been slow to tell their patients about the test. (The Murphys, for example, didn't learn about the test from their doctor; *they* had to tell *him* it was available.) The Cystic Fibrosis Foundation has also stayed away from the test, afraid of the association with abortion, and Gilbert gets no funds from it.

IN EVERY HUMAN CELL, THERE ARE 23 PAIRS OF CHROMOsomes that contain the 50,000 to 100,000 bits of genetic information called genes. The genes are made up of DNA (six feet of it in each cell), a long, two-stranded molecule. DNA, which looks like a tightly coiled rope ladder, is a compound made up of phosphate, sugar, and four nucleic acids: adenine (A), thymine (T), guanine (G), and cytosine (C). Each "rung" of the ladder (and there are 3 billion rungs in each person's DNA) is made up of a pair of nucleic acids, and they always join in the same way: Adenine connects only to thymine, guanine to cytosine. Only in the last few years have scientists been able to read the order of these nucleic acids.

Sequences of these four nucleic acids form genes. Genes, in turn, determine a person's physical characteristics, though no one knows the process that, for example, leads the genes for shape and form to generate the five fingers of the hand. A genetic disease often results from a "typo" in the order of the nucleic acids; like a misspelling in language, a typo in DNA can change the message of a gene. For example, the normal DNA code for glutamic acid is GAG, but in a person suffering from sickle-cell anemia, thymine replaces adenine (so the code reads GTG) and the blood is unable to absorb oxygen correctly. Surprisingly, though, only 10 percent of the base pairs actually form genes. The other 90 percent—mostly repetitive sequences—apparently have no meaning, and typos in that part of DNA seem to have no effect on health.

In the early seventies, researchers found a way to break the DNA ladder into two separate strands. When they exposed these strands of DNA to certain enzymes (substances that break up, or "cut," nucleic acids), they found that the enzymes could "read" DNA and make a cut at a specific sequence of bases. If just one base was changed—if, for example, adenine was substituted for thymine in the sequence GCAT, making it GCAA—the enzymes didn't cut. By looking at the size and number of DNA fragments created by a particular enzyme, researchers could tell when a typo existed.

Looking for the typo that causes a genetic disease like CF is like trying to find a person somewhere in the United States without knowing his name, the town he lives in, or even what part of the country he's in. But there are clues. For one thing, harmless variations in DNA are passed along with ones that cause diseases. Using enzymes, researchers were able to study the DNA from a family with a genetic disease and try to find typos (which they call RFLPs—restriction fragment-length polymorphisms) that traveled with the harmful gene. The RFLPs, then, are genetic flags or markers that let molecular biologists know that a disease gene is present.

Harvard's Dr. James Gusella was the first to have success: Studying a large family with Huntington's (a fatal disease that damages the central nervous system of its victims, usually when they reach mid-life), he cut family members' DNA and then exposed it to a dozen radioactively tagged genetic probes—each a cloned piece of DNA. One, called G8, was able to track the Huntington's gene in affected family members.

That connection showed up again when Gusella looked at other Huntington's families—he'd found a genetic marker for the disease. And, since G8 had been located on chromosome 4 (except for the X and Y sex chromosomes, pairs are numbered from 1 to 22), Gusella reasoned that the Huntington's gene was there as well. A genetic marker is only the first step, and Gusella

THE NEXT STAGE: GENE THERAPY

Prenatal diagnosis using DNA is still in its early stages, but scientists are already working on the ultimate genetic goal—gene therapy, which could render unnecessary the use of abortion after an inherited disease is diagnosed in the womb. Genetically engineered cures seem the stuff of science fiction; they conjure up Aldous Huxley's *Brave New World*, with its Betas, Gammas, and Alphas, all programmed to fit society's requirements. Though the advent of gene therapy is, as Dr. Richard Mulivor of the Coriell Institute puts it, "way down the line somewhere past Buck Rogers's time," it's likely to happen; a handful of researchers are already laying the groundwork for it. Ultimately, genetic cures would mean "correcting all the sperm and eggs as well, so you could never pass it on," says Mount Sinai's Dr. Robert Desnick. "I think that's a long, long way off. What geneticists are interested in doing today is providing therapy using DNA technology for those suffering from these diseases."

Potential gene therapists must do three things: put the new gene into the target cells, get the transplanted gene to replace the function of the harmful one, and show that the new gene doesn't harm the patient. Researchers are concentrating on diseases of the immune system—for example, adenosine deaminase deficiency (ADA, or "boy-in-the-bubble syndrome," whose victims are born without immune defenses)—because they involve white blood cells, which grow in bone marrow and are easily transplanted.

The most successful efforts have come from a team working under Dr. French Anderson of the National Institutes of Health. Collaborating with doctors at Memorial Sloan-Kettering, Anderson is removing bone marrow from monkeys, inserting the human gene for ADA into their bone-marrow cells, and then putting the cells back into the monkeys. Three weeks later, their blood is checked for traces of the human enzyme. So far, a Sloan-Kettering monkey known as Robert has been the star performer: His blood contained 0.5 percent of the human enzyme.

Gene cures won't be easy, though, and a different one will have to be developed for each inherited condition. Anderson's efforts are "good for a genetic disease of white cells," says Dr. Mulivor. "But the more dispersed the effect of the bad gene, meaning the number of tissues or organs involved, probably the less likelihood of gene therapy succeeding." —A.V.

is now trying to find the harmful gene itself. Other researchers are using similar technology to hunt the genes responsible for a host of inherited diseases.

Gusella's lab is just one of the thousands working on the frontiers of genetics. Molecular biology is one of the fastest-growing research specialties, as labs across the country join the search for harmful genes. The National Institutes of Health's yearly allocation for genetic research has increased by $73-million since 1984; in 1987, the NIH will spend an estimated $283 million.

Already, developments are flooding the scientific journals. Last October, Dr. Louis Kunkel of Boston Children's Hospital announced that he had discovered the gene responsible for Duchenne muscular dystrophy, a fatal disorder that strikes mostly boys, progressively destroying the muscles. Duchenne affects one in every 3,300 male children. A screening test for the general population is still several years away.

At Columbia University Medical School, Dr. Charles Cantor and Dr. Cassandra Smith are making the first attempt to map a human chromosome (No. 21); in mid-February, a group of researchers revealed that they had discovered a genetic marker for Alzheimer's disease; just a week later, researcher Janice Egeland announced that

CRITICS OF CVS FEAR THAT IT WILL ENCOURAGE MANY PARENTS TO SELECT THE SEX OF THEIR CHILD.

she'd found a genetic marker for manic depression among the Old Order Amish; a Swedish-American study of adoptees has linked alcoholism to heredity; and a blood test is in the works that will determine, at birth, a person's risk of developing atherosclerosis, a result of research that found a defective gene was a major cause of the disease for more than 2.3 million Americans.

"We're at the same point now in regard to genetic diseases that we were 50 years ago with infectious diseases," says Dr. Gerard McGarrity, president of the Coriell Institute for Medical Research in Camden, New Jersey. "In those days you had to worry about tuberculosis, polio, diphtheria, and everything else. I'm sure then that the concept of having a society virtually free of infectious diseases was mind-boggling. It's not beyond the realm of possibility to start thinking about a day when there's not going to be any genetic disease."

'TWENTY YEARS AGO, WHEN A WOMAN GAVE BIRTH to an impaired child, people would say, 'Oh, what a heartbreak that she has to live with that situation.' But you'd look for reasons for it," says writer Deborah Batterman, 37, who had amniocentesis during her pregnancy last year. "You'd say, 'I have to examine my life and see why this happened and what I can do to make the best of the situation.' That's the martyrdom indoctrination. We don't have to do that now. We're not raised to be martyrs."

Amniocentesis was the first step away from martyrdom. Introduced in the late sixties, it was the solution to a very specific problem: Women over 40 have an extremely high likelihood (one in 50) of having a child with Down syndrome, a condition caused by an extra copy of chromosome 21 in each cell and characterized by mental retardation, heart defects, and what people once called a "mongoloid" appearance. (A 25-year-old has a one-in-1,000 chance of having a child with Down.) As amnio became less risky, it was given to younger women. Today, it is routinely administered to women 35 and older; in this city, it's even used on women who are 32 or 33.

The test, done when a woman is about sixteen weeks pregnant, involves inserting a needle through the abdomen and drawing fluid from the amniotic sac surrounding the fetus. Fetal cells in the fluid are then grown and the chromosomes examined. Amnio is most often used to predict chromosomal abnormalities, like Down. It can also be used to look at DNA and to predict the sex of the child. Getting the results of an amnio takes about three weeks, which means that couples can't make a decision about abortion until the nineteenth week of pregnancy. A late abortion is so physically and emotionally harrowing that few genetic counselors believe amnio is used for frivolous reasons like sex selection.

Chorionic villus sampling can be done much earlier. The procedure has been used for just four years, and fewer than 30 medical centers around the country perform it. CVS is done at the ninth to eleventh week of pregnancy, and can take less than twenty minutes: While watching the position of the fetus on an ultrasound monitor, a doctor inserts a thin tube through the vagina and into the uterus, and then extracts a tiny piece of the placenta. Like amnio, CVS can be used to look at chromosomes and at DNA; it also predicts fetal sex. Often, the results are back in less than a week. Those who decide to abort can do so in the first trimester.

The test's critics fear that CVS will make it so easy to abort that couples will do so even if their child is diagnosed as having only a mildly disabling condition or a nonlethal disease, and they argue that it will encourage parents to use prenatal diagnosis to choose the sex of their child.

If a blood test that can make the same predictions is developed, testing for gender seems even more likely, especially in cultures that have a strong bias in favor of males. If it seems improbable that couples would choose to abort because they're having a child of the "wrong" sex, consider that in just three months ProCare Industries sold 50,000 Gender Choice kits, which promised parents that they could pick the sex of their child by timing conception correctly. According to the FDA, the test is a fraud.

A third test, for alpha-fetoprotein (AFP), which is used to diagnose spina bifida (incomplete formation of the spinal column), a condition that causes birth defects ranging from severe retardation to mild physical disability, recently became the first prenatal screening technique mandated by law: Last April, California began requiring obstetricians to tell expectant mothers about it. The technique measures the level of a fetal protein in a mother's blood, and only those women with abnormal levels need a follow-up amnio. Unlike amnio or CVS, AFP is a blood test, and it's cheap—$50 (compared with about $1,000 for amnio and $1,200 for CVS). AFP screening is strongly recommended by the American College of Obstetricians and Gynecologists' legal committee, because the OB/GYNs believe that the test will help cut down on the growing number of "wrongful birth" malpractice suits.

IN HER TINY SUBBASEMENT OFFICE AT NEW YORK HOSPItal, genetic counselor Nancy Zellers, an energetic woman in her late thirties, sees fifteen couples each week. Her clients tend to be young, urban, and professional. Typical, she says, is "the banker who may be planning a very small family and she is going to live in an apartment in the city. And probably, because she established her career, she is getting started on children much later and she is going to have a limited family.

"Her concern is 'How is this going to fit into my life-style if it is not a normal child? How am I going to handle this? What's it going to do to my relationship?' Maybe both husband and wife have to work. What's it going to do to the family unit as a whole? I don't think they're unreasonable," says Zellers. "I think that they're anxious."

Susan Katz was. At 36, she is the New York editor-in-chief of Holt, Rinehart and Winston. Her dark hair is lightly streaked with gray, and her office is decorated with bright paintings signed by her seven-year-old daughter from her first marriage. Katz's husband, Howard Radin, owns a computer-software business.

During her first pregnancy, Susan had had a blood test for German measles but nothing else. "This time," she says, "I thought the chances of something being wrong—because I'm 36—are greater. My sense was that it was worth going through, and that people would have thought I was crazy not to have it."

Neither Katz nor Radin has a family history of genetic disease, so Susan had three tests: carrier screening for Tay-Sachs disease, which strikes one in every 2,500 Jews of Eastern and Central European ancestry and causes children to wither and die in their first years of life (both Katz and Radin are Jewish); amniocentesis, to look for Down syndrome; and the AFP test for spina bifida. All three were negative. Though they didn't talk about it much, both Susan and Howard think she would have had an abortion if a defect had turned up. "It's not even the time and the money as much as it breaks your heart," explains Susan. "I think part of the goal of raising a happy child is to create an independent adult, and I think of seriously impaired children as never reaching independent adulthood, and something about that is very sad to me."

"WOULD WE HAVE BEEN BETTER OFF WITHOUT WOODY GUTHRIE, WHO DIED OF HUNTINGTON'S?" ASKS AN ETHICIST.

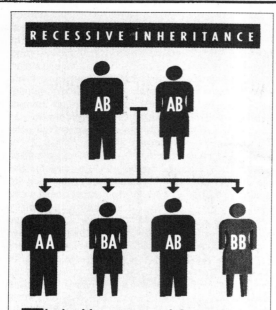

RECESSIVE INHERITANCE

The healthy parents each have one normal gene and one defective, disease-causing one, so they are carriers. A child inherits one gene from each parent. Each child has a one-in-four chance of inheriting two normal genes and being healthy (far left), two chances in four of being a healthy carrier (inheriting one normal gene and one defective one; center), and one in four of inheriting two defective genes and having the disease (far right).

Illustration by Anders Wenngren.

For many women, the decision to use a prenatal test is an automatic one—their friends have done it, their doctor recommends it, and they feel they'd be foolish not to—and among the white middle class, diagnosis in the womb is becoming routine. Deborah Batterman, for example, didn't think much about amnio. "I figured as long as I was going to have a baby at this stage, I might as well go through with it," she says. "Since you have choices today—abortion is available and you don't *have* to have a child—you may as well take advantage of all the options."

Though Batterman and her husband, Lew Dolin, never discussed what they would do if a defect turned up, she assumes that she would have had an abortion. "I guess we've all become a little bit cavalier about these decisions," she says. "It's available, it's easy, we don't have to think about the problems of bringing up a child with Down syndrome or something as terrible. So given those options, we just assume, Let's go for the healthy baby."

"I might be a good person and I love my family and all that, but I am not one of these people who would have been able to give a child I knew to be handicapped the right love and atten-

tion in this world," says Judith Liebman, who chose CVS during her second pregnancy (she had amnio during her first). "It'd be different if I didn't know. I am sure that there are women out there whose opinion is 'It is God's will.' Great! That is terrific. That is not my opinion. I can change it. And I don't think you get struck down for changing it. I think it's a true analysis of what you're ready to deal with."

SINCE GENETIC DISEASE IS RARE, PRENATAL TESTS ARE a reassurance for more than 98 percent of the couples who use them, a fact that genetic counselors emphasize when answering their critics. "I spend every Monday calling people and telling them that their results are normal," says Beth Fine, a genetic counselor at Chicago's Michael Reese Hospital and president of the National Society of Genetic Counselors. "And it's a pleasure.

"People always argue that these tests are a search-and-destroy mission, or encourage abortion," Fine continues, "but we have more pregnancies that result because of the availability of these tests. So there are many more people who can go ahead and get pregnant, as opposed to those who end up testing just to abort. It gives people an opportunity, and that's something you can't take away."

CVS may give that opportunity to even more couples. Melissa Gordon*, a doctor who lost a brother to hemophilia, wasn't sure if she could raise a son with the disease (hemophilia is passed from mothers to sons, and Gordon's child had a fifty-fifty chance of having the disease). When she wanted to become pregnant, none of the DNA probes for hemophilia were applicable to her case (they now are). Gordon consulted genetic counselor Fine and then decided to have CVS; if the child was male, she's decide whether to have an abortion.

"It was a very difficult decision, because we wouldn't know whether we were aborting a normal boy or not," says Gordon. "It would be impossible to go through a pregnancy that long [with amnio] and consider the possibility of an abortion that late. The other reason CVS was important was that if we decided not to abort, my plans to go back to work would have been very different." Happily, Gordon discovered at ten weeks that she was carrying a girl.

For people like Gordon, who have experienced the pain—physical and emotional—of inherited disease, the tests are especially reassuring. "We're allowing people to have healthy families," says genetic counselor Nancy Zellers. "I can't believe that if someone had a baby with Tay-Sachs and watched a child die before their eyes, anyone would expect them to take a risk of going through that again, if they really don't want to. Or can blame that family for having an early abortion rather than going through that again with another affected child. How much tragedy can people take?"

Neither CVS nor amnio can guarantee a healthy baby: Geneticists estimate that even *with* prenatal diagnosis, 2 to 4 percent of all children will be born with serious birth defects. Still, they allay the fears of prospective parents who feel incapable of caring for a child with a problem. "If we're talking about gross birth defects, terrible handicapping conditions, potentially fatal disease, a level of mental retardation that we don't think we can live with, I don't think there's anything wrong with people not wanting to have a child with serious problems," says Zellers. "We have to be willing to accept a certain amount of risk. It doesn't mean we're not sad, and burdened, and angry, and frustrated, and just devastated if a baby's born with something wrong with it. But a certain number of babies are going to be born with something wrong with them. And yet people are still having babies, right?"

EMILY PERL KINGSLEY CONSIDERS HERSELF A FEMInist; a member of NOW, she has belonged to the Abortion Rights Action League. At 47, Kingsley has won four Emmys as a writer for *Sesame Street,* and sometime in the next year, CBS will broadcast a film she wrote. But when she thinks about prenatal testing, Kingsley finds her feminism running up against her feelings about disability rights, her belief in a woman's right to choose to have an abortion colliding with her belief in the value of the lives of handicapped children. Her son, Jason, thirteen, has Down syndrome. "When you lose me," says Kingsley, "is when you say that the world would be better off without people like my son. I can't go along with that. The only drawback to Down is the pain that he will experience. I would do anything to save him that pain—short of killing him."

Though she was offered amniocentesis during her pregnancy, Kingsley passed on it (a decision she says she has never regretted). At 34, she was a year short of the cutoff date, and at the time, amnio was a risky procedure. Kingsley and her husband, Charles, were both on their second marriage, and both were sure they wanted the child. When Jason was born, their obstetrician suggested they might put him in an institution, tell their families he'd died, and try again. Instead, they took him home to Chappaqua and enrolled him in infant-stimulation classes designed to saturate Down children with information and activity.

Infant-stimulation classes were a new concept then, and the Kingsleys had no assurance that they'd be of any real help; today, Jason functions at a high level for a Down-syndrome child. Though he will probably never drive a car, live without some supervision, or go to Yale, Jason can read and write, manage in social situations, and follow complex directions. (He has also appeared on *Sesame Street* and other television shows.) "The idea that you ought to abort a child because he might turn out like my son is *crazy* to me," says Kingsley. "It's crazy."

For Kingsley and other disability-rights activists, their movement is a rerun of the civil-rights and feminist fights of the sixties and seventies, and their goals—acceptance by and access to the mainstream of society—are similar. "No one's entitled to tell us, 'No, you can't. Your kid isn't smart enough. Not smart enough to swim on this beach. Not smart enough to play with the kids in this group.' A lot of the things that they get away with saying to us—if they said, 'You can't get into this class because your skin is black...'"

This time around, though, they find themselves edging toward agreement with people who probably fought on the other side in those earlier battles—right-to-life activists and those on the political right who are opposed not only to prenatal testing but to all abortion. (Many disability-rights supporters don't object to abortion in general, only to abortion to prevent the birth of a handicapped child.)

Still, even within the movement, there's no consensus on prenatal testing. The disabled and their families, after all, are of no particular political persuasion, socioeconomic class, or religion. Instead, they are united by accidents of birth and chance. Though one woman who works with the mentally retarded claims that "the lives of the disabled are debased each time a disabled child is aborted," a large number of activists are unwilling to take a stand on prenatal diagnosis. Their concern, they say, is those who have already been born.

Most, like Carol Levine, an ethics specialist at the Hastings Center in Briarcliff Manor, feel that the severity of the disease must come into play—that a short life followed by a painful death might better be avoided, but that life in a wheelchair is not grounds for abortion. "There's a difference," says Levine, "between being able to test for a lethal disease like Tay-Sachs and a disease like cystic fibrosis that people are living with—not to great old age, but it's not incompatible with life and even productive, happy life. So the choices that people will make are inevitably going to be colored by the differences in the conditions. I think it's pointed up even more in Huntington's disease, where people live to be 50. Would we have been better off without Woody Guthrie, who died of Huntington's? Well, I don't think anybody would say that. But is there a right answer? I don't think so."

Not everyone whose child would be affected by a disease chooses abortion: Of the fifteen pregnancies Fred Gilbert has studied with DNA probes, four have come up positive for CF, and half of those families kept the children. Parents who are told their child has Down syndrome can put him up for adoption; there are waiting lists of families happy to take Down babies. Even some Roman Catholic hospitals are now offering prenatal diagnosis (without abortion), on the theory that parents who know about their child's problems will be able to cope with them better. The numbers they see are dramatically lower than those at secular hospitals: At Creighton University Medical Center in Omaha, about 5 prenatal tests are done a month; New York Hospital, which is three times as large, does approximately 39 a *week*. And for some disabilities the abortion rate is high—91 percent of the women who get a positive diagnosis for Down syndrome terminate their pregnancies (many women who don't feel they could do it simply do not take the test).

"WE'RE ALL GENETIC MESSES," SAYS THE NAtional Institutes of Health's John Fletcher. "There's no such thing as perfection, and there never will be." Yet each year, more people are searching for it. In Manhattan, about 50 percent of the women at risk for chromosomal abnormalities have prenatal diagnosis, while nationwide, the number of such women using the tests has more than doubled since 1977 (to about 20 percent). An Indiana hospital that did ten amnios in 1971 now does fifteen a *week*. In New York State, 40 percent of women 35 and over had some form of prenatal test in 1984, up from about 5 percent in 1977. As the tests become routine, the definition of abortable defects may become wider and, some people fear, parents may reject a potential child for what seem to be frivolous reasons. Ever more genetic hurdles might be set up, and in order to be born a fetus might have to clear them all.

"Anything that has an aspect of mental retardation is already quite unacceptable for most people," says Carol Levine. "I think that the limits of tolerability will be stretched very far. More and more things will be seen as disabilities, and smaller disabilities will be seen less tolerably than they are now. And a society that isn't tolerant of diversity is one that is bereft of imagination and creativity."

Prenatal diagnosis may also change our notion of parental responsibility. "To know that you're bringing forth a child who has to use a wheelchair.... With a few curb cuts, life in a wheelchair is not a tragedy," says Barbara Katz Rothman. "But then you're going to look at this particular kid who's going to say to you, 'I am in this wheelchair because you thought it was a good idea.' And there's going to be an element of truth in that. And that's an incredible responsibility to take on in a society that's not supportive of people in wheelchairs, people with mild retardation, people with any kind of problem. I think there's going to be a certain attitude: 'This isn't an act of God anymore that could happen to anybody; this is your selfish choice, lady.'"

But who can blame parents for wanting their children to be healthy? Though the treatments for diseases like CF and Duchenne muscular dystrophy have improved in recent years, there is still no cure for most of them, and victims may face frequent hospitalization and even early death. And though the lives of the disabled have improved immeasurably in the last

FOR MANY WOMEN WHO WORK, IT IS ALMOST IMPOSSIBLE TO VISUALIZE RAISING A SEVERELY HANDICAPPED CHILD.

decade, as federal regulations ensuring rights of the handicapped have been put into effect, discussions with the parents of handicapped children, and with those who work with them, reveal just how difficult life with a mentally or physically disabling condition can be. They speak of how hard it is to find a class for their child, of bus drivers who won't stop for people in wheelchairs, of New York's lack of simple amenities, like curb cuts, and, most important, of the cruel treatment their children receive, both from other youngsters and from adults.

"The burdens of raising a kid with Down syndrome have practically nothing to do with the child," says Emily Kingsley. "If anything, the child is easy. The burdens are the attitudes and prejudices you meet from people. Having to overcome their queasiness or whatever it is. People are afraid. 'My God, what if that happened to me?' Isn't it neater to keep these people in the closet and not have to think about them, not have to face them, not have to make ramps for them?"

For many women who work, it is almost impossible to visualize raising a severely handicapped child. "I think the real lives of women, especially women who work outside the home, mean that the juggling act implied in motherhood is already very, very tough," says Rayna Rapp, a New School anthropologist who has been studying prenatal diagnosis for the last three years. "And compared with most Western societies, we have fewer social services, fewer maternity benefits, less day care. All of those very large-scale factors go into an assessment at the time of a life crisis. You're not thinking in general about what it means to be the mother of *a* child; you're thinking rather specifically about how your life will change to become the parent of a disabled child, right now and here. And I'm not arguing that if the services were perfect everybody would go ahead and have a disabled baby. But I think some people might have a very different sense of it if the climate around disability and disability services was transformed."

Rapp, 41, is a quick, small woman with shaggy brown hair. She began her anthropological study of prenatal diagnosis after going through the experience herself. Pregnant for the first time at 36, Rapp saw amnio as part of the trade-off she had to make because she had devoted 10 years of her life to her academic and political concerns. As it turned out, she was one of the unlucky 2 percent: The fetus was diagnosed as having Down.

"When Nancy [her genetic counselor] called me twelve days after the tap," Rapp wrote in a *Ms.* article about her amnio, "I began to scream as soon as I recognized her voice.... The image of myself, alone, screaming into a white plastic telephone is indelible." Rapp and her husband, Mike Hooper, decided to have an abortion.

"It was a decision made so that my husband and I could have a certain kind of relationship to a child and to each other and to our adult lives," says Rapp. "It had a lot to do with a sense of responsibility, starting out life as older parents. That was the choice, to have delayed childbearing to do the other things we had done in life. And that meant we had to confront something that was very, very upsetting. But in some senses, I wouldn't have wished away the last ten years of my life in order not to have faced the decision.

"Paradoxically," says Rapp, "there's less choice for people who are better educated to understand prenatal diagnosis. The more you know about this technology, the more likely you are to feel its necessity. But unless the conditions under which Americans view, deal with, and respond to a range of disabling conditions are also put up for discussion about choice—until that larger picture changes—I think it's real hard for many, many people to imagine making a choice other than abortion for something like Down."

In fact, as screening tests for inherited diseases like CF, Huntington's, and the muscular dystrophies become available in the next decade, it's likely that many couples—especially urban, middle-class ones—will consider them a normal part of pregnancy. And as prenatal diagnosis is done earlier in pregnancy, the likelihood that prospective parents will decide to have an abortion for one of those conditions will probably also increase. That attitude is deeply disturbing to opponents of the tests, including Dr. Brian Scully, a Catholic infectious-diseases specialist who works with CF patients at Columbia-Presbyterian Medical Center. Scully, who grew up in Ireland, is opposed to genetic testing—for any condition—that leads to abortion. "The idea that I'm only going to have a child if it's going to be a perfect child, that I'll only accept a baby if it's an acceptable baby—I don't sympathize with that at all," says Scully. "I don't want children to have cystic fibrosis. But to say that if you have cystic fibrosis I'm not going to have you, I think that's wrong."

In 1985, Scully denounced the tests in a letter to *The Lancet*, a widely read British medical journal. In return, he got a note from a London biochemist who, in Scully's words, "rationalized that it was just that they were helping nature. A proportion of babies are lost because of defects naturally—I forget what that proportion is—but he felt that they were just supporting nature, and weeding out the undesirable, imperfect children. And he felt that was fine. I would say, Why not wait until they're born, and you can get everybody."

Like other decisions forced on us by advancing technology, the choice involved in genetic testing can exceed one's moral grasp. And while Scully raises an important argument, the pro-choice position is just as cogent: Can parents who have seen one child suffer with a disease be forced to risk having another? Should people be told they must have a child—even one they don't feel capable of caring for? And, in the case of a disease like Tay-Sachs, does anyone benefit from the child's being born? Parents must make all those decisions for themselves—not lightly, but with awareness of the real moral weight of the final choice. And there's no right answer, a fact that Susan Murphy acutely understands. "I don't think that you should judge people by the decisions they make, whether it's to terminate or not to terminate a pregnancy," she says. "Because you don't know what hell they went through."

The age of the test-tube baby is fast developing. Already science has produced an array of artificial methods for creating life, offering solutions to the growing problem of infertility. In these stories, TIME explores the startling techniques, from laboratory conception to surrogate mothers, and examines the complex legal and ethical issues they raise.

The New Origins of Life

How the science of conception brings hope to childless couples

A group of women sit quietly chatting, their heads bowed over needlepoint and knitting, in the gracious parlor at Bourn Hall. The mansion's carved stone mantelpieces, rich wood paneling and crystal chandeliers give it an air of grandeur, a reflection of the days when it was the seat of the Earl De La Warr. In the well-kept gardens behind the house, Indian women in brilliant saris float on the arms of their husbands. The verdant meadows of Cambridgeshire lie serenely in the distance. To the casual observer, this stately home could be an elegant British country hotel. For the women and their husbands, however, it is a last resort.

Each has come to the Bourn Hall clinic to make a final stand against a cruel and unyielding enemy: infertility. They have come from around the globe to be treated by the world-renowned team of Obstetrician Patrick Steptoe and Reproductive Physiologist Robert Edwards, the men responsible for the birth of the world's first test-tube baby, Louise Brown, in 1978. Many of the patients have spent more than a decade trying to conceive a child, undergoing tests and surgery and taking fertility drugs. Most have waited more than a year just to be admitted to the clinic. Some have mortgaged their homes, sold their cars or borrowed from relatives to scrape together the $3,510 fee for foreign visitors to be treated at Bourn Hall (British citizens pay $2,340). All are brimming over with hope that their prayers will be answered by in-vitro fertilization (IVF), the mating of egg and sperm in a laboratory dish. "They depend on Mr. Steptoe utterly," observes the husband of one patient. "Knowing him is like dying and being a friend of St. Peter's."

In the six years that have passed since the birth of Louise Brown, some 700 test-tube babies have been born as a result of the work done at Bourn Hall and the approximately 200 other IVF clinics that have sprung up around the world. By year's end there will be about 1,000 such infants. Among their number are 56 pairs of test-tube twins, eight sets of triplets and two sets of quads.

New variations on the original technique are multiplying almost as fast as the test-tube population. Already it is possible for Reproductive Endocrinologist Martin Quigley of the Cleveland Clinic to speak of "old-fashioned IVF" (in which a woman's eggs are removed, fertilized with her husband's sperm and then placed in her uterus). "The modern way," he notes, "mixes and matches donors and recipients" *(see chart page 20)*. Thus a woman's egg may be fertilized with a donor's sperm, or a donor's egg may be fertilized with the husband's sperm, or, in yet another scenario, the husband and wife contribute their sperm and egg, but the resulting embryo is carried by a third party who is, in a sense, donating the use of her womb. "The possibilities are limited only by your imagination," observes Clifford Grobstein, professor of biological science and public policy at the University of California, San Diego. Says John Noonan, professor of law at the University of California, Berkeley: "We really are plunging into the Brave New World."

Though the new technologies have raised all sorts of politically explosive ethical questions, the demand for them is rapidly growing. Reason: infertility, which now affects one in six American couples, is on the rise *(see box page 21)*. According to a study by the National Center for Health Statistics, the incidence of infertility among married women aged 20 to 24, normally the most fertile age group, jumped 177% between 1965 and 1982. At the same time, the increasing use of abortion to end unwanted pregnancies and the growing social acceptance of single motherhood have drastically reduced the availability of children for adoption.

At Catholic Charities, for instance, couples must now wait seven years for a child. As a result, more and more couples are turning to IVF. Predicts Clifford Stratton, director of an in-vitro lab in Reno: "In five years, there will be a successful IVF clinic in every U.S. city."

It is a long, hard road that leads a couple to the in-vitro fertilization clinic, and the journey has been known to rock the soundest marriages. "If you want to illustrate your story on infertility, take a picture of a couple and tear it in half," says Cleveland Businessman James Popela, 36, speaking from bitter experience. "It is not just the pain and indignity of the medical tests and treatment," observes Betty Orlandino, who counsels infertile couples in Oak Park, Ill. "Infertility rips at the core of the couple's relationship; it affects sexuality, self-image and self-esteem. It stalls careers, devastates savings and damages associations with friends and family."

For women, the most common reason for infertility is a blockage or abnormality of the fallopian tubes. These thin, flexible structures, which convey the egg from the ovaries to the uterus, are where fertilization normally occurs. If they are blocked or damaged or frozen in place by scar tissue, the egg will be unable to complete its journey. To examine the tubes, a doctor uses X rays or a telescope-like instrument called a laparoscope, which is inserted directly into the pelvic area through a small, abdominal incision. Delicate microsurgery, and, more recently, laser surgery, sometimes can repair the damage successfully. According to Beverly Freeman, executive director of Resolve, a national infertility-counseling organization, microsurgery can restore fertility in 70% of women with minor scarring around their tubes. But for those whose tubes are completely blocked, the chance of success ranges from 20% to zero. These women are the usual candidates for in-vitro fertilization.

Much has been learned about the technique since the pioneering days of Steptoe and Edwards. When the two Englishmen first started out, they assumed that the entire process must be carried out at breakneck speed: harvesting the egg the minute it is ripe and immediately adding the sperm. This was quite a challenge, given that the collaborators spent most of their time 155 miles apart, with Edwards teaching physiology at Cambridge and Steptoe practicing obstetrics in the northwestern mill town of Oldham. Sometimes, when one of Steptoe's patients was about to ovulate, the doctor would have to summon his partner by phone. Edwards would then jump into his car and charge down the old country roads to Oldham. Once there, the two would remove the egg and mate it with sperm without wasting a moment; by the time Lesley Brown became their patient, they could perform the procedure in two minutes flat. They believed that speed was the important factor in the conception of Louise Brown.

As it happens, they were wrong. Says Gynecologist Howard Jones, who, together with his wife, Endocrinologist Georgeanna Seegar Jones, founded the first American in-vitro program at Norfolk in 1978: "It turns out that if you get the sperm to the egg quickly, most often you inhibit the process." According to Jones, the pioneers of IVF made so many wrong assumptions that "the birth of Louise Brown now seems like a fortunate coincidence."

Essential to in-vitro fertilization, of course, is retrieval of the one egg normally produced in the ovaries each month. Today in-vitro clinics help nature along by administering such drugs as Clomid and Pergonal, which can result in the development of more than one egg at a time. By using hormonal stimulants, Howard Jones "harvests" an average of 5.8 eggs per patient; it is possible to obtain as many as 17. "I felt like a pumpkin ready to burst," recalls Loretto Leyland, 33, of Melbourne, who produced eleven eggs at an Australian clinic, one of which became her daughter Zoe.

According to Quigley, the chances for pregnancy are best when the eggs are retrieved during the three- to four-hour period when they are fully mature. At Bourn Hall, women remain on the premises, waiting for that moment to occur. Each morning, Steptoe, now 71 and walking with a cane, arrives on the ward to check their charts. The husband of one patient describes the scene: "Looking at a woman like an astonished owl, he'll say, 'Your estrogen is rising nicely.' The diffidence is his means of defense against desperate women. They think he can get them pregnant just by looking at them."

When blood tests and ultrasound monitoring indicate that the ova are ripe, the eggs are extracted in a delicate operation performed under general anesthesia. The surgeons first insert a laparoscope,

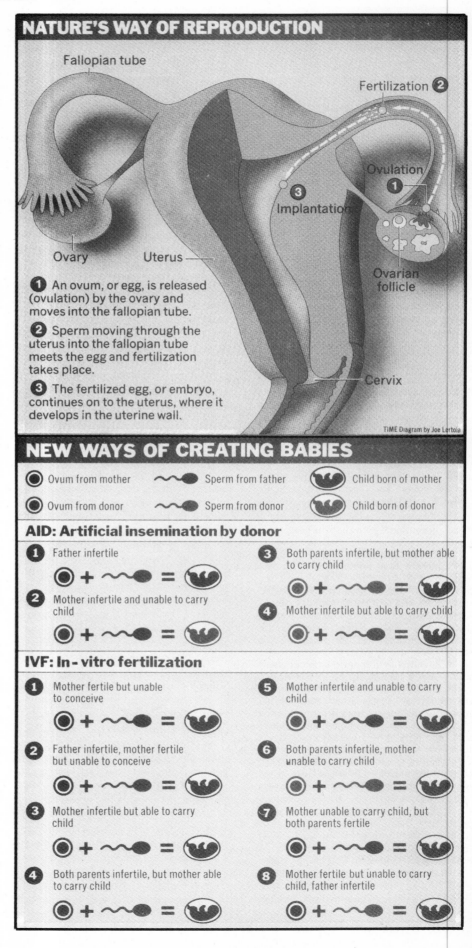

NATURE'S WAY OF REPRODUCTION

Fallopian tube

Fertilization ❷

Ovulation ❶

❸

Implantation

Ovarian follicle

Ovary   Uterus

Cervix

❶ An ovum, or egg, is released (ovulation) by the ovary and moves into the fallopian tube.

❷ Sperm moving through the uterus into the fallopian tube meets the egg and fertilization takes place.

❸ The fertilized egg, or embryo, continues on to the uterus, where it develops in the uterine wall.

TiME Diagram by Joe Lertola

NEW WAYS OF CREATING BABIES

◉ Ovum from mother   〜● Sperm from father   👶 Child born of mother

◉ Ovum from donor   〜● Sperm from donor   👶 Child born of donor

AID: Artificial insemination by donor

❶ Father infertile
◉ + 〜● = 👶

❷ Mother infertile and unable to carry child
◉ + 〜● = 👶

❸ Both parents infertile, but mother able to carry child
◉ + 〜● = 👶

❹ Mother infertile but able to carry child
◉ + 〜● = 👶

IVF: In-vitro fertilization

❶ Mother fertile but unable to conceive
◉ + 〜● = 👶

❷ Father infertile, mother fertile but unable to conceive
◉ + 〜● = 👶

❸ Mother infertile but able to carry child
◉ + 〜● = 👶

❹ Both parents infertile, but mother able to carry child
◉ + 〜● = 👶

❺ Mother infertile and unable to carry child
◉ + 〜● = 👶

❻ Both parents infertile, mother unable to carry child
◉ + 〜● = 👶

❼ Mother unable to carry child, but both parents fertile
◉ + 〜● = 👶

❽ Mother fertile but unable to carry child, father infertile
◉ + 〜● = 👶

which is about ⅓ in. in diameter, so that they can see the target: the small, bluish pocket, or follicle, inside the ovary, where each egg is produced. Then, a long, hollow needle is inserted through a second incision, and the eggs and the surrounding fluid are gently suctioned up. Some clinics are beginning to use ultrasound imaging instead of a laparoscope to guide the needle into the follicles. This procedure can be done in a doctor's office under local anesthesia; it is less expensive than laparoscopy but may be less reliable.

Once extracted, the follicular fluid is rushed to an adjoining laboratory and examined under a microscope to confirm that it contains an egg (the ovum measures only four-thousandths of an inch across). The ova are carefully washed, placed in petri dishes containing a solution of nutrients and then deposited in an incubator for four to eight hours. The husband, meanwhile, has produced a sperm sample. It is hardly a romantic moment, recalls Cleveland Businessman Popela, who made four trips to Cambridgeshire with his wife, each time without success. "You have to take the jar and walk past a group of people as you go into the designated room, where there's an old brass bed and a couple of *Playboy* magazines. They all know what you're doing and they're watching the clock, because there are several people behind you waiting their turn."

The sperm is prepared in a solution and then added to the dishes where the eggs are waiting. The transcendent moment of union, when a new life begins, occurs some time during the next 24 hours,

in the twilight of an incubator set at body heat. If all goes well, several of the eggs will be fertilized and start to divide. When the embryo is at least two to eight cells in size, it is placed in the woman's uterus. During this procedure, which requires no anesthetic, Steptoe likes to have the husband present talking to his wife. "The skill of the person doing the replacement is very important," he says. "The womb doesn't like things being put into it. It contracts and tries to push things out. We try to do it with as little disturbance as possible."

The tension of the next two weeks, as the couple awaits the results of pregnancy tests, is agonizing. "Women have been known to break out in hives," reports Linda Bailey, nurse-coordinator at the IVF program at North Carolina Memorial Hospital in Chapel Hill. Success rates vary from clinic to clinic; some centers open and close without a single success. But even the best clinics offer little more than a 20% chance of pregnancy. Since tiny factors like water quality seem to affect results, both physicians and patients tend to become almost superstitious about what else might sway the odds. Said one doctor: "If someone told us that painting the ceiling pink would make a difference, we would do it."

In recent years, IVF practitioners have discovered a more reliable way of improving results: transferring more than one embryo at a time. At the Jones' clinic, which has one of the world's highest success rates, there is a 20%

chance of pregnancy if one embryo is inserted, a 28% chance if two are used and a 38% chance with three. However, transferring more than one embryo also increases the likelihood of multiple births.

For couples who have struggled for years to have a child, the phrase "you are pregnant" is magical. "We thought we would never hear those words," sighs Risa Green, 35, of Framingham, Mass., now the mother of a month-old boy. But even if the news is good, the tension continues. One-third of IVF pregnancies spontaneously miscarry in the first three months, a perplexing problem that is currently under investigation. Says one veteran of Steptoe's program: "Every week you call for test results to see if the embryo is still there. Then you wait to see if your period comes." The return of menstruation is like a death in the family; often it is mourned by the entire clinic.

Many couples have a strong compulsion to try again immediately after in vitro fails. Popela of Cleveland compares it to a gambling addiction: "Each time you get more desperate, each time you say, 'Just one more time.'" In fact, the odds do improve with each successive try, as doctors learn more about the individual patient. But the stakes are high: in the U.S., each attempt costs between $3,000 and $5,000, not including travel costs and time away from work. Lynn Kellert, 31, and her husband Mitchell, 34, of New York City, who tried seven times at Norfolk before finally achieving pregnancy, figure the total cost was $80,000. Thus far, few insurance companies have been willing to

The Saddest Epidemic

Richard and Diana Barger of Virginia could be a textbook case of an infertile couple. Diana's fallopian tubes and left ovary are blocked with scar tissue, ironically the result of an intrauterine device (I.U.D.) she used for three years. Even if an egg did manage to become fertilized, the embryo might be rejected by her uterus, which has been deformed since birth. Richard has his own difficulties: his sperm count is 6.7 million per milliliter, considerably below the number ordinarily required for fertilization under normal conditions. Says Diana: "I never thought getting pregnant would be so difficult."

The Bargers are victims of what Reproductive Endocrinologist Martin Quigley of the Cleveland Clinic calls "an epidemic" of infertility in the U.S. In the past 20 years, the incidence of barrenness has nearly tripled, so that today one in six American couples is designated as infertile, the scientific term for those who have tried to conceive for a year or more without success. More than a million of these desperate couples seek the help of doctors and clinics every year. Women no longer carry the sole blame for childless marriages. Research has found that male deficiencies are the cause 40% of the time, and problems with both members of the marriage account for 20% of reported cases of infertility.

Doctors place much of the blame for the epidemic on liberalized sexual attitudes, which in women have led to an increasing occurrence of genital infections known collectively as pelvic inflammatory disease. Such infections scar the delicate tissue of the fallopian tubes, ovaries and uterus. Half of these cases result from chlamydia, a common venereal disease, and 25% stem from gonorrhea.

Other attitudes are also at fault: by postponing childbirth until their mid- or even late 30s, women risk a barren future. A Yale University study of 40 childless women found that after 35 years of age, the time it takes to conceive lengthens from an average of six months to more than two years.

Other surveys have found that such athletic women as distance runners, dancers and joggers can suffer temporary infertility. The reason is that their body fat sometimes becomes too low for the production of the critical hormone estrogen. Stress can also suppress ovulation; women executives often miss two or three consecutive menstrual periods.

Infertility is easier to trace in men, but often much harder to treat. The commonest problems are low sperm counts and blocked sperm ducts. Among all men, 15% have varicose veins on the left testicle, which can reduce sperm production. Certain drugs and chemicals such as insecticides can also lower sperm counts. A man's fecundity also decreases with age, although not with the dramatic finality of female menopause. Happily, the source of infertility in couples can be diagnosed 95% of the time, and half of all these cases can be treated.

foot the bill, arguing that IVF is still experimental. But, observes Grobstein of UCSD, "it's going to be increasingly difficult for them to maintain that position."

Second and third attempts will become easier and less costly with the wider use of cryopreservation, a process in which unused embryos are frozen in liquid nitrogen. The embryos can be thawed and then transferred to the woman's uterus, eliminating the need to repeat egg retrieval and fertilization. Some 30% to 50% of embryos do not survive the deep freeze. Those that do may actually have a

better chance of successful implantation than do newly fertilized embryos. This is because the recipient has not been given hormones to stimulate ovulation, a treatment that may actually interfere with implantation.

Opinion is sharply divided as to how age affects the results of IVF. Although most clinics once rejected women over age 35, many now accept them. While one faction maintains that older women have a greater tendency to miscarry, Quigley, for one, insists that "age should not affect the success rate." Curiously, the Joneses in Norfolk have achieved their best results with women age 35 to 40. This year one of

their patients, Barbara Brooks of Springfield, Va., had a test-tube son at age 41; she can hardly wait to try again.

Doctors are also beginning to use IVF as a solution to male infertility. Ordinarily, about 30 million sperm must be produced to give one a chance of penetrating and fertilizing the egg. In the laboratory, the chances for fertilization are good with only 50,000 sperm. "In vitro may be one of the most effective ways of treating men with a low sperm count or low sperm motility, problems that affect as many as 10 million American men," says Andrologist Wylie Hembree of Columbia-Presbyterian Medical Center in New York City.

A Surrogate's Story

Valerie is a New Jersey mother of two boys, age two and three, whom she describes as "little monsters full of mischief." Her husband works as a truck driver, and money is tight. The family of four is living with her mother while they save for an apartment of their own. One day last March, Valerie, 23, who prefers to remain anonymous, saw the following advertisement in a local New Jersey paper: *Surrogate mother wanted. Couple unable to have child willing to pay $10,000 fee and expenses to woman to carry husband's child. Conception by artificial insemination. All replies strictly confidential.*

The advertisement made Valerie stop and think. "I had very easy pregnancies," she says, "and I didn't think it would be a problem for me to carry another child. I figured maybe I could help someone." And then there was the lure of the $10,000 fee. "The money could help pay for my children's education," she says, "or just generally to make their lives better."

The next day Valerie went for an interview at the Infertility Center of New York, a profit-making agency owned by Michigan Attorney Noel Keane, a pioneer in the controversial business of matching surrogate mothers with infertile parents. She was asked to fill out a five-page application, detailing her medical history and reasons for applying. Most applicants are "genuine, sincere, family-oriented women," says agency Administrator Donna Spiselman. The motives they list range from "I enjoy being pregnant" and an urge to "share maternal joy" to a need to alleviate guilt about a past abortion by bearing someone else's child. Valerie's application and her color photograph were added to 300 others kept in scrapbooks for prospective parents to peruse. Valerie was amazed when only one week later her application was selected, and she asked to return to the agency to meet the couple.

Like most people who find their way to surrogate agencies, "Aaron" and "Mandy" (not their real names) had undergone years of treatment for infertility. Aaron, 36, a Yale-educated lawyer, and his advertising-executive wife, 30, had planned to have children soon after marrying in 1980. They bought a two-bedroom town house in Hoboken, N.J., in a neighborhood that Aaron describes as being "full of babies." But after three years of tests, it became painfully clear that there was little hope of having the child they longed for. They considered adoption, but were discouraged by the long waiting lists at American agencies and the expense and complexity of foreign adoptions. Then, to Aaron's surprise, Mandy suggested that they try a surrogate.

Their first choice from the Manhattan agency failed her mandatory psychological test, which found her to be too emotionally unstable. Valerie, who was Aaron and Mandy's second choice, passed without a hitch. A vivacious woman who is an avid reader, she more than met the couple's demands for a surrogate who was "reasonably pretty," did not smoke or drink heavily and had no family history of genetic disease. Says Aaron: "We were particularly pleased that she asked us questions to find out whether we really want this child."

At first, Valerie's husband had some reservations about the arrangement, but, she says, he ultimately supported it "100%." Valerie is not concerned about what her neighbors might think because the family is planning to move after the birth. Nor does she believe that her children will be troubled by the arrangement because, she says, they are too young to understand. And although her parents are being deprived of another grandchild, they have raised no objections.

For their part, Aaron and Mandy have agreed to pay Valerie $10,000 to be kept in an escrow account until the child is in their legal custody. In addition, they have paid an agency fee of $7,500 and are responsible for up to $4,000 in doctors' fees, lab tests, legal costs, maternity clothes and other expenses. In April, Valerie became pregnant after just one insemination with Aaron's sperm. Mandy says she was speechless with joy when she heard the news.

Relationships between surrogate mothers and their employers vary widely. At the National Center for Surrogate Parenting, an agency in Chevy Chase, Md., the two parties never meet. At the opposite extreme is the case of Marilyn Johnston, 31, of Detroit. Johnston and the couple who hired her became so close during her pregnancy that they named their daughter after her. She continues to make occasional visits to see the child she bore and says, "I feel like a loving aunt to her."

Not all surrogate arrangements work so well. Some women have refused to give up the child they carried for nine months. As a lawyer, Aaron is aware that the contract he signed with Valerie would not hold up in court, should she decide to back out of it. "But I'm a romantic," he says. "I have always felt that the real binding force was not paper but human commitment." Valerie, whose pregnancy is just beginning to show, says she is "conditioning" herself not to become too attached to the baby. "It is not my husband's child," she says, "so I don't have the feeling behind it as if it were ours." She does not plan to see the infant after it is born, but, she admits, "I might like to see a picture once in a while." —*By Claudia Wallis.*
Reported by Ruth Mehrtens Galvin/New York

While most clinics originally restricted IVF to couples who produced normal sperm and eggs, this too is changing. Today, when the husband cannot supply adequate sperm, most clinics are willing to use sperm from a donor, usually obtained from one of the nation's more than 20 sperm banks. An even more radical departure is the use of donor eggs, pioneered two years ago by Dr. Alan Trounson and Dr. Carl Wood of Melbourne's Monash University. The method can be used to bring about pregnancy in women who lack functioning ovaries. It is also being sought by women who are known carriers of genetic diseases. The donated eggs may come from a woman in the Monash IVF program who has produced more ova than she can use. Alternately, they could come from a relative or acquaintance of the recipient, providing that she is willing to go through the elaborate egg-retrieval process.

At Harbor Hospital in Torrance, Calif., which is affiliated with the UCLA School of Medicine, a team headed by Obstetrician John Buster has devised a variant method of egg donation. Instead of fertilizing the ova in a dish, doctors simply inseminate the donor with the husband's sperm. About five days later, the fertilized egg is washed out of the donor's uterus in a painless procedure called lavage. It is then placed in the recipient's womb. The process, which has to date produced two children, "has an advantage over IVF," says Buster, "because it is nonsurgical and can be easily repeated until it works." But the technique also has its perils. If lavage fails to flush out the embryo, the donor faces an unwanted pregnancy.

The most controversial of the new methods of reproduction does not depend on advanced fertilization techniques. A growing number of couples are hiring surrogate mothers (see box) to bear their children. Surrogates are being used in cases where the husband is fertile, but his wife is unable to sustain pregnancy, perhaps because of illness or because she has had a hysterectomy. Usually, the hired woman is simply artificially inseminated with the husband's sperm. However, if the wife is capable of producing a normal egg but not capable of carrying the child, the surrogate can be implanted with an embryo conceived by the couple. This technique has been attempted several times, so far without success.

The medical profession in general is apprehensive about the use of paid surrogates. "It is difficult to differentiate between payment for a child and payment for carrying the child," observes Dr. Ervin Nichols, director of practice activity for the American College of Obstetrics and Gynecology. The college has issued strict guidelines to doctors, urging them to screen carefully would-be surrogates and the couples who hire them for their medical and psychological fitness. "I would hate to say there is no place for surrogate motherhood," says Nichols, "but it should be kept to an absolute minimum."

In contrast, in-vitro fertilization has become a standard part of medical practice. The risks to the mother, even after repeated attempts at egg retrieval, are "minimal," points out Nichols. Nor has the much feared risk of birth defects materialized. Even frozen-embryo babies seem to suffer no increased risk of abnormalities. However, as Steptoe points out, "we need more research before we know for sure."

The need for research is almost an obsession among IVF doctors. They are eager to understand why so many of their patients miscarry; they long to discover ways of examining eggs to determine which ones are most likely to be fertilized, and they want to develop methods of testing an embryo to be certain that it is normal and viable. "Right now, all we know how to do is look at them under the microscope," says a frustrated Gary Hodgen, scientific director at the Norfolk clinic.

Many scientists see research with embryos as a way of finding answers to many problems in medicine. For instance, by learning more about the reproductive process, biologists may uncover better methods of contraception. Cancer research may also benefit, because tumor cells have many characteristics in common with embryonic tissues. Some doctors believe that these tissues, with their tremendous capacity for growth and differentiation, may ultimately prove useful in understanding and treating diseases such as childhood diabetes. Also in the future lies the possibility of identifying and then correcting genetic defects in embryos. Gene therapy, Hodgen says enthusiastically, "is the biggest idea since Pasteur learned to immunize an entire generation against disease." It is, however, at least a decade away.

American scientists have no trouble dreaming up these and other possibilities, but, for the moment, dreaming is all they can do. Because of the political sensitivity of experiments with human embryos, federal grant money, which fuels 85% of biomedical research in the U.S., has been denied to scientists in this field. So controversial is the issue that four successive Secretaries of Health and Human Services (formerly Health, Education and Welfare) have refused to deal with it. This summer, Norfolk's Hodgen resigned as chief of pregnancy research at the National Institutes of Health. He explained his frustration at a congressional hearing: "No mentor of young physicians and scientists beginning their academic careers in reproductive medicine can deny the central importance of IVF–embryo transfer research." In Hodgen's view the curb on research funds is also a breach of government responsibility toward "generations of unborn" and toward infertile couples who still desperately want help.

In an obstetrics waiting room at Norfolk's in-vitro clinic, a woman sits crying. Thirty-year-old Michel Jones and her husband Richard, 33, a welder at the Norfolk Navy yard, have been through the program four times, without success. Now their insurance company is refusing to pay for another attempt, and says Richard indignantly, "they even want their money back for the first three times." On a bulletin board in the room is a sign giving the schedule for blood tests, ultrasound and other medical exams. Beside it hangs a small picture of a soaring bird and the message: *You never fail until you stop trying.* Michel Jones is not about to quit. Says she: "You have a dream to come here and get pregnant. It is the chance of a lifetime. I won't give up."

—By Claudia Wallis. Reported by Mary Cronin/London, Patricia Delaney/Washington and Ruth Mehrtens Galvin/Norfolk

All About Twins

They share traits and emotional bonds, communicate with each other in mysterious ways and provide tantalizing insights into human nature

Two years ago engineer Donald Keith was walking down a hall at his office in Rockville, Md., when he suddenly experienced a series of sharp pains like jolts of electricity in his groin. An acquaintance phoned to ask Keith if he knew anything about the "shared pain" that twins sometimes experience. Keith had never thought much about it. But it made him wonder if his identical twin, Louis, a Chicago obstetrician, might have had a flare-up of his bad back recently. He phoned Louis, who said no, but he *had* just injured a groin muscle. "The hair on the back of my neck stood straight up," Keith recalls.

Such uncanny affinities are just part of the mystery and fascination of twins. Twins improve on the miracle of birth by doing it twice, a trick shot that never fails to dazzle us. Their matching looks inspire a certain voyeuristic curiosity, but at the same time their special bond arouses envy. Every lonely adolescent fantasizes a lost twin somewhere, the perfect companion, confidant and soul mate—another self. In their later years twins seem to embody the idyllic, unrivalrous sibling relationship so many of us, in vain, longed for. Phil and Frank Interlandi, 63-year-old identical twins in Laguna Beach, Calif., attended the Chicago Academy of Fine Arts together, landed advertising jobs together and quit about the same time. Well-known cartoonists, they see each other every day at their favorite bar and grill. "It's hard to say how we're different," says Phil, whose work appears in Playboy. Frank, who drew for the Los Angeles Times before he retired, says, "Phil and I have more intimacy in thinking than I have with my wife. He can understand me better than she can."

Twins are equally intriguing for what they may reveal about the rest of us. As scientists study them in an attempt to sort out which qualities of body and mind are shaped by our genes, and which by our upbringing, the answers are forcing revisions in many cherished notions about how personality develops and how much control we have over our own lives.

But twins themselves still pose multiple puzzles. Although scientists have pretty much figured out how the rest of us are made, they remain a bit baffled about the origins of some twins, including how eggs and sperm fuse to create them. The questions scarcely end with twins' entrance into the world. Psychologists are trying to fathom how being half of a biological pair forever stamps a twin's sense of identity. And they are trying to understand how upbringing can influence twins' relationships (page 28). Researchers needn't worry about running out of subjects: there are 2.4 million sets of twins in the United States, and 33,000 more each year.

Researchers are now seeking to explain the strikingly similar choices many identical twins make even when they live far apart. Take the renowned case of Jim Springer and Jim Lewis, identical twins separated just four weeks after they were born in Ohio 48 years ago. Reunited 39 years later in a study on twins at the University of Minnesota, they discovered that they had married and divorced women named Linda, married second wives named Betty and named their first sons James Allan and James Alan, respectively. That's not all: they both drove the same model of blue Chevrolet and they both enjoyed woodworking (and had built identical benches around trees in their backyards). They often vacationed on the same small beach in St. Petersburg, Fla., and owned dogs named Toy.

Researchers are wondering whether these "coincidences" are something more—a clue to the forces that shape beliefs, personality and even the path one chooses to follow in life. For 100 years twins have been used to study how genes make people what they are. Identical twins reared apart are the ideal keys to unlock such mysteries. Because they share precisely the same genes but live in different surroundings under different influences, they are helping science sort out the relative influence of heredity and environment on such traits as shyness and thrill seeking. Or as Luigi Gedda, director of the Gregor Mendel Institute for twin studies in Rome, Italy, says, "Twins are not just a curiosity. They are a doctrine." In short, twins promise to reveal much about what makes us what we are.

'Twince Charming': Far from being self-conscious, twins often exult in their twinness. They have their own magazine—called Twins, of course—and hold an annual summer festival in, inevitably, the Ohio town of Twinsburg (which changed its name from Millsville at the request of twin brothers who donated land and $20 to the town in

The Birds and the Bees of Double Births

Identical Twins

1 Accounting for about 1 in 250 births, these are created when a single egg is fertilized by one sperm.

2 The egg splits into halves. Each develops into a fetus with the same genetic composition.

Fraternal Twins

1 Twice as common as identicals, fraternals arise when two eggs are released at once.

2 If both are fertilized by separate sperm, two fetuses form. Genetically they are just ordinary siblings.

There are more kinds of twins than the well-known fraternals and identicals, and they differ in their degree of genetic similarity.

Half-Identical Twins

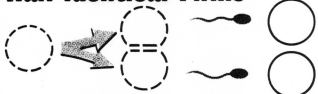 

1 A rare type, half-identicals form when a precursor to an egg splits evenly and is fertilized by two sperm.

2 The fetuses have about half of their genes in common—those from the mother.

Twins of Different Fathers

1 In extremely rare cases, an egg is released even though the previous month's egg was fertilized.

2 If the second egg is fertilized by another man, the fetuses are no more alike genetically than half siblings.

CHRISTOPH BLUMRICH—NEWSWEEK

1817). At the most recent Twinsburg gathering last August, more than 1,300 pairs from around the world sported punny T shirts ("Twince Charming") and competed in contests for most and least alike, oldest and farthest traveled.

The closeness of the twins was obvious. One effusive pair, Karah and Sarah Isom, 11, of Akron, Ohio, have twin Cabbage Patch Kids, twin German shepherds and the same best friend. When Sarah had to take a summer math class, Karah begged to be allowed to join her—and they missed the same eight out of 200 questions on the final exam. When Karah broke her leg, Sarah asked for—and got—an Ace bandage and crutches from the hospital too. One would as soon leave the other as cut off her hand. "We want to get twin houses, twin cars," says Sarah. "We want to have our yards the same and live next door."

John and Buell Fuller, 79-year-old Eagle Scout leaders in Boston, are equally devoted to each other. They have always lived together, except for two years Buell served in the Air Force. Neither ever married. Every morning they race each other out of bed, don matching watches and outfits and plot new ways to fool a world of singletons. "Our hobby is confusing people," says John. While working for an airline, they took breaks at different times in the same cafeteria seat, leading people to think that a fellow named Mr. Fuller never worked.

Private language: Sometimes such intimacy can be destructive. Psychologists are trying to find out where closeness ends and pathology begins. From the time June and Jennifer Gibbons could speak, they seldom did—except for rare simple sentences to adults and some words to other children. Eventually, they spoke only with each other. Around their hometown of Haverfordwest, Wales, they became known as the Silent Twins. They developed a private language that no one else could understand. Jennifer was so jealous of June—whom she thought prettier and more loved by their parents—that she forced her into a childhood vow of silence. One acquaintance said Jennifer seemed to have "possessed" her sister: with eye signals, she told June when to talk, how to move, what to do. Each monitored the other so they could act in perfect unison. They even fell off horses at the same time.

Once Jennifer intoned to her sister, "You are Jennifer. You are me." An agonized June screamed back, "I am June! I am June!" But in some ways the twins flowered—writing diaries of 1 million words and, in June's case, a novel published by a vanity press. Still, they remained deathly afraid of being seen as individuals. After committing petty theft and arson, they were sentenced in 1982 "without limit of time" to Broadmoor, a British institution for the criminal insane.

Such closeness can produce behavior that outsiders find mystifying, even eerie.

Twins are renowned for creating "secret languages," supposedly unique tongues that only they understand. But more careful study has removed the mystery from several of these cases. The Silent Twins of Britain, for example, turned out to be using rapid-fire English with odd stresses. Jeffrey and Kristopher Cardwell of Midland, Texas, 10, used made-up words, much like the ones single children would. But while singletons discard the nonsense sounds after being corrected, twins keep them because each child reinforces the other. "They talked about 'woggies,' which seem to have been airplanes," says the Cardwell twins' mother, Linda. "'Gooden-goodens' were any small animal, especially raccoons. They kept this stuff up for years." By some estimates, 40 percent of twins develop a private language, although they usually drop it by the age of five or six.

ESP events: Other mysteries of twin behavior have been harder to unravel—particularly ESP experiences. Psychologists have heard dozens of such stories over the years, mostly from identical twins. The ESP generally revolves around major events: injuries, births, deaths. Nancy Segal, codirector of an ongoing eight-year study of twins at the University of Minnesota, says she doesn't "doubt the reality of [ESP] events," since the stories are too numerous to be total fabrications. But she is reluctant to label them paranormal. She notes that researchers "never hear of the cases where one twin is sure the other is lying dead in the gutter, and he isn't." And since twins think about each other more than other siblings, experiences labeled ESP may be just coincidence.

Donald Keith, who studies twins with his twin, Louis, thinks there's more to it. He says that, by concentrating, he can make Louis phone him. At one point Donald was successfully sending the "phone home" message several times a week. "I think of there being electrical windows in the mind," he says. "Because twins are closer and more open to each other's thoughts, they are more likely to leave the windows open."

At times the legendary closeness of twins can be a liability. Twins used to sensing each other's moods often have to struggle "to say what they think and feel," notes Barbara Schave, coauthor of the 1983 book "Identity and Intimacy in Twins," "because they are accustomed to their nonverbal twin relationship." But another aspect of being twins can pay off. Their close bonds are attractive to other children, so they readily form other friendships. One reason, suggests Kay Cassill, president of the Twins Foundation in Providence, R.I., is that they are able to empathize with their friends quite early in life—a carry-over from their closeness to their twin. There are few better empathizers than "Dear Abby" (Abigail Van Buren) and "Ann Landers"—identical twins.

Different dads: Most people know of only two types of twins—identical and fraternal. But there seem to be more kinds of twins than scientists thought. Besides identical, who have exactly the same genes, and fraternal, who are born at the same time but are as genetically different as other siblings, there may be "half-identical" twins (chart). Half-identicals arise when a precursor to a true ovum divides into identical halves and is fertilized by two sperm. Thus they are more alike than fraternals, who come from two different eggs, but less alike than identicals, who come from one sperm. Twins can also have different fathers. If an ovum is released after the previous month's has begun developing into an embryo, it can be fertilized by the next act of sexual intercourse. They are, in essence, step-siblings, for they have the same mother but different fathers.

Identical twins, who make up a third of all twin births, seem to arise randomly. The rate among all races is about four sets per 1,000 births. But the long shot may come in more than once, as happened to Susan and Don Taft of Corpus Christi, Texas. Four of their five children are identicals: Josh and Christian, 15, and Edward and Andrew, nine.

Chance plays a lesser role in fraternal twins. For one thing, they run in families. A study of 4,000 Mormon mothers found that they had a 1-in-58 chance of bearing twins if they themselves were twins. Someone with no twins on her family tree has a slightly greater than 1 percent chance of conceiving a double blessing. A woman also has better odds of having fraternal twins if she has already borne children (especially other twins), used fertility drugs or is between 35 and 40. And she is likelier to have twins if she conceives soon after going off the Pill, since that triggers a burst of gonadotropins, the hormones that promote ovulation. Gonadotropins are used in fertility drugs: if they work too well, more than a single egg will be released. One study at Yale University showed that women who had used the Pill and who conceived within two months of quitting had a two-times-normal chance of having twins.

One of the more provocative recent discoveries about twins is that many more are conceived than are born—a phenomenon called the vanishing-twin syndrome. Sonogram studies indicate that up to 70 percent of pregnancies that start with two tiny fetuses end up with only one by about the fifth month. Apparently, many singletons really began life as a twin, which raises the possibility that the vague yearning some single children feel for a supposed long-lost twin may have a biological basis. Dr. Louis Keith, an obstetrician at Northwestern Medical School and the Center for Multiple Birth in Chicago, now believes the actual incidence of multiple conception may be as high as 20 percent, not the 1 in 90 that birth figures show. "We have taken the view that humans are animals that give birth to a single child except in rare circumstances," he says. "That may not be so."

Odd cannibalism: A fetus can vanish in two ways. Its gestational sac can be resorbed by the mother, for reasons no one understands. This has no effect on the surviving twin. But in an odd sort of cannibalism, a fetus can also absorb or envelop its sibling, keeping the vanished twin inside itself long after birth. Such was the case with Nick Hill, a service-station attendant in Idaho. All his life Hill had complained of near-paralyzing headaches. Finally, when he was 21, doctors performed exploratory brain surgery. They found a mass of embryonic bone, skin and hair—thought to be the remains of an unborn twin. Other people have had cysts that contained vestigial teeth, hair or limbs—possibly relics of the twin who didn't make it.

At Rome's Mendel Institute, Gedda and his colleagues see striking peculiarities in how twins develop. "We have found that [identical] twins must be treated simultaneously," says Gedda, who founded the institute in 1953. "If one has a cavity, the other has it in the same tooth, or soon will. Female twins often experience their first menstruation simultaneously, sometimes in the same night." Twins also cut their first teeth at the same time and go bald together. Gedda believes our genes come with built-in clocks that activate such developmental milestones. The idea of a gene clock may explain why the Jim twins—Springer and Lewis—have the same kind of headaches at the same times in their

Twins share physical traits and emotional bonds. They also are able to communicate with each other in ways that present the scientific community with tantalizing insights into human nature.

Do You Really Want Twins? Try Yams

No culture in the world has more experience with twins than the 18 million-member Yoruba tribe of western Nigeria. Twins account for about 3 percent of the tribe's births, compared with 1.7 percent for other blacks, 1 percent for whites and .5 percent for Asians. At one time many Africans viewed double births as bad omens, and the newborns often were left to die. More recently the Yorubas decided that twins are harbingers of good fortune. Now research suggests the double blessings signify something else entirely: twins, it seems, may be caused by yams.

A typical Yoruba consumes huge numbers of local yams, a staple of the tribal diet. The vegetable contains high amounts of a substance similar to the female hormone estrogen, which in turn may stimulate the production of other hormones called gonadotropins. One gonadotropin, called follicle-stimulating hormone (FSH), may trigger the release of more than one ovum from the ovaries, thus paving the way for the conception of fraternal twins. (Yorubas who have abandoned yams in favor of more novel fare have fewer twins.) Although the hormone connection remains murky, there is circumstantial evidence that these chemicals do influence twinning. Yoruba women have higher levels of FSH than Americans, who have higher levels than Japanese women—a pattern that matches the rates of twin births in these groups. Americans who are eager to hear the pitter-patter of four little feet in the near future might thus do well to choose their Thanksgiving vegetables with care.

lives and even at the same time of day.

But what about Bridget Harrison and Dorothy Lowe, identical sisters separated shortly after birth? Reunited at the University of Minnesota twins project, each arrived—not having seen each other for 34 years—wearing rings on seven fingers, a bracelet on each wrist and a watch. Bridget's children were Richard Andrew and Karen Louise; Dorothy's, Andrew Richard and Catherine Louise.

No scientist would seriously argue that somewhere in our chromosomes lies a gene for ring wearing or child naming. But Bridget and Dorothy's similarities nevertheless suggest that behavior patterns seemingly remote from the influence of heredity may in fact be genetically based. The women, for example, might have inherited genes that gave them attractive fingers, eliciting compliments from friends. Then, independently, they might have come to wear numerous rings to show off their hands. And their children's names? The chance of any two people choosing the same names is high enough that statisticans just label it coincidence.

For the Parents, a Delicate Balance

Raising one child at a time is hard enough; parents of twins have an even more difficult job. They must help their youngsters to grow together and apart, to stay friends and still develop separate identities. "It's strenuous physically and mentally," says 39-year-old Elaine Simon of Philadelphia, the mother of Nora and Claire, two. When she first found out she was going to have two children instead of one, Simon was "shocked." After the girls were born, she and her husband, David Crawford, struggled with a double load of diapers and midnight feedings. Now Simon is back at work as assistant director of urban studies at the University of Pennsylvania and the girls are in separate play groups, still enjoying each other. "I can't imagine having just *one* kid," says Simon. "It seems like it would be too easy."

The balancing act never ends, but as the twins get older, the trick is to make sure the children realize they are individuals first and twins second. "We don't want them to be so bonded to each other that one becomes a missing part to the other," says Adele Faber, coauthor of "Siblings Without Rivalry." Says Linda Cardwell, a Texas mother of 10-year-old twin boys: "Most parents are working very hard to give each of our children their own identity."

Competition between twins is almost inevitable, says Lawrence Balter, professor of educational psychology at New York University, because twins are always at the same stage of development with the same set of parents. That may force them to develop different interests in order to maintain their identities, Balter says. Erica Frederick and her twin sister, Sheila Lambert, 40, sometimes felt like "interchangeable parts" when they were young, Frederick says. But, as they grew up, their paths diverged. Sheila was a cheerleader, Erica was not. They went to different colleges. Erica worked for nonprofit agencies; she is now executive director of the Karen Horney Clinic in New York. Sheila, publisher of Moody's Investors Service, says she's the "capitalist" of the pair. "It's hard to ever completely prevent rivalry among siblings," Sheila says, but "if you really encourage them to follow their own interests, there will be less rivalry than if you're trying to make them the same."

Still, it's impossible to ignore that twin connection. "Identical twins who are genetically exactly alike may be predisposed toward certain things," says Terry Pink Alexander, author of "Make Room for Twins" and mother of eight-year-old twins. Larry Silverstein, 26, a New York City lawyer, jokes that he and his identical twin, Lenny, have always had the same hobbies: "sports and sex." Although they went to different colleges, they now live in the same city and see each other twice a week. But the special intimacy that twins share can lead to problems. As they grow older, they may find it hard to develop intimate friendships or romantic relationships because they think no one will ever know them as well as their twin. Barbara Unell, cofounder of Twins magazine and the mother of seven-year-old twins, advises parents to talk to their twins about different kinds of relationships and how each can be rewarding.

Two heartbeats: There are many resources available to parents who are having trouble raising doubles. The National Organization of Mothers of Twins Clubs in Albuquerque, N.M., has free booklets with twin-rearing advice and can refer parents to 300 chapters. TWINLINE, based in Berkeley, Calif., is a social-service agency and hot line that offers advice on multiple-birth care and development. Unell's Twins magazine, based in Overland Park, Kans., is full of articles on such topics as breast-feeding, toilet training and research on twins. Sometimes, the best advice comes from twins and their parents who have been through it all. "Twins are not twice the trouble," says Detroit obstetrician Robert Sokol, father of twin 15-year-old boys. "They're about one and a half times the trouble." Sokol remembers that momentous day when he first heard two heartbeats inside his wife's womb. Years later, he says, "I still think it's neat." And, Sokol adds, "it's a great conversation piece at cocktail parties."

BARBARA KANTROWITZ *and*
KAREN SPRINGEN

By studying twins reared apart, scientists are learning how the forces of nature and nurture interact to make us what we are. Last December researchers led by Thomas Bouchard of the University of Minnesota announced preliminary results of a study of 350 pairs of identical twins reared apart. Running them through a battery of physical and psychological tests that included 15,000 questions, the Minnesota team found striking similarities between the pairs in several major personality traits, suggesting a strong role for genetics. On the basis of the tests, the researchers estimate that 61 percent of leadership ability is inherited. Other characteristics also showed strong genetic influence: the capacities for imaginative experiences, vulnerability to stress, alienation and a desire to shun risks were 50 percent to 60 percent inherited. But environment was a powerful factor in other traits. Aggression, achievement, orderliness and social closeness were all more strongly influenced by upbringing, with genetic components of 48 percent to 33 percent. "There is a very significant genetic influence across the broad range of personality characteristics," says Bouchard.

Water phobia: Fears and phobias seem to have sizable genetic components too. In a study of 15 pairs of identicals reared apart, the Minnesota team found three sets with multiple phobias. Two siblings, for instance, were afraid of water and heights and had claustrophobia. Three other pairs of twins shared a single phobia. One of these pairs, although separated soon after birth, worked out the same solution to their water phobia: they waded into the ocean backward, averting their eyes from the surf. Perhaps, researchers speculate, evolution selected genes for such fears because they conferred a survival advantage on early humans. Avoiding heights and water is a good way to avoid falling off cliffs or drowning.

Like phobias, shyness seems an inborn characteristic. Studies of identical twins of all ages show that "shyness is the most heritable of any trait," says Robert Plomin, professor of human development at Pennsylvania State University. Although there is no shyness gene as such, an interplay of genes does seem to put shy people in a higher state of physiological arousal. "People in a high arousal state would try to avoid a lot of activity and stimulation to keep their arousal level down," explains Laura Baker of the University of Southern California—making them seem shy.

Researchers lately have been trying to understand how genetic influence on the brain can produce a whole spectrum of different behaviors. One nascent theory holds that a person's genetic makeup influences the kinds of environments he seeks, which in turn influence personality. In this way genes would have an indirect effect as well as a direct one—a sort of nature *via* nurture, as Minnesota's David Lykken puts it. A genetic predisposition toward risk taking, for instance, may create a typical urban criminal. So while there is no gene for criminality, there may be one for behavior that, in the wrong setting, produces it. By the same token, genes may make us more sensitive to our environment. In a study of anxiety and depression published in May, researchers at the Medical College of Virginia examined 3,798 pairs of identical and fraternal twins. They found identical twins more susceptible than fraternal twins, clear evidence of a genetic connection. But there do not seem to be separate genes for anxiety and depression. Rather, it appears that there is only a single set of genes. Its function: to make one "particularly sensitive to environmental effects," says MCV's Lindon Eaves. A person with this sensitivity who experiences, say, a death in the family, may plummet into depression. Another who has a stress-filled job may succumb to anxiety.

Long-held notions: The idea that genes strongly influence how we think and act upsets long-held notions about the primacy of free will. But clearly, we are not slaves of our genes. The proof is obvious: identical twins are not necessarily identical in behavior. One twin studied by Minnesota's Lykken was separated from her identical sister at birth. She gew up to be "a professional-caliber pianist," says Lykken, although her adoptive family was not at all musical. Her sister was adopted by a piano teacher—and doesn't play at all. "An environmental difference, different families, reached into the genes and pulled out something," says Lykken.

If identical twins teach us anything, it is that while we may be stuck with what nature deals us, nurture is somewhat negotiable. Says USC's Baker, "Given the same genes, different circumstances—including different educational opportunities—will produce different results."

To identical twins, just being half of one of nature's most delightful sleights of hand can seem the best of circumstances. Karen Braaten and Laura Terheggen, 35-year-old identical twins, couldn't be closer—although they have always hated to be compared. At one point, Karen seriously thought about having her nose changed. When she heard about it, says Laura, "I rushed to the mirror to say, 'What for?'" Both married and residents of South Padre Island, Texas, for the past decade, they are now undergoing a difficult two-year separation while Laura takes a degree in physical therapy. "Twins are wonderful," observes Karen, confidently summing up the feelings of most of her singular breed of doubles. "You have a built-in best friend. If you can't relate to something half of yourself, you can't relate to anything."

SHARON BEGLEY *with* ANDREW MURR *in Chicago.* KAREN SPRINGEN *in Twinsburg.* JEANNE GORDON *in Los Angeles.* JOANNE HARRISON *in Houston and bureau reports*

Development During Infancy and Early Childhood

No age period in human development has received more attention during the past quarter century than infancy. Today we are more certain than ever before that the events of the first several years of life build a foundation for subsequent development. This does not imply, however, that the events of infancy determine later development. For example, high-quality infant care lays the foundation for successful development during the preschool and school-age years, but optimal care during infancy does not make the preschool or school-age child any less vulnerable to inadequate caregiving. Child abuse first experienced during the school-age years can be as devastating to subsequent development as child abuse first experienced during infancy.

Although developmentalists now recognize that the events of one age period do not necessarily predetermine the events of a subsequent age period, it would be a mistake to underplay the importance of infancy. Although it is well established that the fetus is responsive to external sensory stimulation, whether or not the fetus learns from such experiences is a hotly debated topic (discussed in the article "The First 9 Months of School"). As indicated in "Making of a Mind," fundamental changes in the organization of the brain occur during infancy and early childhood, and these changes are intimately related to the quality, quantity, and timing of environmental stimulation to which the infant is exposed. In one sense the newborn's mind is like a blank slate; no one yet has provided evidence to support the belief that the newborn's mind is filled with innate ideas. In another sense, however, the newborn enters the world with the full set of species characteristics that provide the basis for perceptual discrimination of colors and sounds, emotional expressiveness and responsiveness, learning and imitative behavior, communicative behavior and language acquisition, social interaction, and differentiation of personality. In short, the newborn begins life with all that is necessary to construct an active, adaptive fit to the environment. How infants go about constructing this adaptive fit, and how they communicate with their parents before mastering the intricacies of speech, are discussed in "What Do Babies Know?" and "The Art of Talking to a Baby."

"The Child-Care Dilemma" considers a critical problem in the United States: the lack of quality supplemental child-care facilities for children. Fifty percent of working mothers or about 9 million children are involved. Such figures point out the magnitude of the problem and the urgent need for effective solutions. In addition, we have relatively little information about how day-care staffs influence child development. In the article "What is Beautiful is Good," evidence is presented which suggests that something as basic as a child's physical appearance can shape the attributes adults assign to the child.

When reading the selections in this section, students should keep in mind three important points. First, development during infancy is characterized by plasticity in brain-behavior organization; that is, the potential for change over time is relatively great. As a result, it is unlikely that any single time-limited event can have long-lasting consequences for subsequent development. For example, immediate contact between a mother and a newly born infant can be a positive and rewarding experience for both mother and infant. However, constructing a positive, supportive, binding, loving relationship between mother and infant—generally referred to as attachment—requires considerably more time and interaction than the immediate post-natal contact allows. Second, the infant is an active agent in influencing the caregiving environment. The infant's cry is a communicative act that provides feedback about the infant's ability to influence events in the environment. The infant's vocalizations, smiles, laughter, and first unassisted steps are behaviors designed to provoke responsiveness from caregivers. The nature of the caregiver's response is the basis for the third important point: namely, that both the quality and quantity of caregiving are important influences in structuring the match or fit between caregiver and infant. It is the caregiver's responsiveness that provides feedback to the infant about the effectiveness of the infant's behavior. To one extent or another, each of the articles in this unit focuses on these three points: the plasticity of early development, the infant as an active agent, and the adaptive fit constructed by the infant and its caregivers.

Looking Ahead: Challenge Questions

Environmental stimulation is a key ingredient in brain organization. What additional factors must be taken into account when translating this prescriptive statement into

the practical language of the world of infant care? Put another way, what are the relative advantages and disadvantages of quality, quantity, and timing of stimulation for brain growth and development?

Summarize the opinions of infant researchers with respect to the effectiveness of early infant stimulation programs. Explain why you agree or disagree with their opinions. Why do you suppose that it took so long for researchers to discover that infants possess a rich repertoire of abilities?

What effect does work have on the development of attachment relationships between mother and infant, or on the effectiveness of discipline in school-age children? Does society have a responsibility to provide supplementary care for children of working mothers? Does industry have this responsibility?

Do you agree that competent social and emotional behavior seems to derive directly from early experiences provided by caregivers? Is it possible that some children respond appropriately to competent caregiving, whereas others do not because of a mismatch between infant temperament and caregiver temperament? How might peer interactions offset the negative consequences of this type of mismatch?

9 THE FIRST MONTHS OF SCHOOL

William Poole

William Poole is a freelance writer and a registered pediatric nurse.

SURPRISING LESSONS A FETUS LEARNS IN THE WOMB.

ONCE MORE, with the lights blazing and cameras rolling, Rene Van de Carr addresses a row of pregnant bellies. "When the baby kicks," he says, "have Daddy tap the spot with his hand, put the side of his mouth directly on the abdomen, and say, 'Kick Mommy. That's a good baby, kick Mommy here again.'" This activity, called the Kick Game, is what the local television folks are taping today: a gentle, cherubic doctor coaching a squad of recumbent cheerleaders.

Rene Van de Carr is an obstetrician and gynecologist with an unusual calling. He's president of Prenatal University in Hayward, California, a nonprofit institution that seeks to help babies reach their maximum potential by engaging the parents in exercises designed to stimulate the brain of the unborn child. Despite the cuteness factor — special T-shirts for the young graduates, for instance, each inscribed with a smiling bare-bottom baby sporting mortarboard and diploma — Van de Carr is completely serious. He claims not only that fetuses are listening in the womb but that they benefit from an academic head start.

From the Kick Game, started in the fetus' fifth month, the curriculum expands in the seventh month to include the words *pat, rub, squeeze, shake, stroke,* and *tap.* Mommy can also do the talking, using a rolled-up newspaper as a megaphone to direct her voice to her belly. At least twice a day parents speak these words while performing exercises on the mother's swelling abdomen. "Pat, pat, pat," one of them will say, patting Mother's tummy. "Rub, rub, rub. Shake, shake, shake."

"We like to have nice, clear, concise communication," Van de Carr says, "because that's what works." More than a thousand fetuses have now graduated from Prenatal University. Follow-up studies by Van de Carr and his associates demonstrate to their satisfaction that university graduates are more vigorous at birth; smile, sit, walk, and speak sooner; breast-feed longer; and develop stronger emotional bonds with their parents than do the standard run of infants.

Predictably, Van de Carr is not the only one trying to build better babies through fetal education. The notion has always had a certain pull, bolstered as it is by a never-ending supply of anecdotes: a "fetally enriched" baby who speaks complete sentences at a few months of age, a musician who can perform at first sitting and without sheet music a concerto that his mother practiced while he was in the womb.

Half a dozen fetal education programs have cropped up across the country over the last several years, and dozens of books, kits, and videotapes are now available. One company markets "the only portable automated fetal educational communications unit in the world," a tape player with a special speaker belt that is strapped over the mother's abdomen — deluxe model, $244.95. The company's brochure boasts that "many couples world wide have literally created emotionally well adjusted children with genius level I.Q.'s."

Mainstream child development experts view these claims with caution, if not outright hostility. Susan Ludington-Hoe, director of the Infant Stimulation Education Association at the University of California at Los Angeles, employs some of Van de Carr's techniques in her practice, and outlines them in her book *How to Have a Smarter Baby.* Even so, she believes that the method's only proven benefit is that it strengthens the emotional bond between

Reprinted from *Hippocrates*, The Magazine of Health & Medicine, July/August 1987, pp. 68-73. Copyright © 1987 by Hippocrates, Inc.

parents and newborn. Beyond that, she says, "We have unfounded promises all over the place. There is nothing to support cognitive gains."

Michael Meyerhoff, associate director of the Center for Parent Education in Newton, Massachusetts, is another skeptic. "People like to talk to a baby before it's born. It's a natural thing to do. We also talk to dogs and vending machines. But it's irresponsible for a professional to encourage talking to fetuses in this way."

About Van de Carr's pat-and-rub exercise Meyerhoff says, "Everything we know about how the child's mind develops indicates that a fetus would not have the mental capacity to put together the words with the experience." He also questions another of Van de Carr's assertions — that stimulating the fetus' nervous system will save some extra nerve connections that are usually lost before birth. "Nobody to my knowledge has linked those neurons to anything important. The child also loses a lot of hair shortly after birth. Nobody's running around saying you've got to provide the child with a lot of extra hair."

What, then, if anything, do humans learn in the womb? Solid answers to this question have been elusive, but researchers now say that babies do in fact bring memories — primitive impressions, really — with them into the world. This is quiet news compared to tales of sentence-spouting infants and piano-playing prodigies, but it's important news, because it comes from the newborns themselves.

I N A SMALL ROOM down the corridor from the nursery at Columbia-Presbyterian Medical Center in New York City, psychologists William Fifer, Christine Moon, and Lisa Monti prepare to ask a newborn exactly what she knows about the world. This is dark-eyed Maria, almost 48 hours old. The psychologists, dressed in yellow isolation gowns, work quickly to set up their equipment. They have only a brief time before Maria, now alert two and a half hours after feeding, will become dramatically uncooperative in the face of mounting hunger.

The researchers prop Maria in her bassinet and gently fit a set of earphones around her delicate face. In her mouth they tuck a nipple attached by rubber tubing to a cart bearing a pair of tape recorders and a battered computer. During the experiment the computer will signal the tape recorders to send Maria sounds through the earphones, then record how she sucks on the nipple — fast, slow, or not at all — in response to the sounds.

"It's a tough task," Fifer says as he watches Maria get down to work. Through the earphones she hears a voice repeating the syllables *ee, ee, ee, ee,* followed by *ah, ah, ah, ah.* If she sucks on the nipple when she hears the *ee* sound, the computer will stop the recording and maintain silence as long as she continues to suck at least once every two seconds. But if she sucks on the *ah* sound, she will hear a male voice for as long as she sucks. Fifer adds, "It takes most babies the full 18 minutes to

figure out how to hear the voice and to develop a consistent pattern of preferential sucking."

Fifer wants to know what Maria and the other babies in the experiment will have to say about a male voice — whether they will prefer to hear a man speaking to no voice at all. An old wives' tale holds that newborn babies don't like male voices. But Fifer is not much impressed by old wives' tales; what he is impressed by is facts, and so far the facts suggest that babies will readily choose male voices over silence.

Before the 1970s, such an assertion was unimaginable. Newborns were thought to live in a world of "blooming buzzing confusion," to quote psychologist William James. "People thought newborns really couldn't make much sense out of what was out there," Fifer says. But after considering the results of studies conducted throughout the decade, psychologists began to give newborns and fetuses more credit. Researchers already knew that hearing begins late in the sixth month of pregnancy and that fetuses spend many weeks listening to the mother's heartbeat, various gurglings and gut rumblings, the whoosh of blood through the veins and arteries, and human speech — not only the mother's voice but outside voices as well. Years of observation had also shown that newborns tracked and were soothed by sounds, particularly female voices. It seemed natural to wonder if babies form some preferences before birth.

In the late 1970s, Fifer, then a graduate student at the University of North Carolina at Greensboro, and UNC psychologist Anthony DeCasper decided to ask newborns if what they heard in the womb didn't lead them to prefer their mother's voice to that of any other woman. They posed the question by playing the babies recordings of a story.

The researchers asked new mothers at Moses Cone Hospital in Greensboro to tape themselves reading the Dr. Seuss story *To Think That I Saw It on Mulberry Street.* Back at the lab, Fifer edited the tapes, taking out little pauses to produce stories of the same length. Then it was back to the hospital, where he would often find that his work had been wasted. "You'd lug all this equipment in, only to find your baby had gone home, or was too hungry and crying like crazy." In the end it took the researchers ten weeks to ask ten babies, "Which do you prefer, the story as read by Mom, or the story as read by another mother?" In all, nine of the ten babies sucked more often to hear the recording of their own mother's voice.

While many parents went home with a new appreciation for the abilities of their newborns, the researchers were left with a new question: What if this preference for mother's voice had been learned not in the womb but in the first hours after birth?

To rule out this possibility, DeCasper and his student Phyllis Prescott set up a similar experiment using fathers. Earlier studies had shown that by several weeks of age an infant prefers its father's voice to that of another man. But would this preference show up only days after birth? The two-day-

> "**P**eople like to talk to a baby before it's born. It's a natural thing to do. We also talk to dogs and vending machines."

old babies the researchers tested showed no such partiality. It looked as though the bias for mother's voice was learned in the noisy recesses of the womb.

In 1983 DeCasper and psychologist Melanie Spence, who is now at the University of Alabama at Birmingham, decided to find out if fetuses learned even more about voices and sounds. To recruit subjects for the experiment, Spence attended classes for expectant parents at Moses Cone Hospital. "I am looking for pregnant women who would be willing to sit in a quiet place twice a day and read a story aloud to their unborn child," she told each group. This request often brought smiles and a few shakes of the head. "Some people did seem to think it was sort of a ridiculous thing at first," she says.

The women Spence recruited were asked to read portions of three stories into a tape recorder: Dr. Seuss' *The Cat in the Hat*, *The Dog and the Fog* (*The Cat in the Hat* with all the important nouns changed), and *The King, the Mice, and the Cheese*, by Nancy and Eric Gurney. Each woman was then assigned one of the stories to read aloud to her fetus. "Choose a time when you feel your baby is awake," the psychologists said, "and keep a logbook of the amount of reading you get done."

By the end of the experiment, the 16 babies tested had each heard their stories an average of 67 times. After this many readings, the parents were as curious about the results as the psychologists were. Standing in the darkened room out of baby's view, the parents listened through earphones as the researchers played a recording of the story the baby had heard in the womb and one of the others. "They always wanted to know immediately, 'Does he prefer the old story or the new story?'" Spence says. "It's very difficult to determine right off," she told them. "We'll have to wait to see what the computer has to say."

When the results were finally in, it did seem that the little brains had been picking favorites. Tested at one to three days of age, the infants were more likely to suck for a story they had heard in the womb than for a new story. This was true not only for infants who heard a story read by their own mothers but also for those who heard the same story read by other mothers. Infants who hadn't heard any of the stories before birth showed no preference.

Other studies during the last few years have confirmed and expanded on these findings. In one experiment, DeCasper and Robin Panneton, now at the University of Rochester, asked 13 pregnant women to sing "Mary Had a Little Lamb" several times a day using the syllables *la la la* instead of the lyrics. The experiment was designed to test whether it is the melodic qualities of speech that fetuses are really engaged by.

"Our guess was that if fetuses are listening to Mother's speech, they are probably tuning in to the cadence of the voice," Panneton says. She points out that most of the distinguishing sounds of speech have frequencies above 1,000 hertz. But recordings made in the womb show that these high frequencies

MAKING SENSES

When does human sensory ability begin? There is no single threshold. From the second month a fetus may show reflexes that appear to involve the senses; yet it feels things quite differently from the way we do, since the brain is still immature. Most nerve cells are produced during the first few months. But the senses cannot work until these cells make connections, or synapses. Early reflexes and movements appear to play a role in making these connections, molding the senses and training the fetal brain to perceive.

Here is a month-by-month summary of sensory development:

1 Nerve growth begins when a sheet of cells on the back of the embryo folds in the middle to form a tube—the future spinal cord. At one end the tube enlarges to form the brain's major sections.

2 Humans' earliest responses are reflexes, some of which occur even before the sense of touch is developed. Researchers have found that fetuses will flex their heads away from stimulation around the mouth as early as 7½ weeks of age. By the eighth week the ear begins to take shape.

3 Nerve cells have been multiplying rapidly, and synapses are being formed. Touch receptors around the mouth are developed by the 12th week, and elsewhere by the 15th week. The fetus has acquired more reflexes: Touching the palms makes the fingers close, touching the soles of the feet makes the toes curl down, and touching the eyelids makes the eye muscles clench.

4 At 15 weeks the fetus can grasp, frown, squint, and grimace. It may suck its thumb and swallow amniotic fluid. These and other movements probably correspond to development of synapses in parts of the brain.

5 The senses of taste and smell are now formed. An important era for sensory development begins at 20 weeks. Nerve cell production slows down as existing nerve cells grow larger and make more complex connections. The nerve cells serving each of the senses are now developing into specialized areas of the brain.

6 The sense of touch is developed by the end of the sixth month, and the fetus can now feel movement. In some cases researchers have obtained responses to sound as early as 24 weeks.

7 At 25 weeks some healthy babies born prematurely can survive, an indication of how far their nerves and muscles have matured. Response to sound grows more consistent toward the end of the month, when the nerve supply to the ear is complete. A light shone on the mother's abdomen will make the fetus turn its head, implying some functioning of the optic nerve. At 26 weeks brain scans show response to touch, and at 27 weeks to light.

8 At midmonth the eyes open in the womb, and the fetus may see its hand and environment. Nearly all fetuses respond to sound by 30 weeks. Some scientists put the start of awareness at the 32nd week; at this point the neural circuits are as advanced as a newborn's. Toward the end of the month brain scans show periods of dream sleep.

9 During late pregnancy the walls of the abdomen and uterus become thinner and admit more light, and the fetus begins to develop daily activity cycles. At 35 weeks hearing is mature.

At birth (usually 40 weeks) a baby can see colors and shapes within 13 inches of its face, and can distinguish loudness, pitch, and tone. A newborn may even show a preference for sweets and for the scent of its mother's skin.

—*Keiko Ohnuma*

don't get through to the fetus. "Speech would sound very muffled," she says.

Indeed, Panneton's newborns were more likely to change the speed of their sucking to hear the melody "Mary Had a Little Lamb" than to hear "Love Somebody," a tune with the same notes in a different order with a different rhythm. Panneton's work suggests that it isn't necessarily words that are important, and that newborns may simply prefer sequences of sounds they've heard before.

THE NOTION that the human nervous system is forming impressions several months before birth has captured the imaginations of fetal educators. If an infant is born with an impression of mother's voice or a familiar story, they ask, isn't it reasonable to think the infant may have the capacity to recall words, music, or even basic math?

"I don't mind people taking the kind of work that Tony DeCasper, Bill Fifer, and I have done and *speculating* about what kinds of capacities are present in the fetus and young infant," Panneton says, "but it's a whole other thing when they really think that we have concluded, that we have somehow *demonstrated* these higher capacities."

DeCasper has been quizzed on the implications of his work so many times that he keeps a statement ready. "At this point in time, I would caution expectant parents to act as they normally would to promote the health and welfare of their babies. Consistent, loving, wise rearing over the long haul is the real challenge to being a parent."

And William Fifer has proffered his own response to the fetal educators. Tucked in among the bar graphs toward the end of a slide show about his research is a special picture: the cover of a sensational tabloid. "You can see it's a reliable source," he says lightly, this time to a group of New York City public health workers. "Some people think that if babies are learning and capable, a fetus may be learning and capable as well. They say, 'Let's go in and start talking to it and doing all of these things.' Well, we're a long way from that point, I believe, but you couldn't tell that to this gentle-man here." On the screen is a pregnant woman in a Prenatal U sweatshirt. Beside her is a smiling Rene Van de Carr.

This is not to say that the psychologists doing fetal research reject the idea of applying their work. Some hope to use the methods they've developed to test for brain damage in infants deprived of oxygen during birth. And one expert, Norman Krasnegor, chief of the Human Learning and Behavior branch of the National Institute of Child Health and Human Development, says prenatal learning studies help us understand infants' sensory capacities. "They help us to see which babies may be impaired and which may not. If you can make that distinction, you're in a better position to ask, what should I do, and when should I do it?"

It's easy to understand why the psychologists want to emphasize the practical benefits of their research. They're caught between doubters, who consider studies of fetal learning silly and a waste of vital research dollars, and prenatal educators, who sometimes take the psychologists' findings a few steps too far. "Some people won't believe anything you say," says Anthony DeCasper, "and some will believe everything you say and a great deal more."

One who believes more is Rene Van de Carr. Back at Prenatal University, a fetus' curriculum is enriched with a secondary word list three or four weeks before birth. By now, according to Van de Carr, "the baby has learned that words are important," and is ready to hear words it "might need to know in the first few months after birth." This list of 34 words includes *tongue, powder, burp, yawn, ice cream,* and *throw up.*

Next comes Infantspeak, ten special words to be offered as a kind of extra credit for those parents who are "interested in attempting early talking" with their babies. The word list includes such standards as *dada, mama,* and *bye bye,* as well as *din din* (for food other than milk), and *pu pu* (for diaper change).

Thus prepared, the fetus moves out of the classroom — "This is a squeeze," Daddy tells the fetus during labor contractions — and on to a lifetime of graduate work.

*The newborn's brain: registering
every flash of color, caress,
scent, and other stimuli vital to the*

MAKING OF A MIND

KATHLEEN McAULIFFE

"Give me a child for the first six years of life and he'll be a servant of God till his last breath."

—Jesuit maxim

A servant of God or an agent of the devil; a law-abiding citizen or a juvenile delinquent. What the Jesuits knew, scientists are now rapidly confirming—that the mind of the child, in the very first years, even months, of life, is the crucible in which many of his deepest values are formed. It is then that much of what he may become—his talents, his interests, his abilities—are developed and directed. The experiences of his infancy and childhood will profoundly shape everything from his visual acuity to his comprehension of language and social behavior.

What underlies the child's receptivity to new information? And why do adults seem to lose this capacity as they gain more knowledge of the world around them? Why is it that the more we know, the less we *can* know?

Like a Zen koan, this paradox has led scientists down many paths of discovery. Some researchers are studying development processes in infants and children; others search the convoluted passages of the cortex for clues to how memory records learning experiences. Still others are studying the degree to which learning is hard-wired—soldered along strict pathways in the brains of animals and humans.

Another phenomenon recently discovered: Long after patterns of personality have solidified, adults may tap fresh learning centers in the brain, new nerve connections that allow intellectual growth far after fourscore years.

Although much research remains to be done, two decades of investigation have yielded some dramatic—and in some instances unexpected—insights into the developing brain.

An infant's brain is not just a miniature replica of an adult's brain. Spanish neuroscientist José Delgado goes so far as to call the newborn "mindless." Although all the nerve cells a human may have are present at birth, the cerebral cortex, the gray matter that is the seat of higher intellect, barely functions. Surprisingly, the lower brain stem, the section that we have in common with reptiles and other primitive animals, dictates most of the newborn's actions.

This changes drastically in the days, weeks, and months after birth, when the cerebral cortex literally blossoms. During this burst of growth, individual brain cells send out shoots in all directions to produce a jungle of interconnecting nerve fibers. By the time a child is one year old, his brain is 50 percent of its adult weight; by the time he's six, it's 90 percent of its adult weight. And by puberty, when growth trails off, the brain will have quadrupled in size to the average adult weight of about three pounds.

How trillions of nerve cells manage to organize themselves into something as complex as the human brain remains a mystery. But this much is certain: As this integration and development proceeds, experiences can alter the brain's connections in a lasting, even irreversible way.

To demonstrate this, Colin Blakemore, professor of physiology at Oxford University, raised kittens in an environment that had no horizontal lines. Subsequently, they were able to "see" only vertical lines. Yet Blakemore had tested their vision just before the experiment began and found that the kittens had an equal number of cells that responded to each type of line.

Why had the cats become blind to horizontal lines? By the end of the experiment, Blakemore discovered that many more cells in the animals' brains responded to vertical lines than horizontal lines.

As the human brain develops, similar neurological processes probably occur. For example, during a test in which city-dwelling Eurocanadians were exposed to sets of all types of lines, they had the most difficulty seeing oblique lines. By comparison, the Cree Indians, from the east coast of James Bay, Quebec, perceived all orientations of lines equally well. The researchers Robert Annis and Barrie Frost, of Queens University, in Kingston, Ontario, attributed this difference in visual acuity to the subjects' environments. The Eurocanadians grow up in a world dominated by vertical and horizontal lines, whereas the Indians, who live in tepees in coniferous forests, are constantly exposed to surroundings with many different types of angles.

The sounds—as well as the sights—that an infant is exposed to can also influence his future abilities. The phonemes *rah* and *lah,* for instance, are absent from the Japanese language, and as might be expected, adults from that culture confuse English words containing *r* and *l*. (Hence the offering of steamed "lice" in sushi bars.) Tests reveal that Japanese adults are quite literally deaf to these sounds.

Infants, on the other hand, seem to readily distinguish between speech sounds. To test sensitivity to phonemes, researchers measure changes in the infants' heartbeats as different speech sounds are presented. If an infant grows familiar with one sound and then encounters a new sound, his heart rate increases. Although the evidence is still incomplete, tests of babies from linguistic backgrounds as varied as Guatemala's

Spanish culture, Kenya's Kikuyu-speaking area, and the United States all point to the same conclusion: Infants can clearly perceive phonemes present in any language.

The discovery that babies can make linguistic distinctions that adults cannot caused researchers to wonder at what age we lose this natural facility for language. To find out, Janet Werker, of Dalhousie University, in Nova Scotia, and Richard Tees, of Canada's University of British Columbia, began examining the language capabilities of English-speaking adolescents. Werker and Tees tested the subjects to see whether they could discriminate between two phonemes peculiar to the Hindi language.

"We anticipated that linguistic sensitivity declines at puberty, as psychologists have commonly assumed," Werker explains.

The results were surprising. Young adolescents could not make the distinction, nor could eight-year-olds, four-year-olds, or two-year-olds. Finally, Werker and Tees decided to test infants. They discovered that the ability to perceive foreign phonemes declines sharply by one year of age. "All the six-month-olds from English-speaking backgrounds could distinguish between the Hindi phonemes," Werker says. "But by ten to twelve months of age, the babies were unable to make this distinction."

The cutoff point, according to Werker, falls between eight and twelve months of age. If not exposed to Hindi by then people require a lot of learning to catch up. Werker found that English-speaking adults studying Hindi for the first time needed up to five years of training to learn the same phoneme distinctions any six-month-old baby can make. With further testing, Werker succeeded in tracking down one of the learning impairments that thwarted her older subjects. Although there is an audible difference, the adult mind cannot retain it long enough to remember it. "The auditory capabilities are there," Werker says. "It's the language-processing capabilities that have changed."

Even a brief introduction to language during the sensitive period can permanently alter our perception of speech. Werker and Tees tested English-speaking adults who could not speak or understand a word of Hindi, although they had been exposed to the language for the first year or two of life. They found that these adults had a major advantage in learning Hindi, compared with English-speaking adults who lacked such early exposure.

Werker and Tees's studies show that there is an advantage in learning language within the first year of life. But when it comes to learning a second tongue another study has revealed some startling findings: Adults actually master a second language more easily than school-age children do.

For four years Catherine Snow, of the Harvard Graduate School of Education, studied Americans who were learning Dutch for the first time while living in Holland. "When you control for such factors as access to native speakers and the daily exposure level to the language," Snow says, "adults acquire a large vocabulary and rules of grammar more quickly than children do. In my study, adults were found to be as good as children even in pronunciation, although many researchers contend that children have an advantage in speaking like natives."

Obviously not all learning stops when the sensitive period comes to a close. This observation has led some researchers to question the importance of early experiences. What would happen, for example, if a child did not hear a single word of any language until after one year of age? Would the propensity to speak be forfeited forever? Or could later exposure to language make up the deficit?

Because of the unethical nature of per-

❦It's probably fair to say that if you want bright kids, you should cuddle them a lot when they're babies because that increases the number of neural connections.❞

---

forming such an experiment on a child, we may never know the answer to that question. But some indications can be gleaned from animal studies of how early deprivation affects the development of social behavior.

In *An Outline of Psychoanalysis*, Sigmund Freud refers to "the common assertion that the child is psychologically the father of the man and that the events of his first years are of paramount importance for his whole subsequent development." At the University of Wisconsin Primate Laboratory, the pioneering studies of Harry and Margaret Harlow put this belief to the test on our closest living relative—the rhesus monkey.

"Our experiments indicate that there is a critical period somewhere between the third and sixth month of life," write the Harlows, "during which social deprivation, particularly deprivation of the company of [the monkey's] peers, irreversibly blights the animal's capacity for social adjustment."

When later returned to a colony in which there was ample opportunity for interacting with other animals, the experimental monkeys remained withdrawn, self-punishing,

and compulsive. Most significantly, they grew up to be inept both as sexual partners and parents. The females never became impregnated unless artificially inseminated. We don't know whether humans, like Harlow's monkeys, must establish close bonds by a certain age or be forever doomed to social failure. But an ongoing longitudinal study, the Minnesota Preschool Project, offers the encouraging finding that emotionally neglected four-year-olds can still be helped to lead normal, happy lives. To rehabilitate the children, the teachers in the project provide them with the kind of intimate attention that is lacking at home.

Perhaps one of the Harlows' observations sheds light on why the project was successful: During the critical period for social development, the Harlows found that even a little bit of attention goes a long way. During the first year of life, for example, only 20 minutes of playtime a day with other monkeys was apparently sufficient for the animals to grow into well-adjusted adults. L. Alan Sroufe, codirector of the Minnesota Project, tells the story of one four-year-old boy who was constantly defiant—the kind of child who would hit the other children with a toy fire truck. Instead of sending him to a corner, the teacher was instructed to remove him from the group and place him with another teacher. The message they hoped to impart: We are rejecting your behavior, but we're not rejecting you. Within a few months, the antisocial little boy learned to change his behavior.

If children aren't exposed to positive social situations until adolescence, however, the prognosis is poor. Like any complex behavior, human socialization requires an elaborate series of learning steps. So by adolescence, the teenager who missed out on many key social experiences as a child has a tremendous handicap to overcome.

Researchers are finding that each stage of life demands different kinds of competencies. This may be why sensitive learning periods exist. "When a baby is born it has to do two things at the same time," says biochemist Steven Rose, of England's Open University. "One is that it has to survive as a baby. The second is that it has to grow into that very different organism, which is a child and then finally an adult. And it is not simply the case that everything the baby does is a miniature version of what we see in the adult."

For example, the rooting reflex, which enables the baby to suckle, is not a preliminary form of chewing: There's a transitional period in which the child must begin eating solid foods. And then other sorts of skills become necessary—the child must learn to walk, talk, form friendships, and when adulthood is reached, find a sexual partner. "But the child does not have to know all that at the beginning," Rose says. "So sensitive periods are necessary because we have to know how to do certain things at certain times during development."

2. INFANCY AND EARLY CHILDHOOD

During the course of a sensory system's development, several sensitive periods occur. In the case of human vision, for example, depth perception usually emerges by two months of age and after that remains relatively stable. But it takes the first five years of life to acquire the adult level of visual acuity that allows us to see fine details. And during that prolonged period, we are vulnerable to many developmental problems that can cause this process to go awry. For example, a drooping lid or an eye covered by a cataract—virtually anything that obstructs vision in one of the child's eyes for as few as seven days—can lead to a permanent blurring of sight. This condition, known as amblyopia is one of the most common ophthalmological disorders. Treatment works only if carried out within the sensitive period, before the final organization of certain cells in the visual cortex becomes fixed. After five years of age, no amount of visual stimulation is likely to reorganize the connections laid down when the young nervous system was developing.

Like molten plastic, the nervous system is, at its inception, highly pliable. But it quickly settles into a rigid cast—one that has been shaped by experience. Just what neurological events set the mold is not known. Some suggestive findings, however, come from the research of John Cronly-Dillon, a professor of ophthalmic optics at the University of Manchester Institute of Science and Technology, in England.

Working with colleague Gary Perry, Cronly-Dillon studied growth activity in the visual cortex of rat pups reared under normal light conditions. To measure growth, researchers monitored the rate at which certain cells synthesized tubulin, a protein vital for forming and maintaining nerve connections. The researchers found that tubulin production in the visual cortex remained at a low level until day 13, which marks the onset of the sensitive period for visual learning. It coincides with the moment when the animal first opens its eyes. At that time, tubulin production soars, indicating a rise in growth activity.

Cronly-Dillon and Perry found that the rat's visual cortex continues to grow for the next week and then declines. By the end of the critical period, when the pup is roughly five weeks old, tubulin production drops to the level attained before the eyes open.

To Cronly-Dillon the surplus of tubulin at the beginning of the critical period and its subsequent cutback have profound implications. "It means that an uncommonly large number of nerve connections can exist at the peak of the critical period, but only a small fraction of them will be maintained at the end," he says. "So the question, of course, is which nerve connections will be kept?"

If Cronly-Dillon is correct, experience probably stabilizes those connections most often used during the sensitive period. "So

by definition," he says, "what remains is most critical for survival."

Cronly-Dillon's work elaborates on a theory Spanish neurophysiologist Ramon Y Cajal advanced at the turn of the century. According to this view, which has been gaining broader acceptance in recent times, brain development resembles natural selection. Just as the forces of natural selection ensure the survival of the fittest, so do similar forces preserve the most useful brain circuits.

The beauty of this model is that it could explain why the brain is as exquisitely adapted to its immediate surroundings, just as the mouthparts of insects are so perfectly matched to the sexual organs of the flowers they pollinate. The textures, shapes, sounds, and odors we perceive best may have left their imprint years ago in the neural circuitry of the developing mind.

There is also a certain economic appeal to this outlook: Why, for example, should Japanese adults keep active a neural circuit that permits the distinction between r and l sounds when neither of these linguistic components is present in their native tongue?

Yet another economic advantage of the theory is that it would explain how nature can forge something as intricate as the brain out of a relatively limited amount of genetic material. "It looks as though what genetics does is *sort of* make a brain," Blakemore says. "We only have about one hundred thousand genes—and that's to make an entire body. Yet the brain alone has trillions of nerve cells, each one forming as many as ten thousand connections with its neighbors. So imagine the difficulty of trying to encode every step of the wiring process in our DNA."

This vast discrepancy between genes and connections, according to Blakemore, can be overcome by encoding in the DNA the specifications for a "rough brain." "Everything gets roughly laid down in place," Blakemore says. "But the wiring of the young nervous system is far too rich and diffuse. So the brain overconnects and then uses a selection process to fine-tune the system."

The brain of an eight-month-old human fetus is actually estimated to have two to three times more nerve cells than an adult brain does. Just before birth, there is a massive death of unnecessary brain cells, a process that continues through early childhood and then levels off. Presumably many nerve connections that fall into disuse vanish. But that is only part of the selection process—and possibly a small part at that.

According to Blakemore, many neural circuits remain in place but cease to function after a certain age. "I would venture a guess," he says, "that as many as ninety percent of the connections you see in the adult brain are nonfunctional. The time when circuits can be switched on or off probably varies for different parts of the cerebral cortex—depending on what functions they control—and would coincide with the sensitive period of

learning. Once the on–off switch becomes frozen, the sensitive period is over."

This doesn't mean, however, that new circuits can't grow. There appears to be a fine-tuning of perception coinciding with these developmental events. And as the brain becomes a finer sieve, filtering out all but a limited amount of sensory input, its strategy for storing information appears to change.

"Studies indicate that as many as fifty percent of very young children recall things in pictures," says biochemist Rose. "And by the time we're about four or five, we tend to lose our eidetic [photographic] memory and develop sequential methods of recall."

To Rose, who is studying the neurological mechanisms that underlie learning, this shift in memory process may have an intriguing logic. "To be a highly adaptable organism like man, capable of living in a lot of different environments, one must start out with a brain that takes in everything," Rose explains. "And as you develop, you select what is important and what is not important to remember. If you went on remembering absolutely everything, it would be disastrous."

The Russian neurologist A. R. Luria had a patient cursed with such a memory—the man could describe rooms he'd been in years before, pieces of conversations he'd overheard. His memory became such an impediment that he could not hold even a clerk's job; while listening to instructions, so many associations for each word would arise that he couldn't focus on what was being said. The only position he could manage was as a memory man in a theatrical company.

"The crucial thing then," Rose says, "is that you must learn what to *forget*."

Some components of the brain, however, must retain their plasticity into adulthood—otherwise, no further learning would be possible, says neuroscientist Bill Greenough, of the University of Illinois, at Urbana-Champaign. While the adult brain cannot generate new brain cells, Greenough has uncovered evidence that it does continue to generate new *nerve connections*. But as the brain ages, the rate at which it produces these connections slows.

If the young brain can be likened to a sapling sprouting shoots in all directions, then the adult brain is more akin to a tree, whose growth is confined primarily to budding regions. "In the mature brain," Greenough says, "neural connections appear to pop up systematically, precisely where they're needed."

Early experience, then, provides the foundation on which all subsequent knowledge and skills build. "That's why it's extraordinarily difficult to change certain aspects of personality as an adult," says neuroscientist Jonathan Winson, of Rockefeller University. "Psychiatrists have an expression: 'Insight is wonderful, but the psyche fights back.' Unfortunately, one of the drawbacks of critical-period learning is that a lot of misconceptions and unreasonable fears can become frozen in our minds during this very vulnerable period in our development."

Greenough acknowledges that the system isn't perfect; nevertheless, it works to our advantage because you can't build on a wobbly nervous system. "You've got to know who your mother is, and you've got to have perceptual skills," he explains. "These and other types of learning have to jell quickly, or all further development would halt."

Can these insights into the developing brain help educators to devise new strategies for teaching?

"We're a very long way from being able to apply the work of neurobiologists to what chalk-faced teachers are trying to do," says Open University's Rose.

But he can see the rough outline of a new relationship between neurobiology and education, which excites him. "We can now say with considerable certainty that there are important advantages to growing up in an enriched environment," he says. "That does not mean that you should be teaching three-year-olds Einstein's theory of relativity on the grounds that you will be turning them into geniuses later on. But it's probably fair to say that if you want bright kids, you should cuddle them a lot as babies because that increases the number of neural connections produced in the brain."

Although early learning tends to over-shadow the importance of later experience, mental development never ceases. Recent studies indicate that our intellectual abilities continue to expand well into our eighties, provided the brain has not been injured or diseased. Most crucial for maintaining mental vigor, according to Greenough, is staying active and taking on new challenges. In his rat studies, he found that lack of stimulation—much more than age—was the factor that limited the formation of new neural connections in the adult brain.

As long as we don't isolate ourselves as we grow older, one very important type of mental faculty may even improve. Called crystallized intelligence, this ability allows us to draw on the store of accumulated knowledge to provide alternate solutions to complicated problems. Analyzing complex political or military strategies, for example, would exploit crystallized intelligence.

There is a danger in believing that because the brain's anatomical boundaries are roughly established early in life, all mental capabilities are restricted, too. "Intelligence is not something static that can be pinned down with an I.Q. test like butterflies on a sheet of cardboard," says Rose. "It is a constant interplay between internal processes and external forces."

To be sure, many types of learning do favor youth. As violinist Isaac Stern says, "If you haven't begun playing violin by age eight, you'll never be great." But in the opinion of Cronly-Dillon, the best time for learning other types of skills may be much later in life. Although he will not elaborate on this until further studies are done, he believes we may even have sensitive periods with very late onsets. "There's a real need," Cronly-Dillon says, "to define all the different types of sensitive periods so that education can take advantage of biological optimums."

It is said that the ability to learn in later life depends on the retention of childlike innocence. "This old saw," insists Cronly-Dillon, "could have a neurological basis."

What Do Babies Know?

More than many realize, and much earlier, according to new research

"F antastic!" says Michael Lewis, a small, spry man with a gray-flecked beard. "This is great!"

What inspires such glee in Lewis is that two small and curly-haired sisters named Danielle and Stacy, ages twelve and 14 months, are starting to cry. The sound is heart-rending, but not to Lewis.

"Exactly what we expected," he says cheerily as the girls' parents arrive to comfort them. The wailing soon subsides. Lewis, 46, is not a sadistic Scrooge; on the contrary, he is an eminent and kind-hearted psychologist who presides over the Institute for the Study of Child Development at Rutgers Medical School in New Brunswick, N.J. His laboratory is a friendly place filled with dolls and Teddy bears and jigsaw puzzles; blue-red-and-yellow rainbows streak across the walls. Along one of those walls runs a ten-foot-long two-way mirror so that Lewis can study children unobserved and record their activities on two videotape cameras.

Danielle's parents had adopted Stacy. Lewis wanted to observe how two sisters of similar age and upbringing, but totally different genes, would interact with their parents. All four started by playing with toys and puzzles in front of Lewis' mirror. The parents left, first individually, then together. The girls resorted to playing with each other. Then a stranger entered, and that seemed to make the girls more sharply aware of their parents' absence and their own aloneness—hence the outburst of tears. But why is that "great" or "fantastic"? "We're trying to determine exactly what normal behavior is," says Lewis, who sees the large in the small, "in this case the child's developing sense of self, the sense that it is separate from other people."

In a 17th century brick building on Paris' Boulevard de Port-Royal, once the abbey where the Mathematician Blaise Pascal underwent religious conversion, a quite different kind of experiment is taking place. Into a small room of the Baude-locque Maternity Hospital marches a nurse bearing a tiny, wrinkled infant named Géry. He is four days old and weighs 6 lbs. 6 oz. The nurse carefully deposits Géry in a waist-high steel bassinet that stands next to a computer. The computer is attached to an empty nipple. The question to be tested: Exactly what sounds can young Géry recognize?

The nurse pops the nipple into Géry's mouth and then turns on a nearby loudspeaker. A recorded male voice begins to recite a random series of similar syllables: "Bee, see, lee, see, mee, lee, bee, see, lee, mee." Géry's infant fingers clutch at the orange base of the nipple. Whenever he hears a new sound he sucks harder, and his heart beats faster. When he gets used to these sounds, his attention fades, and his sucking slows down. The computer tirelessly counts the number of sucks per minute.

"Da," the loudspeaker suddenly says. Géry sucks harder, then begins to cry. He is hungry, and the empty nipple brings him no food. The nurse comforts him. Even at the age of four days, the lessons of life can be hard.

A ll across the U.S., all over the world, medical and behavioral experiments like these are under way. Each by itself is a small and seemingly inconsequential affair; the results are sometimes inconclusive, sometimes obvious. But taken all together, they represent an enormous research campaign aimed at solving one of the most fundamental and most fascinating riddles of human life: What do newborn children know when they emerge into this world? And how do they begin organizing and using that knowledge during the first years of life to make their way toward the mysterious future?

The basic answer, which is repeatedly being demonstrated in myriad new ways: babies know a lot more than most people used to think. They see more, hear more, understand more, and they are genetically prewired to make friends with any adult who cares for them. The implications of this research challenge some of the standard beliefs on how children should be reared, how they should be educated, and what they are capable of becoming as they grow up. Yale Psychology Professor William Kessen, who has been studying infants for more than 30 years, says in admiration of the newborn baby's zestful approach to life, "He's eating up the world." Harvard Psychology Professor Jerome Kagan, another pioneer, offers only one caveat about the new research: "Don't frighten parents! The baby is a *friendly* computer!"

Many parents do get frightened, of course, particularly when a flood of books and articles keeps telling them what to do and not to do—and above all not to get frightened. The current discoveries about how much a baby sees and hears and knows at the very moment of birth make the parental responsibility seem even more formidable. Most important, in a way, is that these findings are changing the way people actually see their own children, changing how they talk to them, what they expect of them. And these slow and almost imperceptible transformations can hardly help altering, in subtle and equally imperceptible ways, the babies themselves, and thus the adults they will some day become.

The traditional view of infancy was that of Shakespeare, who described the helpless newborn as "mewling and puking in the nurse's arms." Nearly a century later, John Locke proclaimed it as self-evident that the infant's mind was a *tabula rasa*, or blank tablet, waiting to be written upon. William James prided himself on more scientific observations but wrote in *The Principles of Psychology* (1891) that the infant is so "assailed by eyes, ears, nose, skin and entrails at once" that he views the surrounding world as "one great blooming, buzzing confusion." As recently as 1964, a medical textbook reported not only that the average newborn could not fix its eyes or respond to sound but that "consciousness, as we think of it, probably does not exist in the infant."

Such views have been increasingly re-examined and revised during the past two decades, and this research has now grown into a substantial industry. From the Infant Laboratory at M.I.T. to the University of Texas' new Children's Research Center to U.C.L.A.'s Child Study Laboratory, there is hardly a major university without teams of researchers poking and prodding babies. The number of studies of infant cognition has tripled in the past five years, according to Psychologist Richard Held of M.I.T. A conference of experts in Austin last year heard more than 200 research papers ranging from "Sleep-Wake Transitions and Infant Temperament" to "Right-Left Asymmetrics of Neurological Functions in the Newborn Infants." These multitudinous studies do not go unchallenged: researchers in various disciplines fight for their own specialties, psychiatrists differ sharply in their views from neurologists, judgments are often subjective, and babies themselves are as different as snowflakes.

The search for data is being steadily pushed back from childhood to earliest infancy and even before birth. One French obstetrician, for example, inserted a hydrophone into the uterus of a woman about to give birth and tape-recorded what the fetus could hear: the mother's loudly thumping heartbeat, a variety of whooshing sounds, the muffled but distinguishable voices of the mother and her male doctor, and, from a distance, the clearly identifiable strains of Beethoven's *Fifth Symphony*.

The obvious obstacle that long hindered scientific research on babies was that they could not talk,* could not tell what they saw or thought; the consequence was a widespread belief that they saw little and thought less. But that belief was based primarily on adults' dim recollections of their past. As early as the 1950s, a few psychologists were searching for laboratory methods to discover what babies could learn. Case Western Reserve Psychologist Robert Fantz made an important breakthrough in 1958 by demonstrating that babies' fascination with novelty could be turned into a form of silent speech. Specifically, Fantz watched infants move their eyes when he showed them two different objects; he carefully measured what they looked at and for how long. Given a choice, he showed, babies will look at a checkerboard surface rather than a plain one, a bulls-eye target rather than stripes, and in general they prefer the complex to the simple. Says Rutgers' Michael Lewis: "Out of such

---

*The very word infant derives from the Latin *infans*, meaning incapable of speech.

elementary observations, monstrously important consequences grew."

Once the basic approach was discovered, a whole world of previously untried research opened up; new technology made it possible to devise tests that would have been unimaginable a generation earlier. At the most rudimentary level, the videotape machine enables a psychologist to record a baby's wriggling and demonstrate that it often moves in rhythm with its mother's voice. At the most complex levels, surgeons at Prentice Women's Hospital in Chicago can diagnose prenatal hydrocephalus (a brain-damaging excess of cerebrospinal fluid) in a fetus, then introduce a plastic tube into the mother's uterus and into the fetus' head to drain off the surplus fluid inside its brain. Guiding many of these technological innovations is the ubiquitous computer, which can synthesize a mother's voice as easily as it can measure eye movements or count the times that young Géry sucks on his nipple.

The first area to attract a number of researchers was the newborn baby's senses, which were once thought to represent little more than hunger to be fed. Systematic testing soon showed that babies not only perceive a good deal but have distinct preferences in everything. An Israeli neurophysiologist, Jacob Steiner, found that a baby as young as twelve hours old, which has never tasted even its mother's milk, will gurgle with satisfaction when a drop of sugar-water is placed on its tongue and grimace at a drop of lemon juice. More mysteriously, a newborn will smile beatifically when a piece of cotton impregnated with banana essence is waved under its nose, and it will protest at the smell of rotten eggs. Other infant prejudices: vanilla (good), shrimp (bad).

The baby emerges from the darkness of the womb with a rudimentary sense of vision—it would be rated about 20/500, or "legally blind," as one expert puts it, but eyesight develops rapidly. Newborns start by looking at the edges of things, exploring. Even when the lights are turned out, as infra-red cameras show, an infant's eyes open wide to carry on its investigation of its surroundings. At eight weeks, it can differentiate between shapes of objects as well as colors (generally preferring red, then blue); at three months, it begins to develop stereoscopic vision.

Testing such perceptions can be complicated. At M.I.T.'s Infant Laboratory, for example, University of Tokyo Graduate Student Shinsuke Shimojo has programmed a computer to check whether seven-month-old Whitney Warren can differentiate between a straight bar and a slightly indented bar. The computer makes the indented portion of the second bar move slightly. If Whitney can see the indentation, he will see its movement, and Shimojo, crouching behind the computer

screen, can see his eyes move. Most babies spot the movement easily.

Despite their esoteric quality, such experiments can have an immediate practical value: some infants suffer from eye ailments, such as cataracts, severe astigmatism and strabismus, which benefit from treatment much earlier than would once have been possible. No less important, the new research has demonstrated that an impairment of infant vision can damage those parts of the rapidly growing brain that rely on visual information. That brain damage can be permanent unless the eye impairment is treated early.

Unlike the eyes, the baby's ears have been functioning even before birth, and the newborn arrives with a whole set of auditory reactions. As early as the 1960s, tests indicated that babies go to sleep faster to the recorded sound of a human heartbeat or any similarly rhythmic sound. More recent studies indicate that by the time they are born, babies already prefer female voices; within a few weeks, they recognize the sound of their mother's speech.

Many mothers believe they can understand different kinds of crying by their babies (a controlled experiment in 1973 showed they could not), and they believe even more strongly that their babies can understand a parent's murmurings. And perhaps they can. Though children do not ordinarily say anything very elaborate before the age of one year, Psychologist Peter Eimas of Brown University has demonstrated that infants as young as one month can differentiate between sounds in virtually any language. They also have a "very sophisticated" ability to organize sounds into various categories. "A baby already knows which sounds communicate," says Eimas. "I've never heard a baby imitate the sound of a refrigerator, for example. So a child can put all of his energy into learning how to use the rules of the language."

Pursuing the origins of language back into earliest babyhood is an interesting approach to understanding the infant intellect. No less so is the discovery that this intellect is at work long before any language is available as a tool. The key element in that discovery was the baby's desire to imitate its mother's facial movements. Jean Piaget, the celebrated Swiss psychologist who pioneered in this field with extended studies of his own three children, declared that such imitations began only at about eight to twelve months. Earlier than that, he reasoned, the baby could not understand that its own face was similar to that of its mother.

Olga Maratos, a Greek student who was testing seven-week-old infants for her doctorate, went to Piaget's house one snowy day early in 1973 to tell him of her progress. "Do you remember what I am doing?" she said. "I am sticking out my tongue at the babies, and do you know what they are doing?"

"You may tell me," Piaget murmured. "They are sticking out their tongues right back at me! What do you think of that?"

The venerable professor puffed on his pipe for a moment as he contemplated the challenge to his theory. "I think that's very rude," he said.

Maratos' thesis was never published, so the credit for the discovery went mainly to two young psychologists who now teach at the University of Washington, Andrew Meltzoff and M. Keith Moore. Their study, published in 1977, showed that babies only twelve days old could imitate an adult sticking out a tongue. Meltzoff and Moore demonstrated that if a pacifier in the baby's mouth prevented the infant from imitating the adult, it would *remember* what it wanted to do until the pacifier was removed; then the baby would promptly stick out its tongue.

That first study by Meltzoff and Moore aroused considerable skepticism, so they repeated and elaborated it in 1981, eliminating all uncertainties and using still younger children. "We had one baby 42 minutes old, with blood still on its hair," recalls Meltzoff. "We washed it and tested it. We found that even newborns could imitate adults."

These experiments demonstrated the infant's very early capacity for what psychologists call "intermodal perception"—that is, to combine the brain's perceptions of two different activities, in this case vision and muscular action, which is virtually the first form of thinking. Says Yale's Kessen: "The past 15 to 20 years have demonstrated that the child has a mind. The next several years will be used to find out how it works."

Meltzoff pursued his exploration of intermodal perception by a different test of vision and touch. He gave ordinary pacifiers to a group of month-old babies and pacifiers with bumps on them to another group. He then had the babies look at models of the two kinds of nipples. The result, says Meltzoff, was that "they would look at the ones they had felt." Now, with Speech Professor Patricia Kuhl, he has extended those tests to language. The researchers showed infants two films of faces saying "ahh" and "eee," then placed between the two pictures a loudspeaker that could make either sound. The babies invariably looked toward the picture that fit the sound. "This means that babies can detect the relationship between mouth movements and the sounds they hear," says Meltzoff. "Essentially, babies are lip readers."

As they begin to develop this rudimentary capacity for thinking, babies develop an important ability to recognize categories. This was once thought to require language—how can the unnameable be identified?—but babies apparently can organize perceptions without a word. Psychologist Elizabeth Spelke of the University of Pennsylvania showed four-month-old babies a pair of films in which two toys bounced around on a surface in different ways, each with a corresponding sound track. She then played one sound track, and the babies were able to match the correct film to its sound. From the babies' "highly differentiated ability" to decide what goes with what, Spelke went on to deduce that children are born with an innate ability to divide their experiences into categories. Says she: "Obviously, in order to make sense of anything that you're confronted with, you have to bring to bear certain conceptions about the world. Our hope is that we'll learn something about what those initial conceptions are."

It is a puzzle, for babies repeatedly demonstrate a variety of skills and actions that seem to have no basis in their previous experience. Examples:

▶ Bradley Feige, age 11½ months, is sitting on a glass table at U.C.L.A.'s Child Study Laboratory. "Come here, Bradley, come here," his mother coaxes from the other side of the table, about six feet away. At her end, the cloth material under the glass top suddenly drops away to create the illusion that Bradley may plunge several feet if he does what his mother asks. At eight months, and again at ten months, Bradley ignored the illusion of peril and crawled across the table. Now he refuses to budge past the illusionary end of the table, not even when his mother holds out a toy as a lure. "We know that this response is not related to the experiences they've had," says Psychologist Nancy Rader, "but we've found that it relates to the age at which the baby starts crawling, and we're trying to find out why."

At Harvard's Center for Cognitive Studies, infants as young as two weeks were confronted with a cube (or sometimes only the shadow of a cube) that began moving slowly toward them. When it seemed about to hit them, they showed what psychologists call "a strong avoidance-reaction pattern." They turned aside and squirmed and tried to avoid being struck, though they had no previous experience that would make them think that the approaching object would hit them. When such a cube or its shadow approached the babies on an angled path that would miss them, however, the babies followed its motion with their eyes but showed no sign of anxiety. "The consummate skill of these infants in predicting the path of the moving object is astonishing," says Psychologist Jane Flannery Jackson, "and their evident wish to avoid objects on a collision course is even more so."

▶ At the University of Edinburgh, T.G.R. Bower and his associates have been conducting about 1,000 experiments a year on various infant abilities. One of their most startling claims is that babies can tell the gender of other infants they are looking at, and they prefer to look at those of their own sex. Bower made films of an infant boy and girl making various movements, and then deleted from the film all apparent signs of gender and even swapped their clothes. Some adult viewers had difficulty telling them apart, but something about the way the filmed infants moved enabled a group of 13-month-old children to distinguish the boy from the girl. Bower is still trying to figure out how they do that.

How babies do any of the things they do is a matter of considerable complexity. Some theorists, like Thomas Verny, a Canadian psychiatrist who wrote *The Secret Life of the Unborn Child,* believe the infant begins learning behavior patterns while it is still in the uterus. Most experts, however, assume that the genes still carry messages that primitive humans once needed for survival. The so-called Moro reflex,* for example, which makes a newborn infant reach out its arms in a desperate grasping motion whenever it feels itself falling, implies some monkey-like existence at the dawn of time. Says Lewis Lipsitt, director of the Child Study Center at Brown and a pioneer in research on babies: "The human infant is extremely well coordinated and put together for accomplishing the tasks of infancy. These are: sustenance, maintaining contact with other people, and defending itself against noxious stimulation."

One of the oddest elements in their development is that infants soon lose many of the skills they had at birth. A newborn baby that is held upright on a table is nearly able to walk while suspended; immersed in a tub of water, it makes a fairly impressive try at swimming. Those abilities deteriorate within a few months. The same process seems to occur with intellectual skills that are not used. Psychologists Janet Werker of Dalhousie University in Halifax, N.S., and Richard Tees of the University of British Columbia have shown that babies of six to eight months can distinguish sounds that are not used in their native language, but they have much greater difficulty by the age of twelve months. Japanese babies, for example, have no trouble with the "ell" sound that their parents find difficult.

Most experts now think a baby is born with a number of reflexes that are gradually replaced by the "cortical behavior" dictated from the cortex of its rapidly developing brain. Brown's Lipsitt believes that a period of "disarray" during the course of this transition may be an important element in the "crib deaths" that can mysteriously strike during the first year. The struggle to escape from accidental smothering in bedclothes, known as the "respiratory occlusion reflex," is automatic at birth but then needs to be learned. Says Lipsitt:

---

*Named for German Pediatrician Ernst Moro (1874-1951).

"The peak of 'disarray' is right at the point when crib death is most likely to occur, as if the baby doesn't know whether to be reflexive or cognitive. Suppose a child gets into a compromising situation where it has lost the reflex and has not acquired the learned behavior that has to come in to supplant the lost reflex." Lipsitt hopes to devise a specific test that will pinpoint those few children who may be in jeopardy.

Every test for every kind of trouble implies that there is a "normal" time for a baby to demonstrate various abilities. If it does not sit up by six or seven months or stand by nine or ten, a pediatrician may start neurological testing. The disciples of Yale's Arnold Gesell have applied this approach to all phases of childhood ("He wanders from home and gets lost at four," says the latest edition of the Gesell Institute's *Child Behavior.* "He demands to ride his bicycle in the street at eight").

Most current advice givers urge anxious parents not to take such standardization too seriously. Pediatrician T. Berry Brazelton *(see box),* who is publishing a revision of his 1969 bestseller, *Infants and Mothers,* begins by declaring: "There are as many individual variations in new born patterns as there are in infants." Still, though a child's development during its first year is far slower than that of a monkey or even an elephant, it is nonetheless so dramatic—from lying flat on its back to the first creeping across the floor to the first faltering steps around the corner of the kitchen table—that scientists persist in trying to pinpoint when and how it learns each new accomplishment.

Two months, eight months and twelve months seem to mark major periods of change: in brain developments, in various skills and perceptions, in sociability. At about two months, for example, the baby is awake much longer than it was, it smiles a lot and stares with fascination at a new discovery: its own hand. At eight months, the infant is acquiring the important sense of its separate identity, and even an understanding of what Piaget called "object permanence," the realization that an object hidden from sight is still there. It begins to develop fears of strangers and of separation from its parents. At twelve months, the golden age, the baby has begun to walk and talk, and knows that the whole world awaits. Sometimes, clinging to a chair, waving a spoon in a fist, the one year old will throw back its head and crow in sheer delight.

These physical and social achievements have long been obvious: any mother can see them in her own children. What the new research demonstrates is that babies' mental growth can be as early and as striking as the rest of its development. Robert Cooper, a psychologist with Southwest Texas State University, is even testing a group of ten- to twelve-month-old children on their ability to recognize

The New Dr. Spock: "A Great Dad"

Dr. T. (for Thomas) Berry Brazelton, 65, says he is no scientist, which shows a becoming modesty, but he would have a hard time denying that he is the nation's pre-eminent baby doctor. A whole generation of pediatricians has studied and worked with him at Harvard Medical School and the Children's Hospital Medical Center in Boston. Tens of thousands of anxious parents have been reassured by his easygoing guidebooks (*Infants and Mothers, Toddlers and Parents, Doctor and Child, On Becoming a Family*). Millions of infants who never met him have been tested and evaluated by his Neonatal Behavioral Assessment Scale, generally known simply and naturally as the Brazelton.

It is a folksy sort of test, carried out with such implements as a pocket flashlight, an orange rubber ball, a paper clip, some popcorn kernels. The exam starts when the baby is asleep, and it gauges the infant's reactions to a series of stimuli, including light in the eyes, the sound of rattling, a scratch on the foot: 20 reflexes and 26 behavioral responses in all. After 20 minutes or so, a Brazeltonized baby is wide awake and none too happy about all the testing.

Brazelton began devising the exam some 30 years ago to solve a problem that bothered him: babies available for adoption were being kept in institutions until the age of four months because doctors were reluctant to certify that any younger infant was fully normal. "Four months is just too long to deprive anybody of a new baby," Brazelton recalls, with a trace of a Texas drawl that has survived his years in Boston. "That led me to say, 'Well, gosh, anybody can tell whether a new baby's O.K. or not. What is it we're going by?' Then I began to put together all these things that any good clinician uses. Very little about the scale was really new. It was a compilation of a lot of clinical observations that hadn't been documented."

That all sounds rather routine, but Brazelton's tone changes as he starts to talk about his test, about the way a baby's eyes jerkily follow a moving ball. "If you give him a human face to look at instead, his eyes will widen and he'll get more intense and he'll follow you," says Brazelton, "and as he follows, his face gets more and more alert and more and more involved, and you can feel yourself getting more and more involved back. This kind of visual involvement is more than just looking. You've got another component from the baby, which says to the person doing this, 'You're terribly important.' And the person is bound to feel important. What I'm getting at is that the baby's competence will call up competence from parents. We used to see the parents shaping the child, but now we see the child also helping to shape the parents."

Brazelton once wanted to be a veterinarian. At age eight, already an experienced baby sitter, he decided on pediatrics. He went to Princeton, starred in Triangle Club theatricals, even got an offer in 1940 to try out on Broadway for an Ethel Merman musical, *Panama Hattie,* but he held on to the goal of healing infants. His hero, he says, was Benjamin Spock, and although Brazelton is now regarded as the new Spock, he considers himself more a disciple than a rival of the older man.

Brazelton and his wife Christina have three daughters and a son, ages 19 to 32. He still worries about his high expectations and pushing his children too hard. His son, however, calls him "a great dad."

Like Spock, Brazelton makes it a cardinal rule to reassure anxious parents and to encourage them to trust their instincts. "Parents in our culture are so hungry for people to tell them what to do and so vulnerable as a result," he says. "I feel very strongly that telling them what to do is destructive. Supporting them for what they can do is constructive."

Brazelton is in the midst of a project of "intervention research" that involves studying 100 undersized babies and trying to see which of them will need special assistance. Babies that have been undernourished in the uterus are "very scrawny, very hypersensitive to any kind of stimulation, and they become very fussy and difficult for the parents," says Brazelton. "They need help to see their baby as a person. You have to help parents see that you're seeing the same baby they are. And that the baby doesn't need to be like them. And they don't need to be like it. It can be just as exciting to find another kind of person to learn about."

different numbers. They can master up to four, but he adds that "beyond four, there's some controversy." By showing his little subjects various groups of objects, Cooper demonstrates that they can tell the difference between three and five, he says, though the difference between four and five sometimes baffles them.

The idea that infants can start acquiring an education has tempted ambitious parents for centuries. At the age of three, John Stuart Mill learned Greek, and Mozart was playing the harpsichord. Both were taught by their hard-driving fathers. Today, New York City's fashionable nursery schools not only interview two year olds (and charge their anxious parents $1,200 a year for two mornings of schooling a week), but they also report applications outrunning openings by as much as 5 to 1.

The vogue is spreading. Gymboree, a franchise operation that started seven years ago in San Mateo, Calif., now has 61 outposts operating in 14 states that provide educational play for about 10,000 children. "Learning to read begins at birth," says one of Gymboree's brochures, but the $4 classes are mainly physical, ranging from "wee workouts" for beginners up to "gymgrad" for tots as old as four. "We've tried to create a 'yes' environment for the children, to place them in a setting they can master," says Gymboree's founder, Joan Barnes, a former dance teacher.

More strictly pedagogic is a Philadelphia organization called the Better Baby Institute, which offers a training course to enable mothers to "multiply their baby's intelligence." Specifically, the school claims that parents can learn in one week of intense instruction (for a fee of $500) how to teach their infants to swim, to read, to do math, to speak foreign languages and to play the violin at the age of two. You can't make it to Philadelphia? "Better Baby Video," a California-based spin-off, can provide the same lessons in a weeklong course offered primarily in West Coast cities. Some critics believe that all this mainly makes babies learn a few skills by rote, but it is difficult to obtain any scientific assessment of the five-year-old institute.

Many of these ventures in infant education are fueled by eager parents who will try anything to give their children a head start. Similar experiments are arousing interest in those who work among the poor. Dr. Joseph Sparling, for example, has developed and published a series of 100 educational games at the Frank Porter Graham Child Development Center at the University of North Carolina at Chapel Hill. These games, which range from specific subjects like language development to vague concerns like self-image, have been tried out with some success over the past five years in a federally funded program called Project Care. Researchers use the games both in day care centers and in weekly visits to children's homes. They report that the children get "significantly" higher intelligence-test scores at the age of one year than children in a control group who are not exposed to the games.

If nothing else, the push toward earlier education gives infants a valuable chance at making friends. Says Psychologist Colwyn Trevarthen of the University of Edinburgh: "They really have this intrinsic social capacity, and that's what human beings have evolved for, just as giraffes have evolved for eating high leaves."

But is early education itself really desirable? Does the discovery that a young child can absorb large quantities of knowledge require that it be stuffed like a Strasbourg goose? There were social reasons for launching Project Head Start in the 1960s to get poor children into preschool programs. Most psychologists engaged in the new research, however, are strongly opposed to any formal schooling before the age of three or four, even if the child is capable of it. "We know that babies are coming into the world with a lot more sophisticated skills than we had previously thought, but I do not think reading, writing and arithmetic should be in their curriculum," says Psychologist Tiffany Field of the University of Miami School of Medicine. Warns Child Psychiatrist Robert Harmon, director of the Infant Psychiatry Clinic at the University of Colorado School of Medicine: "I think you're going to get children burned out on learning." And University of Denver Psychologist Kurt Fischer says of the baby's first year: "Don't worry about teaching as much as providing a rich and emotionally supportive atmosphere."

As Fischer's statement indicates, much of the new research emphasizes the extreme importance of the infant's relationship with its mother. (And/or its father, and/or what the linguistically liberated call the "caregiver.") She must not only feed it, and love it, but endlessly talk to it, play games with it, show it what is happening in the world. Rutgers' Lewis has tested 100 babies for mental development at three months and recorded their mothers' response to the infants' signs of distress. He was hardly surprised to find that those who had been more warmly cared for had learned more by the time they were retested at the age of one year. This kind of nurturing is essential to both emotional and intellectual growth; indeed, the two are inseparable. "The baby who doesn't smile may be giving us a more reliable indicator than cognitive tests," says Psychiatrist Eleanor Galenson of Manhattan's Mount Sinai Medical Center.

The baby's smile is also a kind of judgment on the care that its mother has been providing. "All these new data about how early the baby can distinguish things should upgrade motherhood, restore some prestige to it," says Dr. Benjamin Spock, 80, who taught a benign form of child rearing to a whole generation of Americans. "Motherhood has had an ever reduced amount of importance placed on it in our strange, overly intellectualized, overly scientific society."

According to traditional wisdom, all mothers know instinctively how to rear their children, but unfortunately that is not always true. Indeed, the instinct has been vehemently denied by Elisabeth Badinter, the French philosophy professor who wrote *Mother Love: Myth and Reality*. But even if a mother's nurturing is an instinct, it requires some experience as well, and if the ability is entirely a learned trait, it is sometimes none too well learned. To check on how consciously mothers interact with their babies, Psychiatrist Daniel Stern of the Cornell University Medical Center has been observing nearly 100 mothers playing with infants eight to twelve months old. "Whenever we notice that the baby has put on an emotional expression that the mother has seen, we look at how she responded to it," says Stern. "Then we ask her why she did it, what she thought the baby was feeling, what she expected to accomplish, and whether she knew what she was doing at the time." His preliminary findings: about one-third of the mothers were fully aware of what Stern calls the attunement with their infants, another third were quite unaware of it, and the rest were essentially unaware but could recall it when it was pointed out to them.

This extremely important emotional interplay, often described as "bonding," is a combination of love and play, but it is now seen as something else, a kind of wordless dialogue. The baby not only understands what the mother is communicating, or not communicating, but it is trying to tell her things, if she will only listen. Says Dr. Bennett Leventhal of the University of Chicago's Child Psychiatry Clinic: "We now know that babies send messages very early. In their first year of life, they are good students. They are also very good teachers, but they have to have someone to interact with them. There are sometimes very competent babies with very incompetent parents."

Many psychologists believe the new research enables them to anticipate future problems in even the youngest children. "We can now document where a baby may be unable to pick up sensory data; we can spot abnormalities in the emotional areas," says Stanley Greenspan, chief of the Clinical Infant Research Unit of the National Institute of Mental Health in Adelphi, Md. "There is no evidence that an infant's emotional problems are self-corrective. The environment that contributed to early damage will continue to contribute if one does not intervene."

One early and important symptom of trouble, says Greenspan, is the failure of mother or child to look at each other.

Greenspan makes videotapes of such cases. Here is Amanda, age four months, who turns her head away and generally shows what Greenspan calls "an active avoidance of the human world." Small wonder. Amanda's mother was raising her alone and suffered bouts of deep depression. Greenspan and his therapists spent four months playing with Amanda and engaging her interest; the videotape taken at eight months shows the baby cheering her mother along. Says Greenspan, with some satisfaction: "She developed coping facilities stronger than those of her mother."

Psychologists talking about "environment" often meant primarily the psychological structure of the family, but the social and economic environment is hardly less important to a child's development. Fully 13.5 million children in the U.S. live below the official poverty line. Nearly 7.5 million children are currently on welfare. More than half a million babies are born every year to American teen-agers.

The effects of such deprivation on infancy are hard to gauge scientifically, but Dr. Gerald Young of Manhattan's Mount Sinai Medical Center says flatly, "If you want to guess what a child will be like at age seven, look first to the socioeconomic background." This is not simply a matter of economic hardship or nutritional deficiency. Says Brown's Lipsitt: "The socioeconomic index is as powerful a predictor of later intellectual prowess as any variable we've got, but it doesn't operate in a vacuum. It is a representation of the way people live and relate toward each other, and the way they behave toward babies."

One interesting demonstration of this theory was undertaken more than a decade ago by a team of psychologists at the University of Wisconsin. Struck by the fact that many of the mentally retarded children in a Milwaukee slum had retarded mothers, they took 40 infants whose mothers had IQs of less than 75 and put 20 of them in special day care centers. From the age of three months on, the children began getting lessons in language and arithmetic as well as various other kinds of stimulation. By the time they reached school age, their average IQ was more than 100 (none was retarded); the 20 children who had received no special treatment had an average IQ of 85, and 60% were judged to be retarded.

The question of child rearing outside the home cuts across all classes. There are currently 4.1 million working women with children under the age of three, and one survey showed that nearly 70% of working women who have babies return to their job within four months. Overall, about 8 million of today's preschool children receive some form of day care (1 million in day care centers, 3.5 million in family day care homes and 3.5 million tended by relatives and baby sitters). If the nurturing mother is as important as psychiatrists say, hired substitutes may seem a poor alternative, but most psychological researchers reject any such conclusion. All a baby basically needs, they say, is at least someone who is consistently there and who really cares. All depends, obviously, on the quality of the day care—and of the home. In the case of the Milwaukee experiment with the potentially retarded, day care was a rescue service. But in one typical Maryland county, 788 regulated day care facilities have room for only 8,560 of the 65,000 children under 14 who have working mothers.

How good the average day care is remains something of a guess. Bernice Weissbourd, who founded the Chicago-based Family Focus groups to provide support and advice for new parents, argues that any day care service that has more than three infants per adult (and that includes most) is inadequate. "Too often," she warns, "the parents' main questions are simply how close to home is it and how much does it cost." But of day care as such, Cornell Psychologist Urie Bronfenbrenner says categorically, "There is no hard evidence that day care has a negative effect."

Whatever the difficulties, the overwhelming majority of parents want very much to do the best for their children, if only they can be sure what that best is, and that is anything but certain. Most experts say the need is great. "Not more than one child in ten gets off to as good a start as he could," says Burton White, author of *The First Three Years of Life*. Harvard's Kagan, on the other hand, urges parents to provide "a nurturant environment" and declares, "It's easy. Oh, it's easy. There's not a lot of witchcraft here."

Important changes come so slowly that they are taken for granted. Children's sheets used to be all white; now they are explosions of color. The mobile over the crib, which first seemed arty and pretentious, has become almost a basic piece of furniture. The backpacks that were once associated with Indian women carrying papooses are now sold everywhere, not only as a convenience for mothers but as an opportunity for the baby to get out of the house and see the world.

Thus the old keeps becoming the new. Much of what modern research is so elaborately documenting is what parents have always known—whether from instinct or from common sense or from the teachings of their own parents—that babies need and respond to love, attention, stimulation, education, in perhaps roughly that order. The research documents not only the importance of such needs but the damage that can occur when they go unanswered. Yet even these blessings of the latest orthodoxy can be overdone. "We are learning that everything will have an impact on an infant, but we still need to know exactly what happens," cautions Psychologist Rose Caron of George Washington University's Infant Research Laboratory in Silver Spring, Md. "It's conceivable that a child's competency might be diminished because of too much early stimulation."

"Do I contradict myself? Very well then I contradict myself," cried Walt Whitman. The creation of a baby is full of paradoxes and illogicalities. The cost of raising a child to 18, approaching $100,000 in the U.S., according to one estimate, would deter any sensible investor. So would the prospect of more than 20 years of anxiety and irritation. But having a baby remains, for most people, an act of faith. It represents a belief in better things to come, not just for themselves but for the world. That is a faith shared by the myriad baby researchers. Says Rutgers' Lewis: "Can we produce a better society with healthier children? The answer is yes." And in the very moment of birth, as a tiny, dark, wet head thrusts out into the world, every baby fulfills that belief. Then comes the first squall.

—By Otto Friedrich. Reported by Ruth Mehrtens Galvin/Boston and Melissa Ludtke/New York, with other bureaus

THE ART OF TALKING TO A BABY

ALICE STERLING HONIG

Alice Sterling Honig, PhD, is professor of Child Development at the College for Human Development of Syracuse University and the author of Playtime Learning Games for Young Children, *published by Syracuse University Press.*

Shoshannah, four months old, began to perk up at the sound of Grandma's voice. From Mother's arms, the baby strained to see the face from which such loving, admiring tones were coming. Grandma was trying to become acquainted with this newest person in the family. Baby beamed and cooed, "Uh! Ahhh! Mmmm!" in animated response to Grandma's greetings and admiring talk. Naturally, Grandma redoubled her praise. Shoshannah cooed vigorously, with a variety of tones. Then both took turns—more than 10 turns—exchanging happy, important communications with each other. Shoshannah had made another conquest. Such a bright child! Such a beautiful baby! Grandma was entranced by this spirited young conversationalist.

Where had it come from—this expressive cooing that mixed open vowels, long vowels and occasional consonants, such as "mmm" or "dll," into the stream of talk? Shoshannah's mom had learned the secret of helping her tiny baby become a special person in the family, a person who could capture and engage others in talk, a person who had interesting feelings to express to others. Both Mama and Papa had been talking to Shoshannah from the moment she was born. They had bundled her like a small papoose on their laps and murmured loving talk to her. At first, she had stared wide-eyed. By three weeks, she smiled up at Papa and made a sound when he did this special talking. As she nursed, she watched Mama's face so intently when Mama talked to her.

Both parents were careful not to overstimulate their baby. When Shoshannah was one month old, just getting through several turns of cooing back and forth with a beloved parent was hard work. Her coos were accompanied by alert gazing and bodily efforts as she struggled to become a partner in conversation. Sometimes the little one just had to avert her eyes or even close them for a while. Sometimes she puckered up her face af-

ter such efforts, but her special grownups were sensitive to Shoshannah's need to be in charge of tuning in and out of these conversations.

Babies delight in emotionally nourishing sights and sounds. But babies know best when they are tired or overstimulated. They need control over the dialogues—whether these are gazing or feeding or cooing dialogues. So Shoshannah's parents were careful to let her decide when she needed a rest from the talking exchanges. An infant who is constantly bombarded with eye contact or jabber will learn to turn away her eyes and head in order not to feel overloaded.

EARLY BODY DIALOGUES

Language is fundamentally a system that empowers humans with the creative ability to generate new ways of speaking with and for and to others about any topic or experience. Yet long before the mastering of vocabulary and grammar and the meaning of words, babies practice communicating games with parents. These interactions set the

From *Baby,* Vol. III, No. 3, 1988, pp. 12-14, 16-17. *Baby,* published by Working Woman/McCall's Group. Copyright © 1988 by WWT Partnership.

All parents are able to master the skill—once they know the secret. Here's a sensitive, step-by-step guide

groundwork for the later ability to communicate with meaningful words. Smiling games, feeding interactions and adjustments in how the infant is held are among the early games that parent and baby play as they get to know each other. Gestures from baby may tell a sensitive adult that the baby needs a hoist to the shoulder for a burp rather than continuation of feeding. Or, baby may mold his body more comfortably on Papa's lap so that they sit together in loving ease. Myriad signals help babies and parents tune into each other's communication patterns and needs.

So adept do some infants and parents become in early prelanguage communication games that even when language is not convenient, babies and parents can play games that enrich intimacy and responsiveness between them:

Heather's mother was talking on the phone while her nine-month-old daughter was playing with scraps of paper on the coffee table two feet away. Heather crumpled one piece of paper into a small ball, looked at her mother and offered the ball to her. Heather's mother continued to talk on the phone but glanced at her daughter's gift. Then she took the ball of paper from her daughter, held it a moment, and gave it back. Heather took the ball of paper, put it in her mouth for a few seconds, and then again handed it to her mother. During the next three minutes, Heather sustained contact with her mother, who remained talking on the phone throughout the exchanges of a crumpled, and increasingly wet, ball of paper. (From Prelinguistic Communication in Infancy, *by A. Ziajka.)*

HOW LANGUAGE GROWS: STEPS AND STAGES

The birth cry is the first vocal signal with which babies greet the world. Over the next weeks, they practice the throaty sounds we call "coos." By four months, babies are producing consonants, such as "mmm" and "ng."

During the first six months of life, infants are biologically primed to produce all the sounds contained in the languages of the world. During the next months, only those sounds will remain in the baby's repertoire that belong to the language of his caregivers.

After the first six months, babies start to combine vowels and consonants and to play with the resulting sounds—"muh, tuh, pa, nah, ma." When doubled, these vowel-consonant combinations sound very much like the words "Papa" and "Mama" and "Nana" that are heard in nursery talk around the world. How clever of adults to seize on the earliest sound-doublings of babies and to rejoice, sure that baby is now calling them!

Between nine and 12 months, many babies produce their first word. Some common beginning words are "li" (light), "ki" (kitty), "ju" (juice), "wa-wa" (water), "ca-kuh" (cracker), "up" and "no." Some parents say this last word themselves far too frequently as baby starts to creep about or play with food or show her developing awareness of adult "toys" that are so attractive to touch. Pointing out pictures of "doggie" or "eye" in response to your request brings great feelings of pride to the baby by the end of the first year. Babies have trouble with the ends of words. "Muh" for "more" may be the best that baby can do. Don't prod or force. Just say the word correctly; nagging will not help. Don't worry about using baby talk. It is fine with babies. Just be sure you switch to correct pronunciation as your toddler becomes more capable with words.

Near the end of the first year, babies begin to use long, murmuring strings of earnest babbling to engage in conversations with loved adults. Decoding babbles is hard and mystifying! Just nod positively, or murmur "Uh-huh" reassuringly. Help your baby to know that you appreciate her communication to you.

Another language skill that baby is developing is the ability to use intonations to suggest questions, commands and declarations. Babies who babble unintelligible syllables often are adept at communication because they use stress, intonation, pitch and the juncture between sounds to express surprise, puzzlement, suspicion, earnest report, angry demand, etc. Thus, even before baby's words are intelligible, we need to respond, to reassure baby that his basic emotional messages are getting through. By valuing baby's babbling, parents give a boost to baby's self-esteem. Here is an incident that shows the importance of baby's talk:

Barbie toddled over to the door behind which the pediatrician and nurse had disappeared with Barbie's older sister, Lucie. When the doctor gave Lucie her shot, she wailed piercingly. Barbie looked troubled, pushed at the door and jabbered excitedly to her mom in long strings of jargon. Lucie's name was heard among the garbled sounds. The mother turned to me in the waiting room and laughed: "Barbie goes on like that all day long. It doesn't mean anything!"

How else could Barbie's mom have responded? She might have listened for the fear and concern in the babbling and soothed her toddler with real respect for toddler feelings: "You are worried about your sister, honey. The doctor just gave her a shot, so she won't get sick. It hurt for a tiny second. Lucie will be OK. She will be out soon. Come here, lovey, and I'll hold you for a while, until your sister comes out of that room."

After the early stage of babbles and of single words, sometime during the second year, toddlers begin to put two or more words together to express universal concepts. I have described these concepts as follows:

Recurrence:	"More juice" "More bouncey-bounce"
Description:	"Hot stove" "Pitty dress"
Want/Wish:	"Me want egg"
Negation:	"No want nighty-night" (I don't want to go to bed)
Agent-Action:	"Daddy read me" "Baby crying" (That baby is crying)
Command:	"Shoe on" (Put my shoe on)
Action-Object:	"Fro baw" (Throw ball)
Agent-Location:	"Book dere"
Classification:	"Dat doggie" (That's a dog)
Possession:	"Mines toy"
Instrument:	"Cut knife" (Cut with the knife)
Question:	"Where juice go?"

Speech at this stage is often called telegraphic. The toddler leaves out little words like "did" or "a" or "is" or "to." Some toddlers may have trouble learning to use "I" instead of their own name. Some will find certain letters like "r" very

Even before a baby's words are intelligible, we need to respond—to reassure him that his basic emotional messages are getting through.

hard to pronounce. Some will confuse "mine" and "yours," both in the words they use and in their actions! Toward the end of the toddler period, connecting words (e.g., "and") appear: "I wanna candy an' a soda."

As toddlers become aware of the rules of language that they are hearing, they begin to abstract those rules and apply them overgenerously. Toddlers who have been saying "I fell down" may now start to say, "I falled down and I hurted myself." Sometimes they say "foots." We call this overgeneralization. If your toddler announces proudly at the table: "Me eated da meat all up," you can respond enthusiastically: "You ate up all your meat. You ate it all up!"

HOW IMPORTANT IS ADULT TALK FOR BABIES?

Sometimes adults believe in old wives' tales: "A good baby is a quiet baby"—an undemanding baby. But what happens when babies are never talked to? Naturally, scientists would never subject infants to such deprivation. But in the 13th century, the Holy Roman Emperor Frederick II set up such an experiment. The chronicler Salimbene gave this account of the findings:

…*He wanted to find out what kind of speech and what manner of speech children would have when they grew up if they spoke to no one beforehand. So he bade foster mothers and nurses to suckle the children, to bathe and wash them, but in no way to prattle with them, or to speak to them, for he wanted to learn whether they would speak the Hebrew language, which was the oldest, or Greek, or Latin, or Arabic, or perhaps the native language of their parents. But he labored in vain because the children all died. For they could not live without the petting, joyful faces and loving words of their foster mothers. (From A Portable Medieval Reader by J.V. Ross and M.M. McLaughlin.)*

Thus, the awesome power of parents (and caretakers) as early language teachers seems startlingly clear. Yet many parents feel they do not have to pay much attention to language. They think that older children will teach baby to talk; that baby will just pick up speech. The truth is, parents are the most important language teachers of their babies.

TIPS FOR BOOSTING BABY LANGUAGE

Nowadays parents are very concerned about how they can best help their babies become bright and intellectually competent children. Parents sometimes worry that their busy life does not give them enough "quality time" to spend helping their baby to learn. Talking with babies is a secret ingredient to early learning that is easy for a parent to carry out. Every diaper change, every pickup in order to shift baby's position, every routine cleaning, bathing, dressing and carrying activity becomes a rich opportunity for talking with baby. Then, in turn, baby becomes an active and important partner in conversing with her mother, father or caregiver. Thus, talking to babies helps them to become distinctive, special persons.

Most adults raise their voice tones when they talk with young babies. Something about raised, loving tones pleases a baby. Babies also love chants and will bounce their behinds in rhythm to the cadenced tones of nursery rhymes.

TALK WHILE PLAYING WITH BABY

Talking can be combined with all sorts of little games. Let baby's fingers grasp yours, as she or he lies on a firm, comfortable surface, such as a rug. Sing "Up and down, up and down, baby is going up and down." As long as you use two different notes, one higher

and one lower, the tune will sound just fine, especially if you accent the upper note. The rule in force, of course, is to watch for baby's signals. Smiles and coos and eager participation in a lifting-to-sitting-position-while-chanting game tell you that the game is OK. Tired or fretful responses tell you that baby needs a rest, or that the game is too strenuous for now. By the way, your baby does not care that you do not have the voice of a Metropolitan Opera star. Babies take pleasure in musical cadences and singsong rhythms as you tend to their needs.

One game that baby will adore is to hear you use plosive sounds. Say words like fabulous, physics, fantastic, fillibuster, fussbudget. Exaggerate the "f" sound in each word. Another time try "p" sounds: pelican, Popeye, pineapple. Babies enjoy play with sounds. "Tsee-tsee bird," repeated forcefully and lovingly, can set a baby chuckling and giggling.

Games that have body actions to go with them are also favorites of young and older babies. Eight-month-olds may enjoy it if you hold their hands and put them through all the motions of "Pat-a-cake, pat-a-cake, baker's man, bake me a cake as fast as you can. Roll it" (slow down as you roll baby's hands), "and pat it and mark it with B" (trace a B on baby's tummy with his own hands), "and put it in the oven for baby and me." End on a triumphant, upbeat note as you bring baby's hands together and pretend he's putting a cake into the oven. Combining body actions with sounds makes language a pleasure for babies.

Older toddlers usually enjoy fingerplay games, such as "Eentsy, beentsy spider"; sociable games like "Ring-around-a-rosy" are also favorites.

TALKING OPPORTUNITIES

How often an adult changes diapers silently, intent on the mechanics of wiping a behind or lifting legs. Yet the diapering table is a language-lesson locale par excellence. You are in a one-to-one situation with baby. Your eyes and mouth are from 12 to 20 inches away from the infant's. This dis-

Each episode during the day provides its own opportunity and scenario for conversing with baby.

Toddlers love a story to sound the same at each reading, and they love to hear it over and over.

tance is ideal for baby to focus on your face and hear you telling and explaining all about how you are changing him, how you are taking off that wet, uncomfortable diaper, how you are cleaning up that bottom, how you are getting baby all clean and comfortable again, how delicious baby's bottom and limbs are, how much you love those toes that wiggle up in the air as you slip a new diaper beneath that dimpled bottom.

During diapering, you need not monopolize the entire time with descriptions of your prowess in the diapering process. This is also a wonderful opportunity to comment on baby's bright eyes, on pretty fingers, on interesting coos and other sounds baby is making. As soon as baby makes a sound, express pleasure. Reward baby's sounds. You may just want to say, "Oh, I heard you saying 'guh.' That is neat—I like the way you say 'guh.'" Or if baby murmurs "Meh," you can repeat the sound and remark, "That's your sound. Yes. I can say meh too. Meh. There! That's a nice sound you taught me!" Much as the sparrow in the Mary Poppins tale was surprised that Mary understood what she called "sparrer" talk, babies are often very pleased when we seem to understand their talk.

Every episode during the day provides its own opportunity and scenario for conversing with baby. Talk while taking a walk over to a window with baby in your arms. Comment on what you are doing as you walk, how you are enjoying your time together. Shared attention, shared experience—language ties them together and gives richer meaning to the time you spend together.

Giving baby a bath, whether in the sink or tub, is a good time to describe the squishing of water on a sponge, the wetness, the delight you feel as baby kicks his legs and splashes the water a bit. Verbal accompaniments to bathtime add flavor to the experience. Such talk is called "self-talk." You talk about what you are doing with and for baby.

O TALKING AND WORKING

Opportunities for talking with babies arise during the homeliest of routine household activities:

"You are really getting hungry. I am cooking this yummy soup for you."

"Your nose has such sniffles. Let me see how gently I can wipe your nose."

"You are hugging dolly so gently." (This kind of talk is called "parallel talk." You give the baby or toddler words that go with her actions or feelings or toys.)

"I know it is hard to wait when you want Mama and Papa and they are not here. You really wish they would come back soon. You are feeling so mad and sad." (This is fine talk for a babysitter who has to deal with a toddler now awake from a nap.)

Babies and toddlers also love it when words are set to a chant: "We're going home from the grocery store, the grocery store, the grocery store." (This may help a toddler find the energy to walk that last block home.)

"Soon we will have our lunch; soon our meat will be on our plate; soon we can eat our carrots." (This is a fine chant to help a hungry toddler accept with patience the moments that it takes to seat her in a high chair, put on her bib and bring the food. Language helps a toddler find internal controls. Reassuring words let her know you are working toward getting food ready. You care about her hunger. You know how hard it is to wait for a meal.)

"You are picking up your toys so that we can get ready to put on your snowsuit and go to the park. You are such a big help. You are a hard worker."

"Can you find me another sock just like this sock with the red stripes? You have to look so carefully in the laundry basket. You are such a careful looker! Thank you! You found the matching sock so nicely." (Such admiration builds a toddler's self-esteem and her image of herself as an important household member.)

R READ TO YOUR BABY

Reading to babies is a powerful booster of language development. In the beginning, the parent should just point to clear, single, colorful pictures that represent familiar items in the baby's world. Pictures of animals, clothing, foods and familiar faces will hold baby's attention. Use your public library and bring home armfuls of books. But beware. Toddlers love a story to sound the same at each reading and love to hear the same story over and over …and over. *Goodnight Moon* is a wonderful book to read at bedtime.

A FINAL THOUGHTS

As you play with, sing with, talk with and read to your baby, remember that some babies speak later than others. Be patient. If your baby seems to understand you, and your toddler can carry out simple requests, then you know that receptive language is just fine. Expressive language takes a little longer for some babies, who might be more spatially oriented than verbally oriented in their early development. As long as your baby is enjoying your language games and reading, then you are providing the secret ingredient for success later on in school.

The Child-Care Dilemma

Millions of U.S. families face a wrenching question: Who's minding the kids?

The smell of wet paint wafts through the house on a tree-lined street on Chicago's North Side. Marena Mc-Pherson, 37, chose a peach tint for the nursery: a gender-neutral color. But the paint had a will of its own and dried a blushing shade of pink. Ah well, no time to worry about that. With the baby due in less than a month, there are too many other concerns. Like choosing a name, furnishing the baby's room, reading up on infant care and attending childbirth classes. Above all, McPherson must tackle the overriding problem that now confronts most expectant American mothers: Who will care for this precious baby when she returns to work?

An attorney who helps run a Chicago social-service agency, McPherson has accumulated two months of paid sick leave and vacation time. She plans to spend an additional four months working part time, but then she must return to her usual full schedule. So for several months she has been exhaustively researching the local child-care scene. The choices, she has learned, are disappointingly few. Only two day-care centers in Chicago accept infants; both are expensive, and neither appeals. "With 20 or 30 babies, it's probably all they can do to get each child's needs met," says McPherson. She would prefer having a baby-sitter come to her home. "That way there's a sense of security and family." But she worries about the cost and reliability: "People will quit, go away for the summer, get sick." In an ideal world, she says, she would choose someone who reflects her own values and does not spend the day watching soaps. "I suspect I will have to settle for things not being perfect."

That anxiety has become a standard rite of passage for American parents. Beaver's family, with Ward Cleaver off to work in his suit and June in her apron in the kitchen, is a vanishing breed. Less than a fifth of American families now fit that model, down from a third 15 years

ago. Today more than 60% of mothers with children under 14 are in the labor force. Even more striking: about half of American women are making the same painful decision as McPherson and returning to work before their child's first birthday. Most do so because they have to: seven out of ten working mothers say they need their salaries to make ends meet.

With both Mom and Dad away at the office or store or factory, the child-care crunch has become the most wrenching personal problem facing millions of American families. In 1986, 9 million preschoolers spent their days in the hands of someone other than their mother. Millions of older children participate in programs providing after-school supervision. As American women continue to pour into the work force, the trend will accelerate. "We are in the midst of an explosion," says Elinor Guggenheimer, president of the Manhattan-based Child Care Action Campaign. In ten years, she predicts, the number of children under six who will need daytime supervision will grow more than 50%. Says Jay Belsky, a professor of human development at Pennsylvania State University: "We are as much a society dependent on female labor, and thus in need of a child-care system, as we are a society dependent on the automobile, and thus in need of roads."

At the moment, though, the American child-care system—to the extent that there is one—is riddled with potholes. Throughout the country, working parents are faced with a triple quandary: day care is hard to find, difficult to afford and often of distressingly poor quality. Waiting lists at good facilities are so long that parents apply for a spot months before their children are born. Or even earlier. The Empire State center in Farmingdale, N.Y., received an application from a woman attorney a week after she became engaged to marry. Apparently she hoped to time her pregnancy for an anticipated opening. The Jeanne Simon center in Burlington, Vt., has a folder of applications labeled "preconception."

Finding an acceptable day-care arrangement is just the beginning of the struggle. Parents must then maneuver to maintain it. Michele Theriot of Santa Monica, Calif., a 37-year-old theatrical producer, has been scrambling ever since her daughter Zoe was born 2½ years ago. In that short period she has employed a Danish au pair, who quit after eight months; a French girl, who stayed 2½ months; and an Iranian, who lasted a week. "If you get a good person, it's great," says Theriot, "but they have a tendency to move on." Last September, Theriot decided to switch Zoe into a "family-care" arrangement, in which she spends seven hours a day in the home of another mother. Theriot toured a dozen such facilities before selecting one. "I can't even tell you what I found out there," she bristles. In one home the "kids were all lined up in front of the TV like a bunch of zombies." At another she was appalled by the filth. "I sat my girl down on the cleanest spot I could find and started interviewing the care giver. And you know what she did?" asks the incredulous mother. "She began throwing empty yogurt cups at my child's head. As though that was playful!"

Theriot is none too sure that the center she finally chose is much better. Zoe's diapers aren't always changed, instructions about giving medicine are sometimes ignored, and worse, "she's started having nightmares." En route to day care on a recent day, Zoe cried out, "No school! No school!" and became distraught. It is time, Theriot concludes, to start the child-care search again.

Fretting about the effects of day care on children has become a national preoccupation. What troubles lie ahead for a generation reared by strangers? What kind of adults will they become? "It is scaring everybody that a whole generation of children is being raised in a way that has never happened before," says Edward Zigler, professor of psychology at Yale and an authority on child care. At least one major survey of current research, by Penn State's Belsky, suggests that extensive day care in the first year of life raises the risk of emotional problems, a conclusion that has mortified already guilty working parents. With high-quality supervision costing upwards of $100 a week, many families are placing their children in the hands of untrained, overworked personnel. "In some places, that means one woman taking care of nine babies," says Zigler. "Nobody doing that can give them the stimulation they need. We encounter some real horror stories out there, with babies being tied into cribs."

The U.S. is the only Western industrialized nation that does not guarantee a working mother the right to a leave of absence after she has a child. Although the Supreme Court ruled last January that states may require businesses to provide maternity leaves with job security, only 40% of working women receive such protection through their companies. Even for these, the leaves are generally brief and unpaid. This forces many women to return to work sooner than they would like and creates a huge demand for infant care, the most expensive and difficult child-care service to supply. The premature separation takes a personal toll as well, observes Harvard Pediatrician T. Berry Brazelton, heir apparent to Benjamin Spock as the country's preeminent guru on child rearing. "Many parents return to the workplace grieving."

New York City Police Officer Janis Curtin resumed her assignment in south Queens just eight weeks after the birth of Peter. The screaming sirens and shrill threats of street thugs were just background noise to a relentless refrain in her head: "Who can I trust to care for my child?" She tried everything, from leaving Peter at the homes of other mothers to handing him over to her police-officer husband at the stationhouse door when they worked alternating shifts. With their schedules in constant flux, there were snags every step of the way. Curtin was more fortunate than most workers: police-department policy allows a year of unpaid "hardship" leave for child care. She decided to invoke that provision.

The absence of national policies to help working mothers reflects traditional American attitudes: old-fashioned motherhood has stood right up there with the flag and apple pie in the pantheon of American ideals. To some people day-care centers, particularly government-sponsored ones, threaten family values; they seem a step on the slippery slope toward an Orwellian socialist nightmare. But such abstract concerns have largely receded as the very concrete need for child care is confronted by people from all walks of life.

Child care is fast emerging as a political issue. At least three Democratic presidential candidates have been emphasizing the need for better facilities and calling for

federal action. Former Arizona Governor Bruce Babbitt has proposed that the U.S. Government establish a voucher system to help low-income parents pay for day care. Delaware Senator Joseph Biden favors federal child-care subsidies for the working poor and tax incentives to encourage businesses to provide day care. If elected, he vows, he will set up a center for White House employees as an example to other employers. Massachusetts Governor Michael Dukakis, who has established the country's most comprehensive state-supported day-care system, would like to see the Federal Government fund similar programs throughout the U.S.

Last week the issue surfaced on Capitol Hill. In the House, Republican Nancy Johnson of Connecticut and Democrat Cardiss Collins of Illinois introduced legislation to establish a national clearinghouse for information on child-care services. A Senate subcommittee began hearings focused on the shortage of good-quality, affordable day care. Says Chairman Christopher Dodd of Connecticut: "It's about time we did something on this critical problem."

Without much federal help, the poorest mothers are caught in a vise. Working is the only way out of poverty, but it means putting children into day care, which is unaffordable. "The typical cost of full-time care is about $3,000 a year for one child, or one-third of the poverty-level income for a family of three," says Helen Blank of the Children's Defense Fund in Washington. As a result, many poor mothers leave their young children alone for long periods or entrust them to siblings only slightly older. Others simply give up on working.

Rosalind Dove, 29, of Los Angeles, is giving it her best shot. A single mother of four, she worked for five years as a custodian in a public high school, bringing home $1,000 in a good month. "I was paying $400 a month for child care," she recalls. "We didn't buy anything." When that failed, she began bringing her children to work with her, hiding them in an empty home-economics classroom while she mopped floors and hauled huge barrels of trash for eight hours a day. "I'd sneak them in after the teacher left and check on them every 30 minutes or so." She finally quit last February and slipped onto the welfare rolls. She applied for state child-care assistance, only to learn there were 3,000 others on the waiting list. Frustrated, she returned to work this month. "Don't ask me how I'm going to manage," she says.

Child care has always been an issue for the working poor. Traditionally, they have relied on neighbors or extended family and, in the worst of times, have left their children to wander in the streets or tied to the bedpost. In the mid-19th century the number of wastrels in the streets was so alarming that charity-minded society ladies established day nurseries in cities around the country. A few were sponsored by employers. Gradually, local

regulatory boards began to discourage infant care, restrict nursery hours and place emphasis on a kindergarten or Montessori-style instructional approach. The nurseries became nursery schools, no longer suited to the needs of working mothers. During World War II, when women were mobilized to join wartime industry, day nurseries returned, with federal and local government sponsorship. Most of the centers vanished in the postwar years, and the Donna Reed era of the idealized nuclear family began.

Two historic forces brought an end to that era, sweeping women out of the home and into the workplace and creating a new demand for child care. First came the feminist movement of the '60s, which encouraged housewives to seek fulfillment in a career. Then economic recessions and inflation struck in the 1970s. Between 1973 and 1983, the median income for young families fell by more than 16%. Suddenly the middle-class dream of a house, a car and three square meals for the kids carried a dual-income price tag. "What was once a problem only of poor families has now become a part of daily life and a basic concern of typical American families," says Sheila B. Kamerman, a professor of social policy and planning at Columbia University and co-author of Child Care: Facing the Hard Choices. Some women are angry that the feminist movement failed to foresee the conflict that would arise between work and family life. "Safe, licensed child care should have been as prominent a feminist rallying cry as safe, legal abortions," observes Joan Walsh, a legislative consultant and essayist in Sacramento.

In the early 1970s, there was a flurry of congressional activity to provide child-care funds for the working poor and regulate standards. But under pressure from conservative groups, Richard Nixon vetoed a comprehensive child-development program in 1971, refusing, he said, to put the Government's "vast moral authority" on the side of "communal" approaches to child rearing. The Reagan Administration has further reduced the federal role in child care. In inflation-adjusted dollars, funding for direct day-care subsidies for low- and middle-income families has dropped by 28%.

California, Minnesota, Massachusetts, New York and Connecticut are among the few states that have devoted considerable resources to improving child-care programs. Most states have done virtually nothing. Thirty-three have lowered their standards and reduced enforcement for licensed day-care centers. As of last year, 23 states were providing fewer children with day care than in 1981.

Nor have American businesses stepped in to fill the void. "They acknowledge that child care is an important need, but they don't see it as their problem," says Kamerman. Of the nation's 6 million employers, only about 3,000 provide some sort of child-care assistance. That is up from about 100 in 1978, but most merely provide advice or referrals. Only about 150 employers provide on-site or near-site day-care centers. "Today's corporate personnel policies remain stuck in a 1950s time warp," charges David Blankenhorn, director of the Manhattan-based Institute for American Values. "They are rooted in the quaint assumption that employees have 'someone at home' to attend to family matters."

There are basically three kinds of day care in the U.S. For children under five, the most common arrangement is "family" or "home-based" care, in which toddlers are minded in the homes of other mothers. According to a Census Bureau report called *Who's Minding the Kids*, 37% of preschool children of working mothers spend their days in such facilities. An additional 23% are in organized day-care centers or preschools. The third type of arrangement, which prevails for older children and for 31% of those under five, is supervision in the child's own home by a nanny, sitter, relative or friend.

Home-based groups are popular primarily because they are affordable, sometimes costing as little as $40 a week. The quality depends on the dedication of the individual mothers, many of whom are busy not only with their paid charges but with their own children as well. Darlene Daniels, 31, a single mother of three in Chicago, has been through four such sitters in six months. Two proved too expensive and careless for Daniels, who was earning $7 an hour as a janitor; another robbed her. "For most people, it's not their own kids, and they're just looking at the dollar sign," she complains.

Only eight states have training requirements for home-based centers. Regulations governing the ratio of attendants to tots vary widely. In Maryland there must be one adult for every two children under age two. But in Georgia each adult is allowed to care for up to ten children under age two and, in Idaho, twelve.

A private nanny or au pair usually assures a child more individual attention. Professional couples, who must work long hours or travel, often find that such live-in arrangements are the only practical solution, though the cost can exceed $300 a week. However, most live-in sitters in the U.S., unlike the licensed nannies of Britain, have no formal training. Many speak English poorly, and agencies frequently do a cursory job of screening them. A Dallas mother who asked an attorney friend to run a check on her newly hired nanny was told the woman was wanted for writing bad checks. "People need a license to cut your hair but not to care for your child," observes Elaine Claar Campbell, a Chicago investment banker. She and her lawyer-husband Ray, armed with five

pages of questions, spent three months interviewing more than 50 people, before settling on Clara Hawkes, 47, an artist from Santa Fe whose own daughter is a National Merit Scholar. "You don't want to gamble with your child," says Ray.

Au pairs, usually European girls between 18 and 25, are less expensive, receiving an average of $100 a week plus room and board. Most stay only a year, and few have legal working papers. The immigration law that took effect this month will make the employers of such workers liable for fines up to $10,000, though the Immigration and Naturalization Service does not plan an aggressive crackdown on domestic help.

Concerns about legality have led more families to hire American au pairs—frequently teenage girls from the Midwest and often Mormons. "We Mormons come from big families, so we have experience with kids," explains Karen Howell, 19, a Californian who is spending a year with a Washington, D.C., family. "We don't drink, and we know the meaning of hard work." Two agencies—the Experiment in International Living and the American Institute for Foreign Study—have Government permission to bring in 3,100 European au pairs a year on cultural-exchange visas. Although the programs are more expensive than traditional au pair arrangements, host families are assured that their helpers are legal.

The professional day-care center is the fastest-growing option for working parents. There are an estimated 60,000 around the country, about half nonprofit and half operated as businesses. Costs vary widely, from $40 a week to as much as $120. In the best centers, children are cared for by dedicated professionals. At the nonprofit Empire State center in Farmingdale, N.Y., teachers make up lesson plans even for infants. Empire, which receives partial funding from New York State, keeps parents closely informed of their child's development. "If a child takes a first step, develops in the least, that parent is called," says Director Ana Fontana.

Not all day-care centers are so conscientious. Day-care staffers rank in the lowest 10% of U.S. wage earners, a fact that contributes to an average turnover rate of 36% a year. Says Caroline Zinsser of the Center for Public Advocacy Research in Manhattan: "It says something about our society's values that we pay animal caretakers more than people who care for our children." Gilda Ongkeko is delighted with the quality of the Hill an' Dale Family Learning Center in Santa Monica, Calif., attended by Jason, 4. In her job as owner of a preschool-supply company, she has come to appreciate how unusual it is. "I've been to over 1,000 child-care centers," she says, "and I'd say that 90% of them should be shut down. It's pathetic."

Experts worry that a two-tier system is emerging, with quality care available to the affluent, and everyone else settling for less. "We are at about the same place with child care as we were when we started universal education," says Zigler of Yale. "Then some kids were getting Latin and Greek and being prepared for Harvard, Yale and Princeton. Other kids were lucky if they could learn to write their own name."

In 1827 Massachusetts led the way to universal education by becoming the first state to require towns with 500 or more families to build high schools. Now it is showing the way to universal child care. Aided by a booming economy, the state has worked out a program with employers, school boards, unions and nonprofit groups to encourage the expansion and improvement of child-care facilities. Small companies and groups can receive low-interest loans from the state to build day-care facilities. Funds are earmarked for creating centers in public housing projects. School systems can get financial aid for after-school programs. A statewide referral network serves both individual parents and corporations looking for child care.

Emilia Davis, 38, of Boston's working-class Roslindale section, is the beneficiary of another of the state's far-reaching programs. After years of dependence on welfare to support herself and her five children, Davis, who is separated from her husband, is now going to college with the ultimate hope of finding a job. The state's E.T. (employment and training) program provides her with vouchers for day care in the public housing complex where she lives. "Child care is an absolute precondition if one is serious about trying to help people lift themselves out of poverty," insists Governor Dukakis. Though the state will spend an estimated $27 million on day care under the E.T. program this year—and a total of $101 million on all child-care related services—it claims to have saved $121 million in welfare costs last year alone. Next month the state will begin a pilot program that will pay 20% to 40% of child-care costs for 150 working-class families.

San Francisco has adopted another innovative approach. It requires developers of major new commercial office and hotel space to include an on-site child-care center or pay $1 per sq. ft. of space to the city's child-care fund. The state of California is spending $319 million this year on child-care subsidies for 100,000 children. It also funds a network of 72 resource and referral agencies.

Because such state programs are the exception, a number of political leaders and lobbying groups are calling for federal intervention. This summer a coalition of 64 groups—including the National Education Association, the American Federation of Teachers and the Child Welfare League of America—will propose a comprehensive national child-care bill, which will probably call for increased support to help low- and moderate-income families pay for child care. Legislation has already been introduced in both houses of Congress to create a national parental-leave policy.

In an era of towering federal deficits, much of the future initiative will have to come from the private sector. By the year 2000, women will make up half the work force. Says Labor Secretary Bill Brock: "We still act as though workers have no families. Labor and management haven't faced that adequately, or at all."

A few companies are in the forefront. Merck & Co., a large pharmaceutical concern based in Rahway, N.J., invested $100,000 seven years ago to establish a day-care center in a church less than two miles from its headquarters. Parents pay $550 a month for infants and $385 for toddlers. Many spend lunch hours with their children. "I can be there in four minutes," says Steven Klimczak, a Merck corporate-finance executive whose three-year-old daughter attends the center. "It's very reliable, and that's important in terms of getting your job done."

Elsewhere in the country, companies have banded together to share the costs of providing day-care services to employees. A space in Rich's department store in downtown Atlanta serves the children of not only its own employees but also of workers at the Federal Reserve Bank of Atlanta, the First National Bank of Atlanta, Georgia-Pacific and the Atlanta *Journal* and *Constitution* newspapers.

Businesses that have made the investment in child care say it pays off handsomely by reducing turnover and absenteeism. A large survey has shown that parents lose on average eight days a year from work because of child-care problems and nearly 40% consider quitting. Studies at Merck suggest that the company also saves on sick leave due to stress-related illness. "We have got an awful lot of comments from managers about lessened stress and less unexpected leave time," says Spokesman Art Strohmer. At Stride Rite Corp., a 16-year-old, on-site day-care center in Boston and a newer one at the Cambridge headquarters have engendered unusual company loyalty and low

turnover. "People want to work here, and child care seems to be a catalyst," says Stride Rite Chairman Arnold Hiatt. "To me it is as natural as having a clean-air policy or a medical benefit."

The generation of workers graduating from college today may find themselves in a better position. They belong to the "baby-bust" generation, and their small numbers, says Harvard Economist David Bloom, will force employers to be creative in searching for labor. Child-care arrangements, he says, will be the "fringe benefits of the 1990s." The economics of the situation, if nothing else, will provoke a change in the attitude of business, just as the politics of the situation is changing the attitude of government. In order to attract the necessary women—and men—employers are going to have to help them find ways to cope more easily with their duties as parents.

—*By Claudia Wallis.*
Reported by Jon D. Hull/Los Angeles, Melissa Ludtke/Boston and Elizabeth Taylor/Chicago

"What Is Beautiful Is Good": The Importance of Physical Attractiveness in Infancy and Childhood

Linda A. Jackson

Ph.D. University of Rochester
Associate Professor, Department of Psychology
Michigan State University

Hiram E. Fitzgerald

Ph.D. University of Denver
Professor, Department of Psychology
Michigan State University

"She's so ugly!" "He's got a big nose!" "She's too fat!" Most parent have heard one or more of these comments from their children and can probably recall making similar comments themselves, both as children and as adults. Yet the typical response from parents overhearing such comments from their children goes something like "That's not nice. It's not what a person looks like but what's inside the person that matters." Good advice! But do children take it? Do adults take it? Decades of research in social psychology on the effects of physical appearance on the perception and treatment of others suggests that this advice goes largely unheeded. In fact, children and adults do "judge the book by its cover" and the tendency to do so has implications not only for the immediate interaction but also for the self concepts of those who are responded to on the basis of their appearance.

Research on people's reactions to attractive and unattractive infants, children, and adults has revealed one striking similarity. Across all of these age groups attractive individuals are reacted to more favorably than unattractive individuals, an observation commonly referred to as the "what is beautiful is good" stereotype. Like it or not, the benefits of beauty are evident at birth. For example, in one study nurses gave better prognosis for the intellectual development of attractive premature infants than unattractive premies. By three months of age it is clear that "all babies do not look alike." There are beautiful babies and beauty counts.

What is beauty in infancy? Research indicates that several clearly distinctive physical characteristics are associated with "cuteness" in infants. Most are facial characteristics. Short and narrow facial features, large eyes and pupils, and a large forehead make an infant appear "cute." Female infants reach a "cuteness peak" at nine months of age while male infants reach their "cuteness peak" at eleven months. This difference in peak cuteness is presumed to reflect the more rapid physical development of female infants compared to male infants. If asked to guess the sex of an infant, adults are more likely to guess "female" if the infant is cute, although they rate infants identified as female as less cute. These findings suggest that more stringent standards of physical attractiveness for females than for males exist as early as infancy. Other factors which influence the cuteness ratings of infants are facial expression and birth order. Smiling infants are rated as cuter than crying infants. First born infants are rated as cuter than later born infants, apparently because first borns tend to be better dressed and groomed than later borns. Not surprisingly, parents rate their infants as cuter than strangers rate them, suggesting that there is some truth to the statement that "beauty is in the eye of the beholder."

Given the evidence that infants do not "all look alike," at least in terms of their cuteness, do differences in their appearance have an effect on adult reactions to them? The answer is an unequivocal "yes." The initial adult response to an unfamiliar infant, the smile, oc-

From *Child International*, February 1988. *Child International*, published by Eltan Limited, 411 London House, 26-40 Kensington High Street, London W8, England. Reprinted by permission of the authors and *Child International*.

curs regardless of the infant's cuteness. However, the subsequent response to an unfamiliar infant, the prolonged gaze, is affected by the infant's cuteness. Cuter infants are looked at longer than less cute infants, at least in situations where the adult has a choice about what to look at. Psychologists have speculated that the greater time spent looking at cute infants may result in more care-giving behaviors directed toward them than toward their less cute age mates. For example, in day care situations where caregivers must continually make choices about how to divide their attention among equally familiar infants, being cute may reap benefits. Day care centers with high turnover, in which infants are unfamiliar to caregivers, may be even more susceptible to the effects of an infant's cuteness on the amount of attention and caregiving provided to him or her.

Results of the research on the effects of infants' cuteness on their parents' reactions to them lead to similar conclusions. Mothers of infants who were rated as cute by strangers maintain more eye contact, more ventral holding contact (tummy to tummy), and kiss their infants more than mothers of infants rated as less cute by strangers. Differences in behavior related to the infants' cuteness have also been observed in fathers. Greater stimulation, as evidenced by touching, kissing and moving, is provided by fathers of cute infants than by fathers of less cute infants. Fathers' ratings of their infants' cuteness at three months of age were related to their caregiving behaviors toward their infants, and to ratings of their relationship with their infants. These relations were not obtained for mothers. Taken as a whole, the research suggests that the benefits of beauty begin in infancy and include the responses of parents as well as strangers. However, it is important to keep in mind that parents rate their infants as cuter than strangers rate them, the "eye of the beholder" effect.

Turning from the effects of attractiveness in infancy to its effects in early childhood, it would be heartening to find that appearance factors become less important as there is more and more of the "person within" to respond to. Such is not the case, however. In keeping with the "what is beautiful is good" stereotype, ratings of unfamiliar children by both adults and agemates indicate that attractive children are perceived as more intelligent, popular, and desirable as friends than unattractive children. These differences are evident among raters as young as three years old. Even teachers are not immune to the effects of a child's attractiveness. In one study teachers were asked to rate an unfamiliar child who had committed a transgression. A photograph of the child attached to the description of the transgression indicated that the child was either attractive or unattractive. Teachers rated the transgression as more serious, rated the child as more antisocial, recommended more severe punishment,

and believed the child was more likely to transgress again when the child was unattractive than when she or he was attractive. Other research supports the conclusion that attractive children have an advantage over unattractive children in first encounters with peers and with adults. These findings have important implications since first encounters often set the stage for the development of friendships and expectations for future behavior.

But what about the attractiveness of familiar children? Commonsense suggests that with greater familiarity there should be less of a tendency to behave toward others in stereotypic ways. While there is some evidence to support this commonsense assumption, there is also evidence to support the equally commonsense assumption that being good looking is an asset even among friends. Attractive classmates are viewed as more popular and as having more socially desirable characteristics than unattractive classmates as early as preschool. However, the benefits of beauty among acquaintances appear to be limited to girls. In fact, the research suggests that attractiveness may be a liability for school-aged boys. In several studies children have rated their attractive male classmates as more incompetent, antisocial, and disliked than their unattractive male classmates. Teachers rated attractive unfamiliar boys as more uncontrolled, impulsive, and unpredictable compared to unattractive boys. The basis for this differential response to attractive boys and girls has yet to be understood. It may be that the strong association between attractiveness and "female" or "femininity" is responsible for the negative effects of boys' attractiveness on other's perceptions of them, but additional research is needed to test the validity of this explanation.

Given the above evidence that physical attractiveness influences perceptions of children, does it also influence behavior toward them? Research suggests that there are differences in the treatment of attractive and unattractive children, although the evidence is not as strong as the evidence for differences in perceptions. In one study in which undergraduate women were asked to role play a child's parent, unattractive children were punished more severely than attractive children. In another study, small but significant relations were found between the actual grades of elementary school children and their physical attractiveness. A third study showed that the attractiveness of seventh graders influenced the amount of time teachers spent interacting with them. Attractive seventh graders were interacted with more than unattractive seventh graders, although a similar relation was not obtained for younger children. It has also been observed that teachers' referrals for remedial help are more likely for attractive children than for unattractive children.

Thus, the bulk of research indicates that attractive

Facts About Physical Attractiveness in Infancy and Early Childhood

1. There are distinctive physical characteristics related to ratings of physical attractiveness in infancy. Short and narrow facial features, big eyes and pupils, and a large forehead make an infant "cute."
2. An infant's "cuteness" is related to the amount of time parents and strangers spend looking at the infant. This difference in visual attention has implications for the amount of caregiving behaviors received by attractive and unattractive infants.
3. The "what is beautiful is good" stereotype is evident in children as young as three years. Attractive age mates are believed to be more intelligent, sociable, and desirable as friends than unattractive age mates.
4. The behavior of teachers is influenced by the child's physical attractiveness. Transgressions by unattractive children are viewed as more serious and as deserving of more severe punishment than the same transgressions of attractive children. Teachers believe that unattractive children who transgress are more antisocial and more likely to transgress again than attractive children.
5. Physical attractiveness may be a liability for school-aged boys, in contrast to its ubiquitous benefits for girls of all ages. Attractive boys are viewed by their peers as more antisocial, and dislikable than unattractive boys. Teachers view attractive boys as more impulsive, uncontrollable, and unpredictable compared to unattractive boys.
6. Small but significant relations have been found between children's attractiveness and their school grades and performance on standardized tests of achievement.
7. Variability in ratings of attractiveness within cultures, differences in what constitutes attractiveness between cultures, and cultural differences in the importance of physical attractiveness must be taken into account in understanding the implications of physical attractiveness.

and unattractive infants and young children are perceived differently, and behaved toward differently by children, parents, and other adults. It would therefore be surprising if attractive and unattractive children did not themselves differ in some important ways. Both commonsense psychology and a wealth of research on the self-fulfilling prophecy suggests that what we are is, at least in part, a reflection of how other people see us. There is some evidence of behavior differences between attractive and unattractive children. Relations between attractiveness and scores on standardized achievement tests have been found. Attractive children score higher than unattractive children and this difference increases with age. On the other hand, no relations between attractiveness and IQ scores has been found. Studies of preschoolers indicate that attractive children spend less time in solitary play than average attractiveness or unattractive children. When playing with same-sex age mates, attractive 5-year-olds are less aggressive, less active, and show a slight preference for feminine toys compared to unattractive 5-year-olds. Although the evidence is not overwhelming in its quantity, it nevertheless points to some important differences in the behavior of attractive and unattractive children, differences which have implications for the development of the self concept and adult adjustment.

Exactly what are the implications of the research on the effects of physical attractiveness in infancy and early childhood for parents of today's young children?

Does this research imply that parents should do their utmost to make their children physically attractive so that they may reap the benefits of beauty? Or should parents simply resign themselves to the empirical proof of the fact that books are sometimes judged by their covers? We believe that neither of these extremes is implied by the research on physical attractiveness and young children for a number of reasons. First, much of this research has compared attractive children to unattractive children when, in fact, most children fall somewhere in between. Those few studies which have included average unattractiveness groups often, but not always, find no differences between average and very attractive children. Thus, children of average attractiveness appear to be just as well off as very attractive children, in terms of other's perceptions of them and other's behavior toward them.

Secondly, physical attractiveness, like other physical attributes, is not a particularly stable characteristic over the span of childhood physical development. While the "ugly duckling turned swan" is an extreme example of the instability of physical appearance, most parents are well aware of how much their childrens' appearance changes from year to year. Taken together with the evidence that beauty is, to some extent, in the eye of the beholder, the effects of appearance on the reactions of others are likely to be quite variable over the course of childhood development.

Third, the effects of others' reactions to the child based on the childs' physical appearance are likely to be swamped by the effects of reactions based on other, more enduring and more important characteristics. For example, teachers' reactions to a child depend more on such characteristics as his or her personality, social skills, and academic performance than on his or her physical appearance. This is not to deny the importance of the effects of physical attractiveness demonstrated in the research. But these effects need to be put in a proper perspective.

Finally, it is important to point out that while there is some consensus about what constitutes physical attractiveness in infancy and early childhood, there is also considerable variability within cultures, and even greater variability between cultures in ratings of attractiveness. Cultures also differ in the emphasis they place on physical appearance compared to other characteristics. Since much of the research demonstrating the "what is beautiful is good" stereotype has been conducted in the United States, the generalizability of the results to other cultures remains an open question.

Development During Childhood

- **Social and Emotional Development (Articles 11-18)**
- **Cognitive and Language Development (Articles 19-22)**

Most of the changes that occur during the transition to childhood—as well as those that occur during childhood itself—involve social, cognitive, and language development. These changes are so dramatic that the preschool to school-age transitional period is sometimes referred to as the "five-to-seven shift." During this transitional time significant changes occur in a child's attention span, memory, learning, and problem-solving skills. Articulation improves, vocabulary size increases, and the child achieves a new understanding of the syntactical features of language. Entrance into the formal world of school is the most significant systemic change in the child's environment during childhood. School not only expands the child's contacts with significant adults and peers, but also expands the child's concept of neighborhood.

Social and Emotional Development. During the school years the child's social network expands. School extends the child's peer group beyond the confines of the immediate neighborhood and exposes the child to a new set of authority figures. The quality of the child's interactions with available role models influences his or her sense of social and emotional competence. Specific characteristics of children help to shape these interactions with others; for example, not all children adapt well to new situations or to new people. Research reviewed in "Born to be Shy" suggests that, although behavioral inhibition may have a foundation in biology, parenting practices can help shy children to deal more effectively with novel situations. "Beyond Selfishness" examines how another human trait, selfishness, is manifested in some situations and not others.

All life transitions create some degree of stress. Most children handle everyday stress fairly well; they are able to draw upon a variety of coping skills learned through past interactions with family members and peers. In "Compliance, Control, and Discipline," parts 1 and 2, Alice Honig argues that disciplinary techniques can facilitate or impede development of self-control, compliance, and affiliation. Self-control and self-regulation originate from the secure attachments of infancy and are especially important for the development of coping strategies. Without self-control and the ability to comply effectively with requests, conflict is a certainty. The extent to which conflicts affect personality development during childhood is a topic of special interest to contemporary developmentalists, particularly with respect to the pervasiveness of

aggression and violence throughout the world; this subject is reviewed in "Aggression: The Violence Within."

Children often resort to physical or verbal aggression in their attempts to resolve conflict or to deal with stress. Rarely are such tactics successful; indeed, aggression usually begets aggression. Other children may retreat from stress, drawing into themselves, becoming apathetic, despondent, and depressed. Although depression once was thought to be an affective disorder restricted to adulthood, it is now evident that it can occur as early as infancy. In "Depression at an Early Age," Joseph Alper reviews evidence pointing to childhood depression, and suggests that affective disorders may become the major health problems of the 1990s.

Although some children react to stress by striking out against the perceived source of the stress, and others react to it by withdrawing and attempting to isolate themselves, there are children who seem to take stress-producing situations in stride. These children are referred to as "invulnerable" or "resilient." Resilient children have developmental histories that may include extreme poverty or chronic family stress. Most people would consider both factors to be potentially damaging to the development of personal and social competence skills. Yet these resilient children do not become victims; instead, they develop effective interpersonal skills. Without question, studies of these children will contribute important knowledge to our understanding of personality development and child-rearing practices. There is some evidence to suggest that the development of a successful attachment relationship with a significant caregiver is characteristic of such children. If so, this would suggest that mastery of social competence skills combined with a sense of self-control and personal worth may contribute to the resilient child's ability to cope with stress. Play provides opportunities for children to learn self-control, to gain self-confidence, and to practice skills for effective social interaction. In "The Importance of Play," Bruno Bettelheim suggests that, when adults formalize rules and control children's games, the advantages of play are lost.

Cognitive and Language Development. The articles in this section focus on developmental changes in cognitive processing, context effects on encoding of information, challenges to IQ test definitions of intelligence, computer simulations of brain structure, and issues related to reading and language acquisition. During the past two decades the study of cognitive development was dominated by Piaget's theory. This theory is reviewed in "Practical Piaget: Helping Children Understand." Equally important, however, are the examples provided by the author as to how Piaget's theory can be translated into practice. Although this theory provides a rich description of what a child can and cannot do during a particular stage of development, it is less adequate for explaining how a child acquires various cognitive skills. Thus, many developmentalists have turned to information-processing models in an effort to integrate cognitive psychology with cognitive developmental theory. Information-processing research has directed attention to individual differences in skill acquisition, challenging, among other things, tradi-

tional concepts of intelligence; this is addressed in "Three Heads Are better Than One." The article "How the Brain Really Works Its Wonders" describes research using computer simulations of neural networks to study perception, information processing, and brain reorganization during recovery from brain injury.

Prior to the 1960s, theories of language development stressed its environmental determinants. Noam Chomsky shifted emphasis to genetic factors. While the issue is far from being resolved, most contemporary developmentalists stress language acquisition models which assume that environmental influences are built onto a biological foundation. The study of dyslexia, a problem generally thought to involve problems in visual perception, has been given new direction by evidence indicating that deficits in linguistic processing are at the heart of children's reading difficulties. Frank R. Vellutino addresses this issue in "Dyslexia."

Looking Ahead: Challenge Questions

In a culture that is as aggressive and violent as America's, how can parents suppress aggressiveness in their children? Why do you suppose males are generally found to be more physically aggressive than females in nearly every human culture?

Children who are described as resilient are an enigma. How would you explain such strength of personality in the face of so many potentially disruptive influences in their lives? What coping mechanisms seem to provide resilient children with such strength of character?

The ability to solve problems ranks high among lay definitions of intelligence. Yet problem-solving ability is not exactly the same as the intelligence measured by IQ tests. Which do you believe is the better measure of intelligence? The fact that all human beings are not equally intelligent, any more than they are equally tall, suggests biological variation in the distribution of intelligence. What characteristics of an individual do you think of when you refer to someone as being intelligent?

Do you believe that the American educational system is capable of changing its instructional models to take into account new views of the causes and definitions of mental retardation and learning disabilities? Or is it more convenient to hold on to traditional IQ definitions of mental retardation and learning disabilities, in order to guarantee continued government support for educational programs?

Born to be Shy?

JULES ASHER

Jules Asher is a freelance science writer living in Silver Spring, Maryland.

We've all met shy toddlers—the ones who cling to their parents and only reluctantly venture into an unfamiliar room. Faced with strangers, they first freeze, falling silent and staring at them. They seem visibly tense until they've had a chance to size up the new scene. Parents of such children are likely to say they've always been on the timid side. "It's just his way," one might say.

Despite parents' observations, psychologists have tended to resist the notion that such traits are inborn, concentrating instead on the importance of early experience in the development of character. Today, however, evidence from several lines of research indicates that at least some behavioral traits, such as extreme shyness, are enduring aspects of personality that are grounded in inborn ways of responding physiologically to the environment. In short, they're part of some people's basic temperament.

Harvard University psychologist Jerome Kagan has found this in his long-term studies of human infant development, and psychologist Stephen Suomi, of the National Institutes of Health, has seen the same thing in his developmental studies of monkeys. Both Kagan's "extremely inhibited" kids and Suomi's "uptight" monkeys often behave in similar ways when faced with novel situations, and they have similar physiological reactions as well.

Kagan, in his classic developmental studies with psychologist Howard Moss and colleagues almost a quarter-century ago, found that for traits such as dependence, aggression, dominance and competitiveness, individual children did not remain consistent from age 2 or 3 to age 20. But they were unchanged over the years in one trait—which he then called "passivity" and now calls "behavioral inhibition." In an unfamiliar situation, the children might be cautious; around strangers they might be shy.

The "inhibited" kids, as adults, had unusually stable and high heart rates in response to mild mental stress. At the time, Kagan recalls, this was an intriguing physiological finding but hardly one he saw as a clue to an underlying biological vulnerability. Steeped as he and his colleagues were in the prevailing behaviorism of the day, they assumed they had simply detected "an acquired fearfulness shaped by parents."

SOME CHILDREN—AND MONKEYS TOO— ARE INHERENTLY INHIBITED AND UPTIGHT, BUT BOTH CAN BE HELPED BY GOOD PARENTING.

But when he and other researchers began finding apparently genetic differences in temperament from infancy between Chinese-American and Caucasian children in the mid 1970s, Kagan started to look for the biological underpinnings of the enduring trait they had come upon. He decided to compare very inhibited and very uninhibited children—the 10 percent at either end of the spectrum—rather than average children, since "Otherwise, the biology is just giving a little push and you lose it all."

In 1979, Kagan and psychologists J. Steven Reznick and Nancy Snidman started to follow the development of extremely inhibited and uninhibited children (using two groups, starting at 21 months of age in one group and 31 months in the other). In this study they tracked the children's heart rates and other physiological measures as well as observing their behavior in novel situations.

The children have now been studied through their 6th year, and though the very shy children no longer behave exactly as they did when they were 2, they still show the pattern of very inhibited behavior combined with high physiological responsiveness to mild stress.

These children, in addition to their continuing timidity, show a pattern of excessive physiological responses to mildly stressful situations that wouldn't faze their more easygoing peers. Their unusually intense physical responses to mental stress—which include more dilated pupils and faster and more stable heartbeats—indicate that their sympathetic nervous system is revved up. Higher salivary cortisol levels and indicators of norepinephrine activity round out the picture of an overactive "stress circuit" under mildly stressful conditions (see "A Tour of the Stress Circuit," this article). These temperamental characteristics are so stable that Kagan, looking at the early indicators of behavioral inhibition at

From *Psychology Today,* April 1987, pp. 56-59, 62-64. Copyright © 1987 PT Partners L. P. Reprinted by permission.

SHY KIDS IN A STRANGE SETTING MAY SAY NOTHING FOR 20 MINUTES; BOLD CHILDREN SPEAK ALMOST IMMEDIATELY.

21 months of age, can predict how children will score on several measures of stress reactivity at 5½ years old.

Kagan believes that very shy children tend to become "uncertain" when placed in an unfamiliar situation—a predisposition that may be related to their easily aroused stress circuit. Inhibited toddlers typically cling to their mothers and are very slow to venture into a strange playroom. "These children case the environment very vigilantly," notes Kagan, adding that they also tend to suffer from "stage fright"; very uninhibited children typically speak within the first minute in the lab, but very inhibited children will wait as long as 20 minutes before making a comment. Often conspicuously tense in posture and voice, the very shy children also perform worse than other children on visual matching tasks, leaning either toward impulsivity—and wrong answers—or too much reflection and late answers.

Judging by mothers' reports, Kagan says that the very timid children were, as a group, more colicky, constipated and irritated as infants than were other children. Many have allergies that have continued into middle childhood, while the uninhibited children in Kagan's study are virtually allergy-free. Since one effect of the stress hormone cortisol is to suppress the immune system, he wouldn't be surprised to find a connection between high reactivity to stress and susceptibility to certain physical illnesses.

While Kagan was conducting his early studies of child development in the 1950s and 1960s, psychologist Harry Harlow at the University of Wisconsin-Madison was doing his own classical research on infant monkeys' reactions to separation from their

A TOUR OF THE STRESS CIRCUIT

When uptight monkeys and shy toddlers are stressed, they have similar abnormal physiological reactions. Most of these responses are also seen in adults with clinical anxiety and depression, although shy children do not necessarily grow up to have these disorders.

The reactions center on a biological pathway known as the "HPA axis" or "stress circuit," the feedback loop connecting the hypothalamus with the pituitary and adrenal glands. A stressful situation triggers a cascade of chemical events:

When emotional centers in the brain are activated, chemical messengers such as serotonin and norepinephrine are sent to the hypothalamus, signaling it to secrete corticotropin-releasing hormone. This causes the pituitary gland, just beneath the brain, to release adrenocorticotropin hormone (ACTH) into the bloodstream. ACTH, in turn, stimulates the adrenal glands atop the kidneys to spill the hormone cortisol into the blood. Cortisol triggers a state of heightened bodily arousal to cope with the challenging situation. This state is reflected in a number of measurable changes, including increased heart rate, blood pressure, muscle tone and pupil dilation.

Patients in the throes of serious depression typically show numerous signs of overactivity or poor regulation of this stress circuit. (Some clinicians theorize that early deprivation and stressful events

may make a genetically overactive stress circuit even more active.)

The physiological similarities between uptight monkeys and depressed people have led Suomi to look at how antidepressant drugs affect uptight infant monkeys. Suomi has found that the drug imipramine, used to treat depression and panic attacks in humans, has virtually the same therapeutic effects in anxious monkeys. Such infants, treated with the medication and then separated from their peers, play and explore more than their untreated uptight peers. Separating monkeys, Suomi says, seems to elevate norepinephrine activity in the brain, and imipramine lowers it.

"We're finding that drugs will have different effects depending upon the conditions under which they're taken," Suomi explains. For example, imipramine affects uptight monkeys' stress circuits more powerfully when they are stressed than when they are not, apparently because the animals metabolize the drug faster under stress.

Suomi says that such studies can reveal how environmental events affect both physiological processes and the actions of drugs themselves. However, he emphasizes, they do not suggest that very shy youngsters should be given medications.

"The jury is still out on the effectiveness of pills—for monkeys or for humans," he says, "but we know that sensitive, nurturant parenting can help."

mothers. Harlow showed that while some baby monkeys became depressed, others emerged relatively unscathed. In the 1960s and 1970s, Suomi, working with Harlow, zeroed in on the depression-prone monkeys—the ones he would later identify as "uptight"—and found that they had behavioral and physiological characteristics closely resembling those of Kagan's timid children.

During the first month of life, the uptight monkeys responded to the

stress of a strange situation by being slow to explore. They also lagged slightly behind in motor-reflex development and had poor muscle tone. During infancy, under the stress of brief separations, they became less playful and showed the sharp rises in blood cortisol and the unusually high and stable heart rate seen in Kagan's shy toddlers. They also showed many signs of anxious behavior, such as clasping and grimacing, and had other physiological signs of stress.

NATURE GIVES THE INFANT A VERY SMALL TEMPERAMENTAL BIAS; THE RIGHT ENVIRONMENT CAN CHANGE IT.

Suomi has found that uptight monkeys, like Kagan's shy children, do not often outgrow their abnormal physiological response to stress. Those who, when 22 days old, had elevated cortisol levels in response to mild stress responded to such stress at 1½ years old with the highest heart rates and cortisol levels found among their age group. When these monkeys were later separated from their peers for periods of a week, they continued to show signs of anxious behavior and to have abnormal physiological responses. Once they were reunited, however, everything again appeared to be normal.

Even as late as adolescence—4 to 5 years old in monkeys—those who were uptight at birth continued to react abnormally to stress, but at this stage they tended to become hyperactive. As adults, they seemed to regress in the face of stress, showing the depressed behavior seen in infancy.

The extreme timidity of Suomi's simians and of Kagan's kids seems to have a genetic foundation, the researchers agree. Since most of the monkeys in Harlow's original colony were reared together in nurseries and peer groups, "parenting" clearly was not influencing their individual ways of reacting to stress. Further evidence that uptightness is probably inherited has been accumulating for years in Suomi's breeding colony, the descendants of Harlow's monkeys, in which about one in four monkeys is born uptight. For example, Suomi found that siblings and half-siblings (rhesus monkeys are rarely monogamous) are more likely to react similarly to stress than are unrelated monkeys.

Evidence that humans may have a special biological predisposition for extreme shyness comes from the studies of behavioral geneticists Robert Plomin of Pennsylvania State University and David Rowe of the University of Oklahoma, who found that identical twin pairs are particularly prone to react alike to strangers.

What does this predisposition mean for the babies who inherit such a legacy? Are they destined to be fearful children, hyperactive teenagers, depressed adults? The answer, so far, seems to be "not necessarily." Suomi's current work suggests that, under the right circumstances, being born uptight need not be a social handicap.

While clearly not a perfect model of human behavior, rhesus monkeys share enough brain systems and patterns of social behavior with humans to make for some compelling comparisons. And because monkeys develop faster than people do (by the time Kagan's youngsters have reached second grade, Suomi's monkeys have already reached reproductive maturity), four monkey generations can be compressed into the same time frame as one human generation, offering a long-range view of how heredity and environment interplay.

Because the monkeys in Suomi's colony have been carefully studied over many generations, he knows their pedigrees, their childhood experiences, their adult behavior and their individual characteristics. For example, he has found that monkey mothers, like humans, have varied styles of child rearing; some are quite nurturant and protect their offspring from stress, while others reject, even punish, their children. Such rearing styles are pretty well fixed after one baby has been raised. In addition, the mother's own temperament—whether she herself is emotionally uptight—affects how she raises her babies.

Although the monkeys' genetic heritage seems to play a considerable role in making them uptight or not, learning may also influence how they develop. For example, independent of her genetic contribution to her infant's behavior, an uptight mother might pass on her own emotional tendencies to her offspring simply by providing a model of overreactive behavior or through her particular caretaking style.

To test the relative influences of genetic endowment, caretaker style and caretaker temperament on infant behavior, Suomi is using the monkey equivalent of foster parenting. He is breeding uptight and calm monkeys, then separating them at birth from their biological mothers and placing them with substitute mothers chosen for various combinations of rearing styles and temperaments. Thus, for example, a naturally uptight baby might end up raised by a nurturant, but uptight, mother or one who is relaxed but punitive.

In one such study, infants—half uptight, half calm—were assigned to foster mothers within a week of their birth. Each caretaker-infant pair lived together except for a 20-minute separation once a week during the first month. While the infants were on their home turf with their foster mothers, the mothers' caretaking style was the dominant influence on their behavior. Regardless of their inherited temperament, infants with punitive mothers behaved more anxiously than those with nurturant mothers.

But when the infants were tested during the mildly stressful separation periods, neither the mother's rearing style nor her temperament influenced the infant's behavior; only inherited temperament predicted how the monkey would react to stress. Uptight infants, even with easygoing, nurturant mothers, showed the abnormal stress-reactivity pattern.

After six months, the baby monkeys were separated from their foster mothers for four-day periods, relieved by three-day reunions in the cage. Under these quite stressful conditions, heredity again asserted itself: During separations, uptight youngsters became noticeably disturbed and passive, and their cortisol levels shot up higher than those of their easygoing peers. By the third or fourth separa-

THE TIMID MONKEY'S FORMULA FOR SUCCESS: A NURTURANT MOTHER AND THE RIGHT CONNECTIONS.

tion, the temperamentally uptight mothers also showed signs of anxiety.

At 9 months, the infants were separated from their caretakers and placed in peer groups, where they are still being studied. While characteristically slow to adjust to these new circumstances, the uptight monkeys are for the most part thriving, according to Suomi. In fact, a few lucky enough to be raised by nurturant mothers are now the dominant monkeys in their groups—a sure sign of simian social success. But, not surprisingly, the one unfortunate monkey who was triply handicapped—born uptight and raised by a foster mother who was both uptight and punitive—is struggling at the bottom of his peer group's dominance hierarchy.

How did the naturally timid monkeys overcome their handicaps and rise to the top of the social ladder? In a sense, they cultivated the right connections. The uptight babies who eventually became dominant did so after Suomi introduced older females into the group to act as "foster grandparents." (Such older monkeys are usually available in natural groups.) Having learned to appreciate good caretaking from their nurturant mothers, the uptight monkeys apparently were more motivated than others to seek out similar relationships with the older females. In the rhesus matriarchy, such relationships translate into power, and the lucky ones got it.

Suomi is guardedly optimistic that, at least for monkeys, nurturant parenting can indeed make up for an inherited vulnerability. But what does this mean for the 10 to 15 percent of human babies who Kagan estimates are born with the potential to become

PARENTS SHOULD PUSH VERY TIMID CHILDREN— GENTLY—INTO DOING THE THINGS THEY FEAR.

TRIUMPHING OVER TIMIDITY

Shyness is a widespread condition. Studies by psychologist Philip Zimbardo and colleagues at Stanford University have revealed that 80 percent of Americans say they were shy at some time in their life, and 40 percent say they're currently shy.

According to psychologist Jonathan Cheek of Wellesley College, about half of the latter group seem to be the temperamentally inhibited types studied by psychologist Jerome Kagan of Harvard University, and the other half became shy in early adolescence. The temperamentally shy, even those who don't show their problem behaviorally, usually continue to have excessive physiological arousal under mild stress. Both types can be helped to

overcome their problem, particularly through relaxation techniques, but Cheek says the temperamentally shy have a tougher, longer road.

Shyness is not necessarily a barrier to success. While some shy people choose to work in low-level, invisible jobs, others may become famous in their fields. According to Kagan, both T.S. Eliot and Franz Kafka were shy from a very early age.

Some shy people even become performers. Zimbardo counts among the "shy extroverts" such celebrities as Johnny Carson, Carol Burnett, Barbara Walters and Michael Jackson. Clearly, some shy people, even those with the genetic dice loaded against them, can come up winners.

very shy? Can they avoid it? Can their parents help them?

Kagan says there's nothing deterministic about the biological vulnerability of very shy children. "Nature gives the infant just a very small temperamental bias," he emphasizes. "The proper environmental context can change it profoundly." In his view, "It takes more than simply a biological vulnerability to produce an inhibited child. You need a stressor plus the vulnerability."

"Most parents are fatalistic," says Kagan, "and if you let a child remain fearful for a very long time, then it may become hard to change." However, there's evidence from his study that even though very timid toddlers tend to stay shy, such children can improve with help from their parents. Kagan found that 40 percent of the originally inhibited children—mostly boys—became much less inhibited by 5½ years, while fewer than 10 percent became more timid.

Based on interviews with the parents, Kagan believes that they helped their children overcome shyness by bringing other children into the home and by encouraging the child to cope

with stressful situations. He cites as an example two boys in his study who were very inhibited at 21 months and showed all the characteristic physiological signs. "At 5½ they were uninhibited and their heart rates had changed. There is plenty of malleability in the system. But these children happened to be born to very sensitive, educated parents who decided to change the kids."

He advises parents of very shy children to recognize their problem early, protect them from as much stress as possible and help them learn coping skills. "Parents need to push their children—gently and not too much—into doing the things they fear," Kagan says.

The work of Kagan and Suomi, whose separate research programs have been closely coordinated since 1981, has helped to clarify the delicate interplay and influence of both heredity and environment on behavior and development. In the case of inborn shyness, biology clearly sets the stage, but learning helps to write the script. Many scenarios are possible, including some with reasonably happy endings.

Compliance, Control, and Discipline

PART 1

Alice Sterling Honig

Alice Sterling Honig, Ph.D., Professor, Department of Child and Family Studies, Syracuse University, Syracuse, New York.

"Time for lunch," called Mama as she entered 20-month-old Jason's playroom. Jason clutched his ball and ran waddling and laughing to the other side of the room, repeating "no, no, no, no, no!" Mama stood calmly and pretended to sniff the air. "Mmmm. Hamburger. Yum, yum. Mmmm carrots, yummy carrots!" she remarked with relish. Jason stopped, turned, and galloped past her and into the kitchen. Climbing into his high chair near the table, Jason eagerly called "Meat, mama, meat!"

* * *

Mr. Sims was pouring milk into the dry cereal bowls set for each of the group of children ready for breakfast. "No milk for me," announced Shana, shaking her head vigorously to reinforce her request. "OK, Shana," agreed Mr. Sims, as he skipped her bowl and poured milk for the other children. Shana struggled with her dry cereal. Then she looked up, just as Mr. Sims was walking back toward the refrigerator, and announced "Milk, now, Mr. Sims." He walked back with a smile and poured the milk cheerfully.

Both Jason and Shana are cooperative children. They follow safety rules and play with others without interfering with their rights. But toddlers and young preschoolers often just *need* to say *no*. During the second and third years of life children learn how to assert their own wishes and express their uniqueness. A certain amount of *no* saying will punctuate the conversations of fairly cooperative children. As adults, we can accept their assertions and appreciate the self that is struggling to express its own desires and coordinate those desires with others' wishes.

Origins of compliance

One of the most perplexing problems for parents and teachers is how to help infants, toddlers, and preschoolers, who are already struggling to master locomotion and language skills, also to master compliance and cooperation in a variety of settings. *Choosing discipline techniques wisely depends on understanding the development of compliance.* All adults who care for young children, therefore, must understand how compliance, and ultimately self-control (internalized compliance), emerge.

What is compliance?

Compliance refers to an immediate and appropriate response by the child to an adult's request. A *control technique,* on the other hand, refers to all attempts to change the course of a child's activity (Schaffer and Crook 1980).

Babies learn to imitate kind, helpful adult behavior.

How does compliance emerge? What adult actions promote compliance? How is compliance related to the situation? How can adults promote children's inner controls?

From *Young Children,* January 1985, pp. 50-58. © 1985 by the National Association for the Education of Young Children. Reprinted by permission.

The earliest compliance that infants show is their self-comforting behavior in response to a loving adult's request to stop crying. A very young infant cuddled in an adult's arms hears a soothing voice and tries to focus on the adult's eyes—and quiets down. Infants may put their fists or thumbs into their mouths and thus find another way to comply with the request for self-soothing.

Babies are born with built-in biological skills for learning to comply. Then during the first year of life, effective adult-infant interactions build reciprocal, mutually-satisfying chains of compliance in feeding, soothing, diapering, gazing, and play interactions (Honig 1982a). These often wordless communication games enhance later patterns of mutual cooperation.

> Heather's mother was talking on the phone while her 9-month-old daughter was playing with scraps of paper on the coffee table 12 feet away. Heather crumpled one piece of paper into a small ball, looked at her mother, and offered the ball to her. Heather's mother continued to talk on the phone but glanced at her daughter's gift. Then she took the ball of paper from her daughter, held it a moment, and gave it back. Heather took the ball of paper, put it in her mouth for a few seconds, and then again handed it to her mother. During the next 3 minutes, Heather sustained contact with her mother, who remained talking on the phone, through the exchange of a crumpled, and increasingly wet, ball of paper. (Ziajka 1981, p. 67)

The development of social cooperation is necessarily, then, a joint enterprise of adult and child (Schaffer 1984).

Piagetian theory implies that preschool children, at a preoperational level of thinking, cannot understand the rationale behind cooperative social skills. Yet research on prosocial behavior shows that young infants *are* capable of empathic responses to others (Honig 1982b). Many one-year-olds are quite cooperative in holding out an arm for dressing, or when asked, can stand *fairly* still to have their overalls snapped.

Between 9 and 12 months, infants comply with simple requests such as "Come" or "Show me the doggie in the book." Hay and Rheingold (1983) describe an 18-month-old boy who said, "Night, night, bear" as he put his bear in a cradle, covered it with a blanket, and kissed it. Babies learn to imitate kind, helpful adult behavior.

These researchers also observed 18-, 24-, and 30-month-old children either with their mothers or fathers in a laboratory suite of homey rooms. All of the children joined in to cooperate in some of the experimental tasks set by the parents, such as putting away groceries, sweeping up scraps, and folding laundry. "On the average, parents were assisted by the 18-month-olds on 63%, by the 24-month-olds on 78% and by the 30-month-olds on 89% of the tasks" (Hay and Rheingold 1983, p. 83). And the children responded to parental requests for help with alacrity. They even helped an *unfamiliar* adult to shelve groceries! Their cooperative intentions were often expressed with statements such as "I gonna clean up mess."

Research on moral development reveals that a large majority of children understand the need for controls and discipline between the ages of four and five-and-one-half. When young children, regardless of age, sex, or social class, were presented with stories of noncooperative children (who were throwing sand, refused to give a toy back, or were rude to grandmother), they judged that the mother was "good" (rather than "not good") when she told the child to stop the misbehavior (Siegal and Rablin 1982). Even very young children seem to realize that people must mutually get along with each other so that their needs are fairly considered.

Adult language and compliance

Compliance thus is an early-appearing social skill. Young children are so primed to comply that they will comply if they understand the *verb* of a request—even though the words of the request are scrambled! Two- and three-year-olds complied with their mother's request in 90 percent of normal word order requests ("Give the ball to Mommy"), 81.5 percent of misplaced word order requests ("Can the you throw ball?"), and 79.5 percent of scrambled word order requests ("You how jump me show") (Wetstone and Friedlander 1973).

In another study, children complied with *implicit* verbal requests ("I can't wash your hands unless you put your Teddy down") as well as explicit requests (Holzman 1974).

Secure attachment: Predictor of compliance

Much research in the past decade confirms the critical importance of secure infant-mother attachment for the development of compliance. Secure attachment is fostered by a positively responsive parent who

- is aware of and accurately interprets infant distress signals
- responds to distress signals promptly and effectively to comfort the baby
- has tender and gentle holding and feeding patterns (Ainsworth, Bell, and Stayton 1971).

Stayton, Hogan, and Ainsworth (1971) studied compliance in the last quarter of the first year of life through in-home naturalistic observations of 25 middle-class mothers and their babies every three weeks for four hours. Three scales were devised to assess the degree of harmony in the mother-infant interaction: sensitivity/insensitivity, acceptance/rejection, and cooperation/interference.

Insensitive mothers are geared almost exclusively to their own wishes, moods, and activities. They tend to their babies only when they so desire, so their actions are rarely contingent upon the baby's signals. The accepting mother resolves conflicting or negative feelings about temporary restrictions that the baby may put on her activities, and accepts the responsibility for care.

> The cooperative mother avoids imposing her will on the baby but, rather, arranges the environment and her schedule so as to minimize any need to interrupt or control him [or her].... The interfering mother ... seems to assume that she has a perfect right to do with him what she wishes, imposing her will on his, shaping him to her standards, and interrupting him arbitrarily without regard to his moods, wishes, or activity-in-progress. (Stayton, Hogan, and Ainsworth 1971, p. 1061)

Infant compliance to commands such as "No, no!" or "Come here" was strongly and positively related to all three indicators of the quality of the mother-infant relationship. Thus, the findings of this study suggest that children are more likely to obey parental signals if they have a positive harmonious affectional relationship with their parents. Additionally, those babies who showed a progression toward self-control such as creeping toward a forbidden object and then not touching it (20 percent), had accepting, cooperative mothers who permitted their babies more floor freedom.

Childrearing style and compliance

Parental childrearing styles are related to the level and quality of child compliance in the early years. Bishop (1951) was one of the earliest investigators to show that when mothers were more nonaccepting and directive with their three- to six-year-old children, then the children were more noncooperative and negative in play with their mother.

The mother's harsh tones and use of physical force were related to greater tendency for toddler disobedience.

Clarke-Stewart, VanderStoep, and Killian (1979) found that two-year-old children's positive responsiveness in play with their mother at home and in semistructured situations was most strongly associated with the mother's nondirective speech, her natural positive interactions with her child, and her responsiveness to her child's social expressions.

In an earlier similar study, Beckwith (1972) had observed that when the mother was critical, suppressive, and interfering, then infants from 8 to 11 months of age played less frequently and responsively with their mother.

In an analysis of 120 mother-infant pairs observed at home, Olson, Bates, and Bayles (1984) found that maternal verbal stimulation and responsivity, positive control, and affection best predicted toddler compliance.

McLaughlin (1983) videotaped one-and-one-half-, two-and-one-half-, and three-and-one-half-year-olds at home with toys and with each parent. Parents told their children what to do far more often than what not to do. For mothers, 61 percent of all controls were action controls and 30 percent were attention controls. Fathers used 72 percent action controls, which is consistent with the idea that fathers are more action-oriented with young children than are mothers.

Compliance was greater for attention controls. The three-and-one-half-year-olds showed the most compliance. They also were *more compliant in response to indirect controls* (in the form of questions or declaratives) than they were for direct controls in the imperative form. The one-and-one-half-year-olds showed a nonsignificant tendency to comply more to direct (52 percent) than indirect controls (45 percent). Compliance was present more when there were also nonverbal supports to help the children obey.

How did parents handle noncompliance? One strategy was to repeat the utterance, which fathers did more than mothers. Repeats were higher for the youngest group. For mothers and fathers, the mean eventual compliance rate in episodes of repeating controls were 43 percent and 41 percent respectively. The rate was highest for three-and-one-half-year-olds (53 percent).

McLaughlin feels that when parents say "Look at the bear" or "Open the book" they are trying to keep a child attentive to objects and keep an interaction going rather than to stay in command, and these positive techniques work.

Lytton and Zwirner (1975) also studied compliance in home living rooms. Parent controls preceding a child's comply/noncomply response were coded for 136 male twins and singletons (mean age 32.4 months). Parent *suggestions were more helpful than command-prohibition or reasoning* in gaining child compliance. For mothers, 53 percent of verbal forms of control were followed by compliance. Father's controls boosted child compliance even more. *Physical control and negative action by parents were particularly likely to be followed by noncompliance.*

Londerville and Main (1981) classified 36 middle-class 12-month-olds on security of attachment with the Ainsworth strange situation paradigm.* At the age of 21 months, these children

*The strange situation consists of eight three-minute episodes where a baby is in a playroom with toys and mothers. Two brief separations from and reunions with mother are staged. In one, the baby is left alone with the stranger. In the second, the baby is left alone. Secure attachment is scored when a baby actively greets mother and seeks and accepts physical contact and comfort on reunion, and can then settle down from distress and resume play with toys. Insecure babies are classified as either avoidant (showing less separation protest and behaving less positively to being held and more negatively to being put down) or ambivalent (very distressed by separation, resistant to comfort, anxious and unable to play well on mother's return).

were videotaped during 30 minutes of free play with their mothers in a play-room. Toddler responses to mothers' verbal commands were coded as compliance, active disobedience, or passive noncompliance.

The average toddler complied with 50 percent of the mother's commands. Internalized controls, such as reaching toward a fan (which the mother had initially asked the toddler not to touch) and then withdrawing a hand were seen in 28 percent of the children. The average toddler actively disobeyed 24 percent of mother's commands. Such active disobedience was correlated (r = .52) with mother's description of her child as "troublesome."

Nonsecurely attached toddlers were almost equally likely to obey (42 percent) or actively disobey (38 percent) a maternal command or prohibition.

In contrast, *securely attached toddlers were four times more likely to obey as actively disobey* (57 percent versus 13 percent). None of the securely attached children showed angry active disobedience to their mothers by screaming in rage, hitting, or throwing objects, but 6 of the 14 insecurely attached toddlers did. Thus, *security of attachment at 12 months was a powerful predictor of compliance for boys and girls at 21 months.*

The number of maternal verbal commands or physical interventions was not related to child compliance. Adult style, however, was significantly related to toddler compliance. Maternal harsh tones were related to greater tendency for toddler disobedience. Warmth of tone was correlated with compliance for males only. Greater physical force used by mothers in intervening in their toddlers' play was not associated with toddler compliance, but was significantly related to disobedience with mother.

Clearly, parenting practices can boost child compliance. Early childhood educators need to reach out to share this knowledge with parents!

Secure attachment, compliance, and problem solving: The magic trio

Mahler's and Erikson's theories of toddler development both predict that the second and third years of life will reflect a struggle to develop autonomy and independent initiatives. Both predict that a child with a secure attachment will later exhibit more competence, flexibility in problem solving, and resourceful ability to recruit adult help when needed.

Matas, Arend, and Sroufe (1978) provide powerful evidence that the quality of early attachment is related to later competence *and* toddler cooperation with parents. Initially, 48 female infants were classified with the Ainsworth strange situation paradigm as securely, avoidantly, or ambivalently attached. Then, at 24 months, toddlers returned to the laboratory for a ten-minute free play period, a six-minute cleanup period, and problem-solving tasks, some

A child with secure attachment will later exhibit more competence, flexibility in problem solving, and resourceful ability to recruit adult help when needed.

of which were too hard for a toddler to solve alone.

When mothers asked them to clean up after play, securely attached toddlers showed as much typical opposition as their insecurely attached peers, but their behavior was radically different during the problem-solving tasks.

Compliance with mother's requests during problem-solving tasks was significantly higher (57 percent) for the securely attached toddlers than for the insecurely attached (39.5 percent). Securely attached toddlers ignored the mother less. They were significantly more enthusiastic about solving the problems, spent less time away from task, exhibited fewer frustration behaviors, and were lower on saying "no," crying, and whining. They tried to cooperate with their mothers' suggestions for solving the problems.

Temperament was not related to toddler compliance, nor was general intelligence. *Secure attachment allowed toddlers to use mother as a positive resource when help was needed and there were important problems to solve.* Helpful mothers had more competent toddlers.

The mother's quality of assistance relates more specifically to the cognitive aspect of second-year adaptation.... This requires sensitivity to the child's cognitive, perceptual-motor and information-processing skills—aspects of more autonomous behavior.... Simply telling the child what to do, or overly controlling him, [or her] was not seen as good quality assistance. The mother of a two year old demonstrates her sensitivity ... by giving the minimal assistance needed to keep the child working and directed at the problem solution without solving it for him and by helping her child see the relationship between actions required to solve the problem. (Matas, Arend, and Srouf 1978, p. 555)

Thus, the quality of adult helpfulness can facilitate toddler compliance at developmentally difficult tasks.

In a subsequent study, Sroufe (1979) found that

securely attached infants showed a particular pattern of behavior across tasks. When they came to [a] more challenging ... problem, they maintained their involvement but sought more help. They increased their compliance and decreased their opposition. Their mothers, in turn, maintained a high level of support and offered more directives.... The resistant, difficult-to-settle group, on the other hand, fell apart completely. They became increasingly oppositional, highly frustrated, angry and distressed, even though they did increase their help seeking. Their mothers increased their directives, but the quality of their assistance decreased markedly.... [Avoidant infants] made little adjustment to the harder problem. (p. 840)

Ego resiliency refers to flexibility of controls. Over-controlled children are rigid, not spontaneous; under-controlled children have difficulty in controlling impulses and cannot delay gratification. At five years of age, the children who had earlier been classified as secure in Sroufe's study were described by their teachers as highly ego resilient, self-reliant, and moderate in self-control. Children who had been avoidant and resistant babies were at five years of age significantly less resilient, and were respectively more over-controlled and under-controlled. Thus, *self-control,* the ultimate goal of socialization for cooperation in homes and classrooms, *has been found to be significantly linked to early attachment patterns.*

High-risk infants and compliance

Erickson and Crichton (1981), working with Sroufe, assessed attachment at 12 and 18 months of age and

used the problem-solving situation at 24 months with 267 high-risk infants. At the age of 2 years, compliance in the problem-solving task was found to be unrelated to the children's temperament, irritability, and alertness as infants. These data suggest that compliance is not predicted by early infant variables. Nor did maternal life circumstances predict compliance. However, toddlers (particularly girls) whose attachment classification remained stable and secure from 12 to 18 months of age were more compliant. Of the 9 extremely noncompliant children, 7 were boys.

Mothers of the more compliant children were found to have a better understanding of the psychological complexity of their children and themselves.

Self-control is significantly linked to early attachment patterns.

When their infants were six months old, those mothers had been more sensitive to their babies' signals in feeding and play situations. Once again, this time in a group of high-risk infants, *the quality of infant-mother attachment proved to be a significant predictor of toddler compliance.*

Mother-child relationship and compliance with other adults

In our modern world where many children are cared for part of the time in out-of-home settings, it is important that children learn to cooperate with adults in addition to parents. Research strongly suggests that the quality of a relationship that a preschooler forms with another adult is partly a function of the quality of the relationship that the child has with the mothering one.

In-home frequency of maternal positive interactions with her toddler and accomodating attitudes toward toddlers was predictive of a two-year-old's willingness to interact with an adult stranger (Clarke-Stewart, VanderStoep, and Killian 1979).

In Londerville and Main's study referred to earlier, the greater the force that mothers used with toddlers during free play, the less cooperation those toddlers showed with the Bayley examiner during developmental testing.

In the midst of the free play situation with mothers in that research, a strange play lady entered with a toy baby doll in a carriage and spent 20 minutes engaging the toddler. The play lady pretended to make a toy dog bark and approach the doll. She then left the dog near the doll and requested the toddler "Don't let the doggie bite the baby." Toddlers who complied with this request usually moved the toy dog away or brought it to their mothers. If the mothers had used greater force with their toddlers during free play, then the toddlers were less cooperative with the adult playmate. "Both the Bayley examiner and the adult playmate consistently made requests in a warm and gentle manner, but the nonsecurely attached toddlers were nonetheless uncooperative" (p. 299). Thus, gentleness of strange adults was not the critical variable. Prior relationship with mother was more important in predicting cooperation with a strange adult.

The stranger in another study was a clown (Main and Weston 1981). A masked actor in clown costume played Peek-a-Boo and other games with one-year-old children, who were tested for attachment to mother and father in the Ainsworth strange situation one week later.

Infants were rated for their readiness to establish a positive social relationship with the clown actor and to return his friendly overtures rather than show indifference, fear, or conflicted, disordered behavior. Infants who were later rated as secure were judged from videotapes to show significantly higher relatedness with the clown and lower incidence of conflict behavior.

The most impressive finding of this study was that one-year-old infants who were securely related both to mother and father had the highest mean cooperative score (6.04) with the clown stranger. The other mean scores were 4.87 for infants securely attached to mother but insecurely to father; 3.3 for infants insecure with mother but secure with father; and 2.45 for children insecurely attached to both parents.

Thus, an enduring relationship with each parent in the first year of life has implications for the infant's ability to be affiliative and cooperative with a strange adult. Main and Weston conclude that the effects "of an insecure relationship can be mitigated by a se-

cure relationship" (p. 939). These findings are important for child care staff, home visitors, and other professionals who may work with babies with one or more insecure attachments.

Abused toddlers are peculiarly noncompliant with adults when the children are enrolled in group care. Abused toddlers avoid eye contact, do not respond to positive adult initiatives, sidle up to adults rather than move directly to them, and are more likely than normal toddlers to exhibit aggression against adults as well as other children (George and Main 1979).

Child care staff need to guard against feelings of frustration that can hook them into coercive cycles of mutual noncompliance with abused children. The caring adult must try hard to remain friendly and caring with abused young children, even in difficult discipline situations.

Toddlers were quite compliant when their mothers used distractions or proposed a new activity rather than directly prohibiting a current one.

Compliance as a function of timing and setting

Child compliance may well be related to the particular circumstances under which adults are trying to get children to obey. Asking a child to come in from a playground just before her long-awaited turn to ride the new tricycle will probably result in less compliance than if the request were made just after the ride.

In a laboratory playroom

Control techniques adopted by mothers of 15- and 24-month-olds in a laboratory play situation were timed in such a way that the probability of eliciting requested actions from toddlers was high (Schaffer and Crook 1979). Generally, the mothers used controls in

sequences. Thus, action controls were used at the optimal moment—*after* the mother had used attention controls and had ensured that the toddler's attention was properly focused. Each mother in the playroom had been instructed to make sure that her toddler played with each of eight suitable toys. The mothers of younger toddlers spent one-half their time on attention-focusing devices. Mothers of two-year-olds spent one-quarter of their time this way. Mothers said, "Look," "See," or asked "Where is . . . ?" to focus toddler attention. When the mothers asked for actions, they fully specified the behavior they wanted: "Can you spin the top?" "Now put on the orange ring."

Fewer than six percent of the mothers' utterances were prohibitions. These mothers were far more likely to propose a new activity rather than directly prohibit a current one. They used distractions frequently. These toddlers were quite compliant.

In the supermarket

The supermarket is a setting in which child management techniques are sometimes inadequate to what Holden (1983) calls the "triple threat to a mother: there is food shopping to be done, a child to be managed who is afforded a diverse array of enticing objects, and all the while both mother and child are in the public eye" (p. 234).

Holden observed 24 mother-child dyads (mean child age 31.5 months) on two trips to a supermarket and also interviewed the mothers. Requests for objects (seen as undesirable interruptions to the mother) or gross motor behaviors (such as standing up in the shopping cart) were coded as *child elicitors*. The child was considered to have been compliant with maternal responses to these elicitors if the child terminated an elicitor within 20 seconds following the mother's response.

Mothers refused children's requests 86 percent of the time, usually by responding with reasons or with power assertions. Generally, children complied (69 percent). Power assertion and reasons were used to gain compliance more (about 70 percent) when children acted out motorically rather than made requests.

After a request, the probability of child compliance depended on the type of maternal response. If mothers gave reasons or distracted the child, compliance was 68 percent. Compliance after power assertion was 54 percent. Ignoring or just acknowledging the child lead to only 25 percent compliance. Mothers who consented to the elicitor received compliance 92 percent of the time.

All of these mothers employed proactive controls such as initiating conversations with the child or providing the child with a cracker or toy to keep her or him content. *Proactive controls avoided power battles by engaging the child's attention.*

Holden notes, that with proactive controls

> a child learns what is acceptable behavior in the supermarket under the mother's direction.... The mother, through indicating what behavior is appropriate, facilitates the process of translating the child's view of the supermarket from a place where there are tempting items to play with to one where merchandise is purchased.... Mothers often discuss which items to buy and why.... [They] may be socializing their children into concordant relationships. (p. 239)

Thus, by using disciplinary techniques that reduce the frequency of conflicts in a shopping situation, parents may teach their young children a complying rather than a test-the-parent orientation.

These data are in strong agreement with Hoffman's (1975) position that *inductive discipline works best.*

> Induction includes techniques in which the parent gives explanations or reasons for requiring the child to change ... behavior. Examples are pointing out the physical requirements of the situation or the harmful consequences to the child's behavior for himself or others. (Hoffman 1970, p. 286)

A much less effective discipline technique is *power assertion,* which includes physical punishment, deprivation of material objects or privileges, and use of force or love withdrawal, such as refusing to speak to the child. Hoffman notes that love withdrawal, even though it is not a physical threat to the child, is typically more prolonged and has a highly punitive quality. "It may be ... devastating emotionally ... because it poses the ultimate threat of abandonment or separation" (p. 285).

In reviewing research studies, Hoffman notes that the "frequent use of power assertion by the mother is associated with weak moral development" (p. 292) quite strongly, whereas induc-

tion discipline and affection are associated with advanced moral development.

In the home

In an in-home study of the long term effects of power assertion, the most positive predictors of child compliance were mother's consistent enforcement of rules, amount of play with the child, and use of psychological rewards and reasoning (Lytton 1980). Maternal use of psychological punishment and father's physical punishment were negatively associated with compliance. These findings support Hoffman's prediction that inductive methods are more likely to achieve compliance than are power assertion techniques.

Eimer, Mancuso, and Lehrer (1981) point out that if the parent consistently makes demands that are too difficult or incongruous with the cognitive capacities of the child, then quite possibly the child will not want to be around the parent. This avoidance behavior is likely to be interpreted as noncompliance, sneakiness, or unresponsivity to adult controls. The authors suggest that teachers ask the noncompliant child to think about her or his conception of a rule-related situation in a new way. "I want you to tell me a story. Tell me the story of what I asked you to do. Then tell me what you did. And tell me why I asked you to do what I asked" (p. 10).

Parents and caregivers may find the following books useful in discovering ways to help young children become more compliant and cooperative:

Forehand, R. L. and McMahon, R. J. *Helping the Non-Compliant Child: A Clinician's Guide to Parent Training.* New York: Guilford, 1981.

Smith, C. A. *Promoting the Social Development of Young Children: Strategies and Activities.* Palo Alto, Calif.: Mayfield, 1982.

Wolfgang, C. H. *Helping Aggressive and Passive Preschoolers Through Play.* Columbus, Ohio: Merrill, 1977.

References

Ainsworth, M. D. S.; Bell, M. V.; and Stayton, D. J. "Individual Differences in Strange-Situation Behavior of One Year Olds." In *The Origins of Human Social Relations,* ed. H. R. Schaffer. London: Academic Press, 1971.

Beckwith, L. "Relationships Between Infants' Social Behavior and Their Mothers' Behavior." *Child Development* 43, no. 2 (June 1972): 397–411.

Bishop, B. M. "Mother-Child Interaction and the Social Behavior of Children." *Psycho-*

logical Monographs 65, no. 11 (1951). No. 328.

Chapman, M. and Zahn-Waxler, C. "Young Children's Compliance and Non-Compliance to Parental Discipline in a Natural Setting." *International Journal of Behavioral Development* 5 (1982): 81–94.

Clarke-Stewart, K. A.; VanderStoep, L. P.; and Killian, G. A. "Analysis and Replication of Mother-Child Relations at Two Years of Age." *Child Development* 50, no. 3 (September 1979): 777–793.

Eimer, B. N.; Mancuso, J. C.; and Lehrer, R. "A Constructivist Theory of Reprimand As It Applies to Child Rearing." Paper presented at the Eleventh Annual Interdisciplinary UAP-USC Piagetian Theory and the Helping Professions Conference. Los Angeles, California, February, 1981.

Erickson, M. F. and Crichton, L. "Antecedents of Compliance in Two-Year-Olds from a High-Risk Sample." Paper presented at the Biennial Meeting of the Society for Research in Child Development. Boston, Massachusetts, April 1981.

Forehand, R. L. and McMahon, R. J. *Helping the Non-Compliant Child: A Clinician's Guide to Parent Training.* New York: Guilford, 1981.

George, G. and Main, M. "Social Interactions of Young Abused Children: Approach, Avoidance and Aggression." *Child Development* 50, no. 2 (June 1979): 306–318.

Gordon, T. *Parent Effectiveness Training.* New York: Wyden, 1970.

Hay, D. F. and Rheingold, H. L. "The Early Appearance of Some Valued Social Behaviors." In *The Nature of Prosocial Development: Interdisciplinary Theories and Strategies,* ed. D. L. Bridgeman. New York: Academic Press, 1983.

Hetherington, E. M.; Cox, M.; and Cox, R. "Stress and Coping with Divorce: A Focus on Women." In *Psychology and Transition,* ed. J. Gullahorn. New York: Winston, 1978.

Hoffman, M. L. "Moral Development." In *Carmichael's Manual of Child Psychology,* ed. P. H. Mussen. 3rd ed. New York: Wiley, 1970.

Hoffman, M. L. "Moral Internalization, Parental Power, and the Nature of Parent-Child Interaction." *Developmental Psychology* 11 (1975): 228–239.

Holden, G. W. "Avoiding Conflict: Mothers As Tacticians in the Supermarket." *Child Development* 54, no. 1 (February 1983): 233–240.

Holzman, M. "The Verbal Environment Provided by Mothers for Their Very Young Children." *Merrill-Palmer Quarterly* 20 (1974): 31–42.

Honig, A. S. "Infant-Mother Communication." *Young Children* 37, no. 3 (March 1982a): 52–62.

Honig, A. S. "Prosocial Development in Children." *Young Children* 37, no. 5 (July 1982b): 51–62.

Honig, A. S. "Discipline Tips for Teachers." *Forum,* in press.

Honig, A. S. and Wittmer, D. S. "Caregiver Interaction and Sex of Toddler." Paper presented at the Biennial Conference of the Society for Research in Child Development. Boston, Massachusetts, April 1981.

Honig, A. S. and Wittmer, D. S. "Teachers and Low-Income Toddlers in Metropolitan Day

Care." *Early Child Development and Care* 10, no. 1 (1982): 95–112.

Honig, A. S. and Wittmer, D. S. "Toddler Bids and Teacher Responses." *Child Care Quarterly,* in press.

Howes, C. and Olenick, M. "Family and Child Care Influences on Toddler's Compliance." Paper presented at the Annual Meeting of the American Educational Research Association, New Orleans, Louisiana, April 1984.

Kopp, C. B. "Antecedents of Self-Regulation: A Developmental Perspective." *Developmental Psychology* 18 (1982): 199–214.

Kuczynski, L. "Reasoning, Prohibitions, and Motivations for Compliance." *Developmental Psychology* 19 (1983): 126–134.

Landauer, T. K.; Carlsmith, J. M.; and Lepper, M. "Experimental Analysis of the Factors Determining Obedience of Four-Year-Old Children to Adult Females." *Child Development* 41, no. 3 (September 1970): 601–611.

Londerville, S. and Main, M. "Security Attachment, Compliance, and Maternal Training Methods in the Second Year of Life." *Developmental Psychology* 17 (1981): 289–299.

Lytton, H. "Disciplinary Encounters Between Young Boys and Their Mothers and Fathers: Is There a Contingency System?" *Developmental Psychology* 15 (1979): 256–268.

Lytton, H. *Parent-Child Interaction: The Socialization Process Observed in Twin and Singleton Families.* New York: Plenum Press, 1980.

Lytton, H. and Zwirner, W. "Compliance and Its Controlling Stimuli Observed in a Natural Setting." *Developmental Psychology* 11 (1975): 769–779.

Main, M. and Weston, D. R. "The Quality of the Toddler's Relationship to Mother and Father: Related to Conflict Behavior and the Readiness to Establish New Relationships." *Child Development* 52, no. 3 (September 1981): 932–940.

Martin, J. A. "A Longitudinal Study of the Consequences of Early Mother-Infant Interaction: A Microanalytic Approach." *Monographs of the Society for Research in Child Development* 46, no. 3 (1981). No. 190.

Matas, L.; Arend, R. A.; and Sroufe, A. L. "Continuity of Adaptation in the Second Year: The Relationship Between Quality of Attachment and Later Competence." *Child Development* 49, no. 3 (September 1978): 547–556.

McLaughlin, B. "Child Compliance to Parental Control Techniques." *Developmental Psychology* 19 (1983): 667–673.

Minton, C.; Kagan, J.; and Levine, J. A. "Maternal Control and Obedience in the Two-Year-Old." *Child Development* 42, no. 6 (December 1971): 1873–1894.

Olson, S. L.; Bates, J. E.; and Bayles, K. "Mother Infant Interaction and the Development of Individual Differences in Children's Cognitive Competence." *Developmental Psychology* 20 (1984): 166–179.

Rubenstein, J. L.; Howes, C.; and Boyle, P. "A Two-Year Follow-Up of Infants in Community-Based Day Care." *Journal of Child Psychology and Psychiatry* 22 (1981): 209–218.

Schaffer, H. R. *The Child's Entry into a Social World.* Orlando, Fla.: Academic Press, 1984.

Schaffer, H. R. and Crook, C. K. "Maternal Control Techniques in a Directed Play Situation." *Child Development* 50, no. 4 (December 1979): 989–996.

Schaffer, H. R. and Crook, C. K. "Child Compliance and Maternal Control Techniques." *Developmental Psychology* 16 (1980): 54–61.

Shure, M. B. and Spivak, G. *Problem Solving Techniques in Child Rearing.* San Francisco: Jossey-Bass, 1979.

Siegal, M. and Rablin, J. "Moral Development As Reflected by Young Children's Evaluation of Maternal Discipline." *Merrill-Palmer Quarterly* 28 (1982): 499–509.

Smith, C. A. *Promoting the Social Development of Young Children: Strategies and Activities.* Palo Alto, Calif.: Mayfield, 1982.

Stayton, D. J.; Hogan, R.; and Ainsworth, M. D. S. "Infant Obedience and Maternal Behavior: The Origins of Socialization Reconsidered." *Child Development* 42, no. 4 (October 1971): 1057–1069.

Sroufe, L. A. "The Coherence of Individual Development." *American Psychologist* 34 (1979): 834–841.

Wetstone, H. S. and Friedlander, B. Z. "The Effect of Word Order on Young Children's Responses to Simple Questions and Commands." *Child Development* 44, no. 4 (December 1973): 734–740.

Wolfgang, C. H. *Helping Aggressive and Passive Preschoolers Through Play.* Columbus, Ohio: Merrill, 1977.

Wolfgang, C. H. and Glickman, C. D. *Solving Discipline Problems.* Boston: Allyn & Bacon, 1980.

Ziajka, A. *Prelinguistic Communication in Infancy.* New York: Praeger, 1981.

Compliance, Control, and Discipline

PART 2

Alice Sterling Honig

Alice Sterling Honig, Ph.D., Professor, Department of Child and Family Studies, Syracuse University, Syracuse, New York.

Alice Honig concludes her review of research by discussing the influence of programs for young children and recommending adult techniques to encourage cooperation and compliance.

In an extensive analysis of research on problems of beginning teachers over the past quarter of a century, *classroom discipline* was perceived by teachers as by far their most serious problem (Veenman 1984). Teachers often find it difficult to manage the twin tasks of keeping classroom control while at the same time helping children maintain their autonomy and self-esteem.

Three-year-old Barney murmured to his teacher about a kitty in the picture book lying on his lap. She was reading another book to the group of preschoolers. Barney tried to say something about the kitty twice more as the teacher read her story. He seemed to have an urgent need for an adult response. The teacher ig-

nored him. Finally she finished her story and said, "It's time for nap now, children." Barney promptly said, "No!" "Don't you think the other kids are taking naps now?" the teacher asked. "I don't think so!" answered Barney and then a few other children in chorus.

Cooperation with adults in group care may not come "naturally!" Respect for others must be modeled and taught.

Effects of child care on compliance

An important issue for working parents is whether or not child care from

early infancy will result in preschool behaviors that are socially less desirable.

Peer aggression in day care. Finklestein's (1982) findings from the Abecedarian preschool project make us aware that caregivers must actively teach cooperation and teach conflict resolution without aggression. Well-defined cognitive curricular goals are not enough to ensure positive peer interaction. In this study, low-income kindergarten children, who had attended preschool eight hours per day since infancy, carried out 15 times as many aggressions (mostly hitting), toward peers on the playground as did non-preschool control children. Galvanized by this finding,

From *Young Children*, March 1985, pp. 47-52. © 1985 by the National Association for the Education of Young Children. Reprinted by permission.

the project systematically implemented a positive social curriculum, *My Friends and Me* (Davis 1977). As a result, positive social skills were successfully enhanced and peer aggression was significantly reduced. Preschool teachers need to build prosocial/positive cooperation goals into their curricula. Not only peer aggression is involved. Children in group care may be more defiant or noncompliant with adults too.

Day care and compliance with parents. Rubenstein, Howes, and Boyle (1981) looked at compliance in three-and-one-half-year-old children, half of whom had been in child care during the first three years of life. Mothers were asked to have their child do two boring tasks. From the taped verbal interactions, maternal strategies (such as making the task into a game or telling the child what to do) were coded. Mothers were interviewed, and children were separated from their mothers for 60 minutes of language testing.

Children who had been in child care since infancy had no higher overall level of behavioral problems, nor did they differ in greeting or physical contact after their test. But they were less likely to show attachment behavior to people outside the family, according to mothers. In a sorting task, the child care children were verbally less compliant (50 percent) compared to the children who had not been enrolled in child care (8 percent). Mothers of children in child care made more active efforts (90 percent) to induce compliance compared to the other mothers (38 percent). During testing, child care children were more assertive to the examiner, and declared "I don't know" more often (60 percent) than the children primarily reared at home (15 percent).

What were the complying children like? They manifested less separation distress and fewer behavior problems such as fears or sleep disturbances. Their mothers used more positive reinforcements.

Since temper tantrums and noncompliance were found significantly more frequently in the child care group, the researchers speculate that there are subtle differences in mother-child interactions around issues of control. Daily separation may increase the working mother's desire to make limited time available with the child as pleasant as possible. As a result, she may be more permissive about noncompliance.

Uniform sex differences have not been found when compliance is studied in the home, laboratory, or in child care.

Type of child care has been related to child compliance. Howes and Olenick (1984) compared toddler compliance for families using either low quality or high quality child care (determined by adult-child ratio, staff training, and continuity of staff). Children in high quality child care were more demanding in child care, more skillful negotiators over compliance issues at home, and resisted temptation better in a laboratory task. *Low quality center attendance predicted more child noncompliance at home.* The more hours a child spent alone per week with father, the less noncompliant the child was at home and in the center.

Sex of child and compliance

Uniform sex differences have not been found when compliance is studied in the home, laboratory, or in child care. Compliance was equally typical of one-and-one-half-, two-and-one-half-, and three-and-one-half-year-old boys and girls in interaction with either mother or father in a playroom (McLaughlin 1983).

Preschool boys and girls who were brought into a special playroom with either their own mother or another mother showed no differences in disobedience scores to their own mother or the stranger mothers (Landauer, Carlsmith, and Lepper 1970).

However, Minton, Kagan, and Levine (1971) reported that "Girls were more likely than boys to obey immediately. Boys were likely to resist initially and obey later or be forced to obey" (p. 1885). In general, maternal commands were obeyed about 60 percent of the time, compromised about 15 percent, and disobeyed about 20 percent of the time.

The single most striking difference reported by Minton, Kagan, and Levine was related to mothers' educational

level. Less well educated mothers were likely to scold for petty infractions and noncompliance, used more physical punishment, and gave commands rather than requests. "The better-educated mothers were less authoritarian and less intrusive with their children, more tolerant of mild misdemeanor, and less prepared for mischief" (p. 1887). Almost half of the two-year-old boys in less well educated families were never observed to ask their mothers to play, while more than two-thirds of the boys in families with higher educational levels asked mother to play. In the laboratory playroom, boys whose mothers were physically punitive spent the *least* amount of time near their mothers.

An atmosphere of adult warmth, tolerance, and respect fosters child affiliation and cooperation (Ginott 1965; Greenspan 1983). Inappropriate discipline techniques and harsh responses to noncompliance may negatively change the interaction climate for children and adults. Young children thrive in intimate, interesting engagements with caregivers. Training for parents and caregivers is critical so that they can learn positive techniques to engage young children of both sexes in pleasurable, mutually cooperative interactions that can enhance self-esteem.

Boys are particularly vulnerable. Studies of divorce show that boys receive less positive feedback after compliance than girls do (Hetherington, Cox, and Cox 1978).

Male infants seem to need more control over interactions. Martin (1981) traced individual differences in the development of boys' and girls' compliance from 10 to 42 months. At 10 months, a combination of high mother responsiveness and low infant demandingness was associated with toddler compliance at 22 months. Maternal responsiveness did not influence girls' compliance at 22 months. However, boys who were more demanding and coercive at 42 months had mothers who had not been responsive to them as infants. Martin feels that maternal responsiveness is crucial for the development of compliance in male infants. "Coerciveness in a child develops when he senses that his mother fails to cooperate with him and fails to provide what he wants and needs" (p. 36).

Child care compliance and sex of child

Sometimes males are perceived in

group programs to be more boisterous and less obedient. In a study of low income urban child care centers, 50 toddlers were found to be compliant, regardless of sex of child, about 75 percent of the time (Honig and Wittmer, in press). Nevertheless, the teachers responded to significantly more incidents of male noncompliance. In these centers, teachers received more requests for help and more distress bids from boys, and more positive overtures from girls (Honig and Wittmer 1981, 1982).

Compliance flourishes in a climate of attentive, caring, and affectionate relationships. Male toddlers especially may need a helping hand (and arms!) to discover that such affectionate interchanges are easy, natural, and frequent.

Middle-and-low-SES five- to six-year-olds and eight- to nine-year-olds who had been in child care since age three were rated by their public school teachers on task persistence, talkativeness, and classroom cooperation (Robertson 1981). No differences were reported for the first two factors. But first grade boys who had attended day care were rated by their teachers as significantly more disobedient, more demanding toward adults, more quarrelsome, and jealous of attention paid to other children in comparison with first grade boys reared primarily at home. Boys in group care may require even more teacher ingenuity and care to create a prosocial classroom climate.

Self-regulation: The goal of compliance

Self regulation has been variously defined as the ability to comply with a request, to initiate and cease activities according to situational demands, to modulate the intensity, frequency, and duration of verbal and motor acts in social and educational settings, to postpone acting upon a desired object or goal, and to generate socially approved behavior in the absence of external monitors. (Kopp 1982, pp. 199–200)

When toddlers and preschoolers become able to use representational thinking and have gained some knowledge of, and can recall, social rules and the demand characteristics of certain situations ("At naptime we lie quietly on cots"), then they are able to delay or wait to a limited extent. They may still bang on the table or move their feet from side to side as they try to hold on to self-control if lunch is late. An "in-ternally generated monitoring system" (Kopp 1982, p. 207) has begun to function rather than just compliance to external requests. Self-regulation is the final step in this process. Only when children can use reflection, thinking strategies, and contingency rules can they maintain appropriate, adaptive monitoring of their own behaviors in line with adult requirements.

Adult behaviors are not always conducive to enhancing this capacity that matures as the child develops. Many adults seem to expect compliance and self-regulation from young children. Yet, when it happens, it is accepted but not particularly rewarded.

Lytton (1979) reports that one-third of the time mothers and fathers ignored both compliance and noncompliance. Parental responses, on the whole, were neither particularly reinforcing for compliance nor were they punishing for noncompliance. Lytton dryly observes that "Mothers and fathers do not carry a probabilistic calculus in their heads, adjusting their response to the probability of obtaining eventual compliance. The control system appears on the surface to be erratic" (p. 267).

Self-control and self-regulation will be far easier for young children to learn if adults are *contingent* and *positive* in their responses to child compliance.

Requests must be part of a sequential strategy, the first step of which is to ensure that the child is appropriately oriented.... [The adult] must monitor [the child's] current state of involvement and adjust her behavior accordingly.... By successfully manipulating [the child's initial] state, the parent can avoid the clash-of-wills that is so often portrayed as typical of all socialization efforts. (Schaffer and Crook 1980, p. 60)

Reasoning facilitates long term self-regulation. Reasoning has been proposed as an adult discipline technique that provides motivation for compliance and self-control. Kuczynski (1983) provided one of three kinds of utterances in order to get nine- and ten-year-olds to work instead of look at attractive toys: a prohibition, an other-oriented rationale ("You'll make me unhappy if you look at them now. If you don't work hard enough, I'll have to do some of this work later."), and a self-oriented rationale ("You'll be unhappy if you look at them now. If you don't work hard enough you'll have to do some of this work later and you'll have little time to play with those toys.").

The other-oriented rationale kept children working most consistently even on trials when the experimenter left the room! "These results suggest that other-oriented reasoning aroused internal motivations for compliance and that the self-oriented rationale and the prohibition alone aroused external motivation for compliance" (p. 132).

Thus, for self-regulation in the absence of adult authority figures, inductive reasoning again appears to be an effective discipline technique. However, reasoning may be difficult to use when children are very young and when adults are severely stressed in their lives.

In following the stormy history of four-year-old children for two years after their parents were divorced, Hetherington, Cox, and Cox (1978) found

> For self-regulation in the absence of adult authority figures, inductive reasoning appears to be an effective discipline technique.

that mothers gave far more commands than fathers in that first year. Boys were far less compliant to their mothers than girls one year after divorce. At that low point, reasoning and explanation were less effective for getting compliance than were commands. But two years after divorce, mothers markedly increased their use of reasons and explanations. Then, appeals for the rights and feelings of others were *more* effective in eliciting boys' compliance.

For toddlers, other discipline techniques, such as focusing attention, use of delighted admiration, and provision of action supports may be more effective. Chapman and Zahn-Waxler (1982) have even reported, in a study of the development of prosocial behaviors in toddlers, that verbal prohibition and sometimes love withdrawal produced short-term prompt compliance, but for the long term achievement of self-control and self-regulation, reasoning is a powerful technique.

Conclusions

Achieving inner controls is a long and often uneven developmental process in very young children. They need all the help they can get from perceptive adults in order to achieve self-regulation. Slavish compliance with authority is not the goal. Inner self-regulation and the ability to make judicious choices about compliance with rules and regulations in order to maximize one's own and others' peaceful and harmonious social interactions takes much skill and many helpful experiences with adults and peers. These brief ideas, adapted from Honig (in press) may enable adults to more effectively assist young children to become ego-resilient, self-disciplined people.

- Catch children being good and express pleasure at appropriate behaviors.
- Model considerateness, patience, courtesy, and helpfulness.
- Admire efforts and tries of a less-skilled child, so that you build self-esteem sincerely and consistently.
- Use hugs, shining eyes, and a loving voice tone to build children's basic trust in adults as helpful people whose love and approval are important and pleasurable for the child.
- Hold a child on your lap in front of a mirror and initiate games of facial and body imitations to establish an awareness of body boundaries and self-image.
- Convey to children clearly that hurting others is *not* acceptable. Having and expressing angry or exasperated feelings with words is acceptable. Give children the words for their frustrated feelings ("I'm so mad at Daniel!"). Gordon (1970) calls this active listening.
- Discuss feelings in a group circle, so that children can share occasions when others have made them happy or angry, and experiences where their actions have made another person distressed or happy. Use one such topic per session.
- Encourage children to think of an *alternative* to the unacceptable behavior (Shure and Spivack 1979). For example, "Hitting and grabbing is one way to try to get a toy. What *other* way can you think of to get Jessica to give you the toy?"
- Pair a shy or fearful child with a gregarious, kind child when it seems appropriate.
- Play sociodramatic games with children to give them success in role playing. In this way a less-preferred child can learn how to enter into classroom play productively, rather than in inappropriate ways such as always playing *baby* or *dog* in dramatic play.
- Encourage children to think of the *consequences* of their inappropriate behaviors: "If you hit Mary, then how will she feel? What will she do?"
- Say what you *do* want when possible, rather than what you don't want. "Walk slowly!" is easier to carry out than "Don't run!"
- Physically hold a child who is out of control. Tell her or him "I know you are feeling very upset. I will help you calm down so that you don't get hurt or hurt someone else. I'll help you be safe until you can get back control."
- Give a child with angry feelings a pillow to pound or a pounding peg-board.
- Find a place to be in the classroom so that an angry, acting out child perceives you as a secure safety beacon. You radiate what Wolfgang calls a "zone of safety and control" (Wolfgang and Glickman 1980).
- If a child is doing a fine motor activity and has poor skills, provide bodily help as unobtrusively as possible. For example, steady a shape sorter box, so that the child does not get too upset while working and is able to insert the shapes successfully.
- *Refocus* children from inappropriate or aimless activities to appropriate constructive activities. Redirection can include suggestions, bringing a child into a constructive play situation, or arranging materials for a constructive experience.
- Arrange the environment to promote more harmonious interactions. Book shelves near the block corner will result in frustrated readers. Tricycle riding very close to the fine motor table may irritate or frighten a child who is stringing beads.
- Provide sufficient materials so that children do not wander aimlessly. For example, the creative activities area can provide many choices: water play table, markers and paper, easels, clay, finger paint, and a sand box. Variety can help children to choose comfortably and become involved with materials without infringing on the rights of others.

Such an array of discipline techniques, when used judiciously may help prevent hurtful or distressed behaviors and may help children learn more appropriate ways to settle interpersonal conflicts. Do not be afraid to move from more directive to more democratic procedures as young children develop more skills in interpersonal problem solving and more ability to express their discomforts with statements of their needs rather than with actions that hurt or defy others.

> Catch children being good and express pleasure at appropriate behaviors.

References

Ainsworth, M. D. S.; Bell, M. V.; and Stayton, D. J. "Individual Differences in Strange-Situation Behavior of One Year Olds." In *The Origins of Human Social Relations,* ed. H. R. Schaffer. London: Academic Press, 1971.

Beckwith, L. "Relationships Between Infants' Social Behavior and Their Mothers' Behavior." *Child Development* 43, no. 2 (June 1972): 397–411.

Bishop, B. M. "Mother-Child Interaction and the Social Behavior of Children." *Psychological Monographs* 65, no. 11 (1951). No. 328.

Chapman, M. and Zahn-Waxler, C. "Young Children's Compliance and Non-Compliance to Parental Discipline in a Natural Setting." *International Journal of Behavioral Development* 5 (1982): 81–94.

Clarke-Stewart, K. A.; VanderStoep, L. P.; and Killian, G. A. "Analysis and Replication of Mother-Child Relations at Two Years of Age." *Child Development* 50, no. 3 (September 1979): 777–793.

Davis, D. E. *My Friends and Me.* Circle Pines, Minn.: American Guidance, 1977.

Eimer, B. N.; Mancuso, J. C.; and Lehrer, R. "A Constructivist Theory of Reprimand As It Applies to Child Rearing." Paper presented at the Eleventh Annual Interdisciplinary UAP-USC Piagetian Theory and the Helping Professions Conference. Los Angeles, California, February, 1981.

Erickson, M. F. and Crichton, L. "Antecedents of Compliance in Two-Year-Olds from a High-Risk Sample." Paper presented at the Biennial Meeting of the Society for Research in Child Development. Boston, Massachusetts, April 1981.

Finkelstein, N. W. "Aggression: Is It Stimulated by Day Care?" *Young Children* 37, no. 6 (September 1982): 3–9.

Forehand, R. L. and McMahon, R. J. *Helping the Non-Compliant Child: A Clinician's Guide to Parent Training.* New York: Guilford, 1981.

George, G. and Main, M. "Social Interactions of Young Abused Children: Approach, Avoidance and Aggression." *Child Development* 50, no. 2 (June 1979): 306–318.

Ginott, H. G. *Between Parent and Child: New Solutions to Old Problems.* New York: Macmillan, 1965.

Gordon, T. *Parent Effectiveness Training.* New York: Wyden, 1970.

Greenspan, S. "A Unifying Framework for Educating Caregivers About Discipline." *Child Care Quarterly* 12, no. 1 (1983): 5–27.

Hay, D. F. and Rheingold, H. L. "The Early Appearance of Some Valued Social Behaviors." In *The Nature of Prosocial Development: Interdisciplinary Theories and Strategies,* ed. D. L. Bridgeman. New York: Academic Press, 1983.

Hetherington, E. M.; Cox, M.; and Cox, R. "Stress and Coping with Divorce: A Focus on Women." In *Psychology and Transition,* ed. J. Gullahorn. New York: Winston, 1978.

Hoffman, M. L. "Moral Development." In *Carmichael's Manual of Child Psychology,* ed. P. H. Mussen. 3rd ed. New York: Wiley, 1970.

Hoffman, M. L. "Moral Internalization, Parental Power, and the Nature of Parent-Child Interaction." *Developmental Psychology* 11 (1975): 228–239.

Holden, G. W. "Avoiding Conflict: Mothers As Tacticians in the Supermarket." *Child Development* 54, no. 1 (February 1983): 233–240.

Holzman, M. "The Verbal Environment Provided by Mothers for Their Very Young Children." *Merrill-Palmer Quarterly* 20 (1974): 31–42.

Honig, A. S. "Infant-Mother Communication." *Young Children* 37, no. 3 (March 1982a): 52–62.

Honig, A. S. "Prosocial Development in Children." *Young Children* 37, no. 5 (July 1982b): 51–62.

Honig, A. S. "Discipline Tips for Teachers." *New York State Reporter,* in press.

Honig, A. S. and Wittmer, D. S. "Caregiver Interaction and Sex of Toddler." Paper presented at the Biennial Conference of the Society for Research in Child Development. Boston, Massachusetts, April 1981.

Honig, A. S. and Wittmer, D. S. "Teachers and Low-Income Toddlers in Metropolitan Day Care." *Early Child Development and Care* 10, no. 1 (1982): 95–112.

Honig, A. S. and Wittmer, D. S. "Toddler Bids and Teacher Responses." *Child Care Quarterly,* in press.

Howes, C. and Olenick, M. "Family and Child Care Influences on Toddler's Compliance." Paper presented at the Annual Meeting of the American Educational Research Association, New Orleans, Louisiana, April 1984.

Kopp, C. B. "Antecedents of Self-Regulation: A Developmental Perspective." *Developmental Psychology* 18 (1982): 199–214.

Kuczynski, L. "Reasoning, Prohibitions, and Motivations for Compliance." *Developmental Psychology* 19 (1983): 126–134.

Landauer, T. K.; Carlsmith, J. M.; and Lepper, M. "Experimental Analysis of the Factors Determining Obedience of Four-Year-Old Children to Adult Females." *Child Development* 41, no. 3 (September 1970): 601–611.

Londerville, S. and Main, M. "Security Attachment, Compliance, and Maternal Training Methods in the Second Year of Life." *Developmental Psychology* 17 (1981): 289–299.

Lytton, H. "Disciplinary Encounters Between Young Boys and Their Mothers and Fathers: Is There a Contingency System?" *Developmental Psychology* 15 (1979): 256–268.

Lytton, H. *Parent-Child Interaction: The Socialization Process Observed in Twin and Singleton Families.* New York: Plenum Press, 1980.

Lytton, H. and Zwirner, W. "Compliance and Its Controlling Stimuli Observed in a Natural Setting." *Developmental Psychology* 11 (1975): 769–779.

Main, M. and Weston, D. R. "The Quality of the Toddler's Relationship to Mother and Father: Related to Conflict Behavior and the Readiness to Establish New Relationships." *Child Development* 52, no. 3 (September 1981): 932–940.

Martin, J. A. "A Longitudinal Study of the Consequences of Early Mother-Infant Interaction: A Microanalytic Approach." *Monographs of the Society for Research in Child Development* 46, no. 3 (1981). No. 190.

Matas, L.; Arend, R. A.; and Sroufe, A. L. "Continuity of Adaptation in the Second Year: The Relationship Between Quality of Attachment and Later Competence." *Child Development* 49, no. 3 (September 1978): 547–556.

McLaughlin, B. "Child Compliance to Parental Control Techniques." *Developmental Psychology* 19 (1983): 667–673.

Minton, C.; Kagan, J.; and Levine, J. A. "Maternal Control and Obedience in the Two-Year-Old." *Child Development* 42, no. 6 (December 1971): 1873–1894.

Olson, S. L.; Bates, J. E.; and Bayles, K.

"Mother Infant Interaction and the Development of Individual Differences in Children's Cognitive Competence." *Developmental Psychology* 20 (1984): 166–179.

Robertson, A. "Daycare and Children's Responsiveness to Adults." Paper presented at the Biennial Meetings of the Society for Research in Child Development, Boston, Massachusetts, April 1981.

Rubenstein, J. L.; Howes, C.; and Boyle, P. "A Two-Year Follow-Up of Infants in Community-Based Day Care." *Journal of Child Psychology and Psychiatry* 22 (1981): 209–218.

Schaffer, H. R. *The Child's Entry into a Social World.* Orlando, Fla.: Academic Press, 1984.

Schaffer, H. R. and Crook, C. K. "Maternal Control Techniques in a Directed Play Situation." *Child Development* 50, no. 4 (December 1979): 989–996.

Schaffer, H. R. and Crook, C. K. "Child Compliance and Maternal Control Techniques." *Developmental Psychology* 16 (1980): 54–61.

Shure, M. B. and Spivak, G. *Problem Solving Techniques in Child Rearing.* San Francisco: Jossey-Bass, 1979.

Siegal, M. and Rablin, J. "Moral Development As Reflected by Young Children's Evaluation of Maternal Discipline." *Merrill-Palmer Quarterly* 28 (1982): 499–509.

Smith, C. A. *Promoting the Social Development of Young Children: Strategies and Activities.* Palo Alto, Calif.: Mayfield, 1982.

Stayton, D. J.; Hogan, R.; and Ainsworth, M. D. S. "Infant Obedience and Maternal Behavior: The Origins of Socialization Reconsidered." *Child Development* 42, no. 4 (October 1971): 1057–1069.

Sroufe, L. A. "The Coherence of Individual Development." *American Psychologist* 34 (1979): 834–841.

Veenman, S. "Perceived Problems of Beginning Teachers." *Review of Educational Research* 34, no. 2 (1984): 143–178.

Wetstone, H. S. and Friedlander, B. Z. "The Effect of Word Order on Young Children's Responses to Simple Questions and Commands." *Child Development* 44, no. 4 (December 1973): 734–740.

Wolfgang, C. H. *Helping Aggressive and Passive Preschoolers Through Play.* Columbus, Ohio: Merrill, 1977.

Wolfgang, C. H. and Glickman, C. D. *Solving Discipline Problems.* Boston: Allyn & Bacon, 1980.

Ziajka, A. *Prelinguistic Communication in Infancy.* New York: Praeger, 1981.

DOING GOOD

BEYOND SELFISHNESS

We start helping others early in life,
but we're not always consistent. What makes us
helpful sometimes and not others?

Alfie Kohn

Alfie Kohn, a contributing editor, is the author of No Contest: The Case Against Competition *(Houghton Mifflin).*

You realize you left your wallet on the bus and you give up hope of ever seeing it again. But someone calls that evening asking how to return the wallet to you.

Two toddlers are roughhousing when one suddenly begins to cry. The other child rushes to fetch his own security blanket and offers it to his playmate.

Driving on a lonely country road, you see a car stopped on the shoulder, smoke pouring from the hood. The driver waves to you frantically, and instinctively you pull over to help, putting aside thoughts of your appointments.

Despite the fact that "Look out for Number One" is one of our culture's mantras, these examples of "prosocial" behavior are really not so unusual. "Even in our society," says New York University psychologist Martin Hoffman, "the evidence is overwhelming that most people, when confronted with someone in a distress situation, will make a move to help very quickly if circumstances permit."

Helping may be as dramatic as agreeing to donate a kidney or as mundane as letting another shopper ahead of you in line. But most of us do it frequently and started doing it very early in life.

Psychologists have argued for years about whether our behavior owes more to the situations in which we find ourselves or to our individual characteristics. Prosocial behavior seems to be related to both. On the situation side, research shows that regardless of your personality, you'll be more likely to come to someone's aid if that person is already known to you or is seen as similar to you. Likewise, if you live in a small town rather than a city, the chances of your agreeing to help increase dramatically. In one experiment, a child stood on a busy street and said to passersby, "I'm lost. Can you call my house?" Nearly three-quarters of the adults in small towns did so, as compared with fewer than half in big cities. "City people adjust to the constant demands of urban life by reducing their involvement with others," the researcher concluded. You are also more likely to help someone if no one else is around at the time you hear a cry for help. The original research on this question was conducted by psychologists Bibb Latané and John Darley. They offer three reasons to account for the fact that we're less apt to help when more people are in the area: First, we may get a case of stage fright, fearing to appear foolish if it turns out no help was really necessary. Second, we may conclude from the fact that other people aren't helping that there's really no need for us to intervene either. Finally, the responsibility for doing something is shared by everyone present, so we don't feel a personal obligation to get involved.

But some people seem to be more other-oriented than others regardless of the situation. People who feel in control of what happens in their lives and who have little need for approval from others are the most likely to help others. Similarly, people in a good state of mind, even if only temporarily, are especially inclined to help. "Feel good, do good" is the general rule, researchers say, regardless of whether you feel good from having had a productive day at the office or, say, from finding money in the street. In one study, people got a phone call from a woman who said the operator had given her their number by mistake, and she was now out of change at a pay phone. The woman asked if the person who answered would look up a number, call and deliver a message for her. It turned out that people who had unexpectedly received free stationery a few minutes before were more likely to help out the caller.

But some investigators aren't satisfied with knowing just when prosocial acts will take place or by whom. "Why should we help other people? Why not help Number One? That's the rock-bottom question," says University of Massachusetts psychologist Ervin Staub, who's been wrestling with that problem since the mid 1960s.

 From *Psychology Today*, October 1988, pp. 34-38.

Obviously we do help each other. But it's equally obvious that our motives for doing so aren't always unselfish. Prosocial behavior, which means behavior intended to benefit others, isn't necessarily altruistic. The 17th-century political philosopher Thomas Hobbes, who believed that we always act out of self-interest, was once seen giving money to a beggar. When asked why, he explained that he was mostly trying to relieve his own distress at seeing the beggar's distress.

His explanation will ring true for many of us. But is this always what's going on: helping in order to feel good or to benefit ourselves in some way? Is real altruism a Sunday school myth? Many of us automatically assume so—not because there's good evidence for that belief but because of our basic, and unproved, assumptions about human nature.

New research describes how we feel when helping someone, but that doesn't mean we came to that person's aid in order to feel good. We may have acted out of a simple desire to help. In fact, there is good evidence for the existence of genuine altruism. Consider:

■ **Do we help just to impress others?** "If looking good were the motive, you'd be more likely to help with others watching," says Latané. His experiments showed just the opposite. More evidence comes from an experiment Staub did in 1970: Children who voluntarily shared their candy turned out to have a lower need for approval than those who didn't share. "If I'm feeling good about myself, I can respond to the needs of others," Staub explains. So helping needn't be motivated by a desire for approval.

■ **Do we help just to ease our own distress?** Sometimes our motivation is undoubtedly like that of Hobbes. But the easiest way to stop feeling bad about someone else's suffering is "just to ignore it or leave," says Arizona State University psychologist Nancy Eisenberg. Instead we often stay and help, and "there's no reason to believe we do that just to make ourselves feel better."

When people are distressed over another person's pain they may help—for selfish reasons. But if they have the chance simply to turn away from the cause of their distress, they'll gladly do that instead. People who choose to help when they have the opportunity to pass by, like the biblical Good Samaritan, aren't motivated by their own discomfort. And these people, according to C. Daniel Batson, a psychologist at the University of Kansas, describe their feelings as compassionate and sympathetic

Older children can feel for another person's condition, understanding that distress may result from being part of a class of people who are oppressed.

rather than anxious and apprehensive.

Batson explored this behavior by having students listen to a radio news broadcast about a college senior whose parents had just been killed in a car accident. The students who responded most empathically to her problem also offered the most help, even though it would have been easy for them to say no and put the whole thing out of their minds.

■ **Do we help just to feel pleased with ourselves or to avoid guilt?** The obvious way to test this, Batson argues, is to see how we feel after learning that "someone else" has come to a victim's aid. If we really cared only about patting ourselves on the back (or escaping twinges of guilt), we would insist on being the rescuer. But sometimes we are concerned only to make sure that the person who needs help gets it, regardless of who does the helping. That suggests a truly altruistic motivation.

Pretend you are one of the subjects in a brand-new study of Batson's. You are told that by performing well on a game with numbers, you might be able to help someone else (whose voice you've just heard) avoid mild but unpleasant electric shocks. A little later, you're informed that the person won't be receiving shocks after all. How do you feel? Batson found that many subjects were pleased even though they personally didn't get the chance to do the good deed.

Batson, incidentally, used to assume that we help others primarily to benefit ourselves. But after a decade of studying empathic responses to distress, he's changed his mind. "I feel like the bulk of the evidence points in the direction of the existence of altruism," he says.

■ **If we're naturally selfish, why does helping behavior start so early in life?** At the age of 10 to 14 months, a baby will often look upset when someone else falls down or cries. Obviously made unhappy by another person's unhappiness, the child may seek solace in the mother's lap. In the second year, the child will begin comforting in a rudimentary way, such as by patting the head of someone who seems to

be in pain. "The frequency (of this behavior) will vary, but most kids will do it sometimes," says Eisenberg.

By the time children are 3 or 4, prosocial behavior is common. One group of researchers videotaped 26 3-to-5-year-olds during 30 hours of free play and recorded about 1,200 acts of sharing, helping, comforting and cooperating. Children can be selfish and mean, too, of course, but there's no reason to think that these characteristics are more common or "natural" than their prosocial inclinations.

Psychologist Hoffman points to two studies showing that newborns cried much more intensely at the sound of another baby's cry than at other, equally loud noises. "That isn't what I'd call empathy," he concedes, "but it is evidence of a primitive precursor to it. There's a basic human tendency to be responsive to other persons' needs, not just your own."

Hoffman rejects biological theories that claim altruism amounts to nothing more than "selfish genes" trying to preserve themselves by prompting the individual to help relatives who share those genes. But he does believe "there may be a biological basis for a disposition to altruism. Natural selection demanded that humans evolve as creatures disposed toward helping, rescuing, protecting others in danger" as well as toward looking out for their own needs.

According to Hoffman, the inborn mechanism that forms the basis for altruism is empathy, which he defines as feeling something more appropriate to someone else's situation than to your own. The way he sees it, empathy becomes increasingly more sophisticated as we grow. First, infants are unable to draw sharp boundaries between themselves and others and sometimes react to another's distress as if they, themselves, had been hurt.

By about 18 months, children can distinguish between "me" and "not-me" but will still assume that others' feelings will be similar to their own. That's why if Jason sees his mother cry out in pain, he may fetch his bottle to make her feel better. By age 2 or 3, it is possible to understand

It appears that caring about others is as much a part of human nature as caring about ourselves. Which impulse gets emphasized is a matter of training.

Raising a Helping Child

HOW DO YOU BRING up a child who will be more inclined to help than to hurt? For starters, if you believe aggressive behavior is natural or even desirable—or find it amusing even while telling a child to stop—you are unlikely to curb such behavior. But if you communicate a deeply felt disapproval of hurting, you will be far more effective at discouraging children from doing so.

The intensity of the disapproval is not enough, however. Rather than simply restraining children or punishing them or yelling "No!" you should make the consequences of hurting clear. Children who are plainly told about the effect of aggression on the victim are more likely to refrain from such acts. Of course, helping is more than just the absence of hurting. Those who study child development recommend the following for encouraging specifically prosocial behavior:

■ **Focus on the positive.** Help a child to understand how (and why) to help; telling him or her what not to do isn't enough. "A focus on prohibition may also promote self-concern, thereby diminishing attention to others' needs," Ervin Staub says.

■ **Explain the reason.** Just as children ought to be told why aggression is bad, so they should hear why altruism is desirable: "When you share your toys, Diane gets to play, too, and that makes her feel good."

■ **Set an example.** Particularly before the age of 3, children are powerfully influenced by how adults conduct their own lives. A parent who normally responds to another's distress is teaching a child to do likewise. In general, showing seems to be more effective than telling.

■ **Let them help.** Give children the opportunity to try out what they've learned about being sensitive to others. A child who has taken care of a younger sibling or a pet, for example, has experienced first-hand what it means to be prosocial.

■ **Promote a prosocial self-image.** Children should be encouraged to think of themselves as caring people even though their prosocial tendencies were initially shaped by a parent. "You want them to start thinking of themselves as the kind of people who help," says psychologist Nancy Eisenberg.

■ **Be a warm, empathic parent.** Children who can form a secure attachment to Mom or Dad feel that the world is a basically safe place. They also feel good about themselves and well-disposed toward other people. Such children are more likely to respond to the needs of others than are those raised by parents who emphasize power and control. Responsiveness to children's needs — which includes respecting their occasional preference for distance from you—is also recommended.

that others react differently and also to empathize with more complex emotions.

Finally, older children can feel for another person's life condition, understanding that his or her distress may be chronic or recognizing that the distress may result from being part of a class of people who are oppressed.

Other psychologists, meanwhile, believe that you are more likely to help others not only if you feel their pain but also if you understand the way the world looks to them. This is called "role-taking" or "perspective-taking." "When people put themselves in the shoes of others, they may become more inclined to render them aid," according to Canadian researchers Dennis Krebs and Cristine Russell.

When they asked an 8-year-old boy named Adam whether that seemed right to him, he replied as follows: "Oh yes, what you do is, you forget everything else that's in your head, and then you make your mind into their mind. Then you know how they're feeling, so you know how to help them."

Some people seem more inclined than others to take Adam's advice—and, in general, to be prosocially oriented. Staub has found that such people have three defining characteristics: They have a positive view of people in general, they are concerned about others' welfare and they take personal responsibility for how other people are doing.

All these, but particularly the first, are affected by the kind of culture one lives in. "It's difficult to lead a competitive, individualistic life"—as we're raised to do in American society—"without devaluing others to some extent," says Staub. So raising children to triumph over others in school and at play is a good way to snuff out their inclination to help (see "Raising a Helping Child," this article).

It appears, then, that caring about others is as much a part of human nature as caring about ourselves. Which impulse gets emphasized is a matter of training, according to the experts. "We fundamentally have the potential to develop into caring, altruistic people or violent, aggressive people," says Staub. "No one will be altruistic if their experiences teach them to be concerned only about themselves. But human connection is intrinsically satisfying if we allow it to be."

AGGRESSION

THE VIOLENCE WITHIN

MAYA PINES

Maya Pines has been writing on the brain since research in the field really took off in the 1970s. Her book The Brain Changers: Scientists and the New Mind Control, *which described research on the two hemispheres and on mind-altering drugs, won the National Media Award of the American Psychological Association in 1974. She has also written a book on retarded children and one on the enormous potential of the very young child.*

When his supervisor made a sarcastic comment about the latest production delay, the young engineer said nothing. But he could feel his blood boil. On the way home he had a few drinks to calm down.

That didn't help much. He was not bearing the blame well. At home, he took some Valium. Then he got into an argument with his wife. Suddenly he exploded and punched her. He then smashed a chair, stormed out of the house and got into his car, taking off at top speed. Going out of control, he crashed into another car and wrecked it. The collision broke the other driver's neck.

As the incident illustrates, aggression is not confined to the New York City subways (see box, next page). It lurks in suburban streets and sometimes invades our homes. The so-called civilized world is riddled with violence, both sanctioned and unsanctioned. At a time when more and more people have access to atomic weapons, a single person's aggressive impulses—or perhaps a nation's—could be the most dangerous force on Earth.

Today, we are all at risk of becoming the targets or accidental victims of some kind of violence. Cars driven by aggressive people can be the instruments of suicide or murder. Stabbings, shootings and rapes are now so commonplace that most newspapers don't bother to report them. Law-enforcement agencies reported a total of 1.2 million violent crimes in 1983; probably at least as many other episodes went unrecorded. More than 50,000 Americans are murdered or commit suicide annually.

In the hope of finding ways to prevent or reduce aggression, psychiatrists, brain researchers and behavioral scientists are now making a determined effort to understand its causes. They are not concerned with the kind of drive that fuels ambition and makes people stand up for their rights, nor with aggressive thoughts, but with physical aggression—outright attacks that result in injuries or death. Among some recent findings:

■ Harsh punishment produces aggressive behavior in children. So does the example of violence, at home or on television.

■ Extremely impulsive and aggressive people of both sexes have unusually low levels of a brain chemical that inhibits the firing of nerve cells.

■ Men who are highly aggressive have higher levels of the male hormone testosterone.

■ Treatment with lithium reduces aggressive behavior in highly impulsive and violent people.

While much of the research on the biology of aggression has been carried out on animals, it is known that human beings can resist their biological drives more efficiently than other creatures, because their neocortex—the thinking part of the brain—is more highly developed.

"By the time the human brain matures, the normal individual is controlled very largely by social norms," says Estelle Ramey, a professor of physiology and biophysics at the Georgetown University School of Medicine. "Men don't urinate on the living-room floor, even when they're in agony; they wait until they get to the bathroom. Although urination is a normal, instinctive biological drive, it's very quickly brought under control. Similarly, young men learn quickly who it's safe to be aggressive toward. A man may be a meek and mild Caspar Milquetoast in the office, and kowtow to his boss, and then go home and beat up his wife. Is he an aggressive male?" She answers her own question: "He's aggressive when it's safe for him to be aggressive."

Much depends, therefore, on the level of aggression that is acceptable in a particular society at a particular time. Cultural change is possible, argued John Paul Scott of

EXTREME AGGRESSION TENDS TO RUN IN FAMILIES—A FINDING THAT HAS SET OFF A DEBATE OVER THE CAUSE.

Bowling Green State University at a recent session on evolutionary theory and warfare. Warlike habits can give way to very peace-loving ones. Although Scandinavian culture spawned the Vikings, for instance, it is now represented by people who are "among the most pacific in the world."

Much depends on one's early training. Extreme aggression tends to run in families—a finding that has set off a furious debate over the cause. In the nature-versus-nurture controversy, strong proponents cite genetic causes while others claim children learn from their parents' example.

Last winter, a team of researchers reported on an extraordinary study of aggressive behavior that was made by tracking three generations. It revealed how much children are influenced by their parents' aggression. Led by L. Rowell Huesmann and Leonard Eron of the University of Illinois at Chicago, the team began in 1960 by testing 870

third-graders, who were asked to rate one another on such questions as "who pushes and shoves children?" Their parents were interviewed as well. The researchers then followed up more than 600 of these children and their parents for 22 years—by which time many of the original children had children of their own, who then also became subjects of the study.

The researchers found that the children who most frequently pushed, shoved, started fights and were considered more aggressive by their classmates at age 8 turned into the more aggressive adults. These men were very likely to have criminal records by the age of 30. If their behavior did not land them in jail, they were apt to get into fights, smash things when angry, drive while drunk and abuse their wives. Many of their own children already showed signs of the same type of aggressive behavior.

At the beginning of their study, the team had learned that the more aggressive children had parents who pun-

VIGILANTISM

When Bernhard Goetz drew his gun and shot four youths who, he claims, were harassing him on a New York City subway, he captured the nation's imagination. Not since the days of the Wild West and lynch mobs in the South had people been so preoccupied with a so-called vigilante: A *Washington Post*-ABC News poll found that nearly half of those interviewed supported Goetz. Was his personal crime-fighting crusade an isolated incident, or did it presage a national trend?

"There's a lot of rage and frustration in our society about the ineffectiveness of law enforcement," says Ralph Slovenko, a professor of law and psychiatry at Wayne State University, in Detroit. "Crime is a very safe profession. Only two percent of the people who commit serious crimes are actually sentenced."

A prior mugging victim who had seen his assailants get off easy, Goetz was hardly a vigilante in the traditional sense: Rather than pursuing a specific target in order to right perceived wrongs, he was ready to lash out at the next person who went after him.

"I believe strongly that Mr. Goetz was a man whose emotional state deteriorated in the time after his previous victimization," says psychologist Morton Bard, of the Center for Social Research at the City University of New York. "He was acting out a revenge fantasy that virtually all crime victims have."

What is particularly disturbing is that Bernhard Goetz may be a harbinger of things to come. Everyone agrees that conventional law enforcement is becoming less and less effective; the court system is slow, trial costs are high, and overcrowded prisons mean an increasing reliance on what's known as turnstile justice.

As yet, vigilantism is a predominantly urban phenomenon, fueled by the crowded, alienating environment. The self-styled Army of God is systematically bombing abortion clin-

ics nationwide. A Massachusetts man wounded a teenager as revenge for hitting his car. Abroad, crime-fighting citizens patrol Amsterdam's streets at night, and squads in Dublin combat a growing drug problem. But rural areas can be affected, too: Two men in backwoods Arkansas recently castrated a man charged with rape.

"The American public is fed up," notes James Turner, a clinical psychologist at the University of Tennessee Center for the Health Sciences, in Memphis. "The system no longer protects them. It protects the guilty and punishes the innocent—the victims." He is now writing a book, called *Victims to Vigilantes*, that will document this progression in public opinion.

Researchers have only just begun to consider the effects of crime on the victim. Bard, who recently chaired an American Psychological Association task force on the victims of crime and violence, believes these studies are long overdue. "Victims should have services available to help them deal with the emotional and physical trauma of crime," he says. "Until now, there has been little sensitivity to victims and their 'invisible wounds.' "

Sociologist Emilio Viano, at American University's School of Justice, thinks better treatment of victims will forestall future Bernhard Goetzes. "Victims have been ignored and abused, used only as sources of information. The failure of the court system has weakened the fabric of society. People have nowhere to turn for help. If they want anything done, they feel they must do it themselves."

James Turner is investigating various nonlethal means of self-defense. "We must teach people how to deal with violence, or they will become victims," he says. "Basic police classes are moderately effective; karate needs to be practiced in order to be useful. Mace can make an attacker even angrier, and it's illegal in many major cities." Turner is now reviewing a "stun gun" that uses electronic pulses to temporarily override an attacker's neuromuscular system, causing him to collapse in a daze.

–Andrea Dorfman

CHILDREN ARE COPYCATS, AND WHEN THEY SEE REPEATED VIOLENCE ON TELEVISION, THEY TEND TO IMITATE IT.

ished them far more severely than the less aggressive children. Now the pattern was repeating itself as these aggressive adults severely punished their own children. The most likely explanation for this repetition, the researchers concluded, was that children learn to be aggressive by copying what their parents do to them and to others.

In their report, they said it was most impressive that "the children who are nominated as more aggressive by their third-grade classmates on the average commit more serious crimes as adults." They also found that the degree of aggressiveness in the third-graders was, 22 years later, even more strongly related to the aggressiveness of their children than it was to their own aggressiveness as adults. In short, the study concluded that the aggressive child is father to the aggressive child.

Once aggression is established as a child's "characteristic way of solving social problems," it becomes a relatively stable and self-perpetuating behavior, the researchers emphasized. And by the time this behavior comes to the attention of society, "it is not readily amenable to change."

The idea, then, is to prevent such behavior from becoming fixed before adolescence. One approach is to limit the amount of violence to which children are exposed. Children are copycats, and when they see repeated violence on television, and tend to imitate it. The evidence for this is so strong that the American Academy of Pediatrics is now warning parents about the effects of TV violence and urging them to limit—as well as monitor—what their children watch.

Another approach is to put clear limits on children's aggressive behavior at an early age. Psychologist Gerald Patterson, of the Oregon Social Learning Center in Eugene, and others have found that one of the most effective means of stopping excessive aggression is to isolate children in a room for about five minutes before the child's behavior becomes extreme. Patterson calls this a time-out procedure—a nonpunitive way of saying "that's not acceptable behavior."

If the pattern of violence is not broken before adulthood, little can be done by psychological means. Highly aggressive adults generally resist psychotherapy, according to Gerald Brown, a psychiatrist at the National Institute of Mental Health (NIMH), although behavior therapy is sometimes helpful. Anybody who thinks he or she can reform one of these volatile and violent men through love "is in for a lot of trouble," Brown says. "That idea has brought a lot of grief into marriages. These people can idealize you one minute and attack you the next."

Generally these highly aggressive people are men. Men, in fact, commit about 90 percent of all violent crimes in the United States. This is in part the result of social conditioning. As children grow up, they learn that "it's socially unacceptable for women to become aggressive, but if a man throws his weight around, that's *manly!*" points out Estelle Ramey. But there are also biological reasons for the higher level of aggression in males, as indicated by studies involving certain brain chemicals and male sex hormones.

One chemical that seems to play a key role in preventing or releasing aggression is serotonin, which carries inhibitory messages from cell to cell in the brain. Serotonin is difficult to measure directly, but a substance called 5-HIAA, which is a breakdown product of serotonin, can now be measured in spinal fluid. Several experiments have shown that highly aggressive people have lower levels. This holds true for both men and women, but on the average, men have less 5-HIAA.

Because research involving 5-HIAA requires sometimes painful spinal taps, few studies have been conducted. At the National Naval Medical Center in Bethesda, Maryland, 26 marines and sailors, aged 17 to 32, who had come to the attention of psychiatrists because of their histories of repeated assault, agreed to undergo taps.

"They were the kind of people who'd had temper tantrums as small children and lots of fights in grade school and who would go into bars and tear up the place," says Brown, the NIMH psychiatrist who directed the spinal-fluid study. "They had a very short fuse—they'd be provoked by things others would not find provoking."

Predictable Aggression

Although aggression is highly valued in the military, Brown points out that "it must be controlled and predictable aggression. If people are too unpredictable and keep getting into trouble, they're not suitable for the service." In fact, the young men under study were being examined by a board of officers who were to decide whether they should be discharged from the Navy.

The laboratory that analyzed their spinal fluid found that 14 of the young men had low levels of 5-HIAA, while the rest had nearly double the amount. The first group included the men who had the worst records for impulsive acts of aggression, and 12 were discharged.

"We could have looked at their spinal fluid and predicted with eighty-five percent accuracy which people would be removed," declares Brown.

The results of this study complement earlier research in which scientists lowered the level of serotonin in animals' brains through chemicals or brain surgery and saw a dramatic increase in the animals' aggressive behavior—at least in the kind of aggression characterized by explosive attacks.

Brain researchers have known for decades that there are at least two unrelated kinds of aggression. In the 1920s, the Swiss physiologist Walter Hess, who later won a Nobel prize, described the cat's characteristic "bad-tempered aggression": With dilated pupils and bristling hair, the cat hisses, spits and growls. By contrast, "predatory aggression" is more cold-blooded and deliberate, as when the cat stalks a mouse, kills it quietly and eats it. Only the first variety seems to involve low levels of serotonin.

MOST OF THE OUTRIGHT VIOLENCE IN THE UNITED STATES COMES FROM PEOPLE WHO ARE CHRONICALLY AGGRESSIVE.

Testosterone has long been associated with aggression. A time-honored method of making male animals less aggressive is to castrate them—a procedure that eliminates the source of testosterone. When male mice are castrated at birth, for instance, they do not begin to fight each other at one or two months of age, as normal mice do upon reaching sexual maturity, but giving these castrated mice injections of testosterone makes them fight as if they had not been altered.

"Testosterone increases the biological intensity of stress, so that more adrenaline is released," explains Ramey. This produces a state of anxiety and irritability but also damages the lining of blood vessels and may lead to heart disease. That's why male animals generally die earlier than females, she says—unless they have been castrated, in which case they tend to live as long as females.

In one study, blood from hockey players rated by coaches and teammates as particularly aggressive was found to contain relatively high levels of testosterone. In another study, involving prisoners jailed for violent crimes, the men who were most aggressive toward other prisoners had twice as much testosterone in their blood as those who were not.

While the effects of such biological differences should not be underestimated, researchers emphasize that biology can only set the stage for aggressive acts or make such acts more likely. Biological forces can produce rage or the urge to attack, but they cannot dictate whether an attack will actually take place.

Even dogs react differently when annoyed by their masters, whom they seldom bite, and when annoyed by strangers, whom they will attack with much less provocation. They are particularly lenient toward young children.

In certain circumstances, such as self-defense, any of us may become violent. But when previously unaggressive people suddenly become violent for no apparent reason, they may have had too much alcohol or taken drugs such as PCP (angel dust), both of which often trigger aggression. Psychologist Claude Steele, of the University of Washington, recently found that alcohol accelerates aggressive behavior to an extreme level because, when in a state of conflict, a person using alcohol tends to lessen inhibitions by blocking thoughts of negative consequences.

Most of the violence in the United States comes from people who are chronically aggressive. Some of them simply earn their living from crime. Others have such a short fuse that aggression has become a way of life. For these impulsive people, a new kind of treatment now appears possible in some cases: the use of lithium, a drug that is generally prescribed to treat manic-depressives.

Lithium seems to affect the activity of several brain chemicals, including serotonin, preventing both highs and lows. Scientists don't yet understand exactly how it works, but it has been shown to reduce aggression in rats, mice and fish. On this basis, several psychiatrists have tried it on prison inmates with histories of repeated impulsive assaults. Joe Tupin, a professor of psychiatry at the University of California, Davis, gave lithium to 27 particularly aggressive male prisoners in a maximum-security prison in California. "More than two-thirds of them responded to the lithium," he reports. "It removed the explosive quality of their violence."

The men who responded best were those who would become extremely angry after trivial provocations. Their violence was not only inappropriate but very rapid, Tupin explains. "Between the provocation and the violence, these people didn't think—as if they didn't have the capacity to," he says. "They didn't stop for half a second of internal review during which they could think 'Gee, it was just an accident'; they didn't look at the possible consequences of their acts; they went totally out of control, with no in-between stages."

During the nine months that they were treated with lithium, these men became more reflective and had fewer violent episodes, Tupin says. The drug did not stop them from attacking their fellow prisoners deliberately from time to time, but it did prevent many hair-trigger explosions.

A number of psychiatrists are now giving lithium to patients of this impulsive type. According to Tupin, lithium is "a good choice" for this kind of patient; however, it would be ineffective with people who commit calculated, predatory violence or with psychotics, who are aggressive because of their delusions.

Controlling Behavior

Ideally, psychiatrists agree, highly aggressive behavior should be dealt with in childhood, by psychological means and by improving the economic and social conditions in which children grow up. But inevitably—either because of their brain chemistry or the way they were raised—a small percentage of people will go on being extremely aggressive and dangerous throughout their adult lives. If further research shows that lithium or other drugs are truly effective in such cases, does society have the right to prescribe them to control these people's behavior?

"I certainly wouldn't try to answer that question," says Gerald Brown. "But it is something the public will have to start thinking about, just as it is now thinking about the wisdom of putting mechanical hearts into people." Sooner or later, he says, society will have to decide what is the most humane and rational way to deal with people who keep on hurting others through uncontrolled aggression.

DEPRESSION AT AN EARLY AGE

It strikes in childhood,
and it's on the rise.

JOSEPH ALPER

Joseph Alper is a contributing editor to Science 86.

JANET WAS A BRIGHT GIRL, well liked by her teachers and friends. When she entered junior high and started having trouble paying attention in class, no one made much of it. After all, kids can get pretty restless at that age. In high school she did rather poorly, but since she was quiet and didn't make trouble, her teachers left her alone and assumed she just wasn't much interested in school. Her parents were a bit worried, but there didn't seem to be anything they could do. And the school counselor reassured them that Janet's attitude was just a normal part of adolescence, that she would eventually outgrow her problems.

Instead they got worse. The summer before her senior year, Janet started having trouble getting up in the morning, and when she did make it out of bed, she couldn't get motivated to do anything; it was as if she were stuck in low gear. She lost her appetite, slept poorly, and couldn't stop crying. Her parents decided this was no longer just part of growing up and took her to their family physician. He referred them to the affective disor-

ders clinic of their local hospital, where a psychiatrist who had treated dozens of kids like Janet talked to her and her parents. He told them that Janet's lack of energy, motivation, and appetite were signs of depression and prescribed amitriptyline, an antidepressant.

Within a month, Janet was feeling better than she had for as long as she could remember. She started making new friends, and her schoolwork improved dramatically. The following year she made the dean's list at college. And today, several years later, she continues taking her medication, does well in school, and has a good outlook on life.

"Until maybe 10 years ago, we believed that severe depression was solely an illness of adults," says psychiatrist Frederick K. Goodwin, scientific director of the National Institute of Mental Health in Bethesda, Maryland. "Adolescents didn't develop 'real' depression—they just had 'adolescent adjustment problems,' so most psychiatrists didn't and still don't think to look for it in kids. Now, however, we know that idea is dead wrong. Adolescents, even children, suffer from major depression as much as adults do."

Research is only just beginning, so estimates vary of how many young people suffer from major affective disorders, which include depression and manic-depression, an illness characterized by elation, hyperactivity, or irritability, alternating with depression. Depending on the age of the youngster and how his illness is defined, the estimates can range from one to six percent. There is no question, however, that the problem is serious and that it is growing. One study indicates that the percentage of older teenagers with major affective disorders has increased more than fivefold over the past 40 years.

"The chilling fact is that we may be on the verge of an epidemiclike increase of mania, depression, and suicide," says Elliot S. Gershon, chief of the clinical psychogenetics branch at NIMH. "The trend is rising almost exponentially and shows no signs of letting up. I would go so far as to say this is going to be *the* public health problem of the 1990s and beyond if the trend continues."

Buttressing Gershon's concern are studies revealing that many adults with affective disorders showed the first signs of their illness when they were teenagers or children and that the earlier the onset, the more severe the disease. "So if we are seeing more depression in kids today," he says, "we could be in for real trouble when these kids hit their 30s," the prime time for showing the classic swings of depression.

Not everyone agrees with Gershon's gloomy outlook. Some claim the dramatic rise is due to better reporting and diagnosis, not to rising incidence of disease. But they don't disagree with the trend.

"Depression is a crucial problem," says G. Robert DeLong, a pediatric neurologist at Massachusetts General Hospital, "because this illness clouds a child's or adolescent's perceptions at such a critical time in his social and psychological development. As a result, young people often develop long-lasting problems aside from the original depression or mania. So the earlier we diagnose these kids and treat them, the better chance they have of developing into adults who can enjoy a more normal life."

These findings have important clinical implications. Some psychiatrists believe that if young people can control their affective disorders with drugs during the particularly stressful adolescent years, their biochemistry may stabilize so they need not continue taking drugs for a lifetime as many adults now must.

Furthermore, there is mounting evidence that a constellation of harmful behaviors that accompany depression—suicide attempts, drug abuse, anorexia, bulimia, and juvenile delinquency—may be methods that young people use to try to cope with the anguish they feel. So if depression can be curbed, many of these disorders might disappear as well.

"We're talking about a whole spectrum of problems to which affective disorders seem to be linked," says Joseph T. Coyle, head of child psychiatry at Johns Hopkins. "This is not to say that every kid with these problems has a major affective disorder. But the odds are good he has."

Ben, for example, was eight when his parents brought him to Robert DeLong in 1974. Ben was smart, with an IQ of 128, a nice kid who had become aggressive and nasty. His teachers complained that the boy was extremely disruptive, and neighbors kept calling his folks to report that he had broken a window or beaten up their kid.

DeLong, one of the few nonpsychiatrist physicians to take an interest in childhood mental illness, talked extensively with Ben, who reported feeling

When depression starts in childhood or adolescence, it is likely to have a genetic link.

very sad and confused at times. "He was a classic manic-depressive—wild, aggressive, manic behavior followed by profound sadness," DeLong says. He put the boy on lithium, which is often used to treat manic-depression in adults but rarely in children. "Everyone who knew the child was stunned by his change for the better." Today, thanks to early diagnosis, lithium, and supportive psychotherapy, Ben is a well-adjusted college student who no longer needs lithium.

Ben was fortunate that he was sent to a physician who hadn't swallowed the orthodox view that children can't be manic-depressive. Grace, who is now in her mid-20s, was not as lucky. When she was about 13, she began feeling a bit down in the dumps. Then, at 15, though a promising musician, she started doing poorly in her musical studies, and she began vomiting after meals. It improved her mood, she said.

When her parents started finding jars of vomit hidden around the house, they took her to a psychiatrist, who made a diagnosis of borderline personality disorder, a serious illness involving self-destructive, impulsive behavior that usually does not respond to medication and has a poor prognosis. Grace had a lot of trouble over the next nine months, but then she spontaneously got better. That lasted until she was 18, when she again became depressed. This time she started cutting her wrists, explaining that it made her feel better. She was hospitalized. Again she recovered nine months later. She went to a prestigious college on a music scholarship and did well for another two years but once again became depressed and wound up at the Johns Hopkins Hospital. This time, Raymond DePaulo Jr., director of the affective disorders clinic, diagnosed Grace as manic-depressive and treated her with an antidepressant and lithium. Within a month she was feeling good.

But Grace can't cope with the idea of being seriously ill, so she has stopped taking her medicine, believing that she can control her moods with vomiting and wrist slitting. The years of turmoil have torn her family apart and seriously damaged her musical career. "This is a good example of how early diagnosis and

treatment of an affective disorder would have probably alleviated many of the psychosocial problems this girl and her family now have," DePaulo says. "The odds are she'd be in much better shape today if she'd been diagnosed correctly at 15."

But diagnosing affective disorders is not easy, especially since young people show a more diverse set of symptoms than do adults. And it can be hard to tell which kids are showing typical signs of rebellion and which are depressed. Before diagnosis can become routine, psychiatrists may need a fundamental change in their understanding of mental illness.

Such a change took place in the 1950s, when drugs were discovered that alleviated many of the symptoms of severe psychiatric illnesses, raising the possibility that affective disorders and other mental diseases could be understood in terms of biochemistry and genetics. As researchers learned more about biochemical abnormalities, they hoped to develop better tests to detect psychiatric problems and differentiate among them.

But there still is no perfectly reliable diagnostic test for any psychiatric illness, and just as they've always done, doctors must rely heavily on subjective criteria. "It's a probability game," says Ray DePaulo. "We talk to the patient and to their relatives."

"To gauge depression," says Coyle, longtime friend and collaborator of DePaulo, "we determine if people are sleeping poorly or if they've had a big weight change recently. We try to find out if they have a poor self-image, if they believe they are deficient and feel guilty about their shortcomings, if they feel empty and sad."

To judge manic-depression, DePaulo says, "we look to see if there have been any sudden mood changes. If so, do they take place for a while and then disappear, and then reappear later?"

"A big problem," Coyle says, "is getting this information from kids, because, unlike adults, they have a hard time connecting their unusual behavior to the way they feel. And even when we do get the kids to talk about their feelings, it's rare to find one who satisfies all these criteria."

One important factor in making a diagnosis is a family history of affective disorders. The evidence is compelling that when depression starts in childhood or adolescence, there is likely to be a genetic link.

Mary, for example, was diagnosed last

Between 1960 and 1980,
the suicide rate among 15- to 19-year-olds
increased by 136 percent.

year as a manic-depressive. She was 26 years old. As early as age eight, she had showed episodic symptoms—times when she just wasn't interested in her classes alternating with periods when she wanted to answer every question the teacher asked. Alarmed, her parents took her to a psychiatrist who blamed Mary's troubles on a too-close relationship with her mother.

In a way, he was right. Last year, Mary's mother, who was being treated for manic-depression, realized that Mary's problems were probably related to her own, and she and her husband brought Mary to Johns Hopkins. Looking into the family history, DePaulo learned that Mary's sister had a mood disorder and that Mary's father had been suffering from serious depression for years. The father's brother and an uncle had symptoms that sounded manic-depressive, and two of his relatives had committed suicide. "Mary's symptoms were very mild, and they could be overlooked," DePaulo says. "But that family history was too strong to overlook." Mary is now taking medication, and her prognosis is good.

Family histories are playing an increasingly important role in diagnosing affective disorders in young people. Studies have shown that a youngster with one parent who suffers from manic-depression stands at least a 25 percent chance of developing an affective disorder as an adult. If both parents are ill, the odds rise to between 50 and 75 percent. But genes merely contribute to the vulnerability to depression and manic-depression, and some psychological, social, or biological factor is needed to bring the illness to the fore.

"I think the rise in depression, suicide, and eating disorders is mostly environmental in nature," says Fred Goodwin. "Sure, genetics provides the vulnerability, but over the past 20 years there has been an erosion of sources of an external sense of esteem—religious identity, family cohesiveness, patriotism, etc. The many traumas that have rocked our society since the 1960s have forced people to turn inward, analogous to the grief process, so if there is a genetic vulnerability, it could touch off the problems of affec-

tive disorder. That is why it is so important to find some predictive genetic marker."

At Yale University School of Medicine, epidemiologist Myrna Weissman and geneticist Ken Kidd are looking for such markers by rounding up large families with many members who have affective disorders. Their goal is to use DNA samples from family members to identify genetic markers for affective disorders, in much the same way that researchers two years ago isolated a genetic marker for Huntington's disease. "It will be more difficult than the Huntington's marker because the affective disorders probably involve more than one gene," says Weissman. "We're a long way from the success of the Huntington's marker, but at least it's a beginning."

Such a diagnostic test would, among other things, save lives. "Suicide is the third leading killer in adolescents, and one of the most important risk factors for suicide in this age group is untreated affective disorder," says psychiatrist Susan Blumenthal, chief of behavioral medicine at NIMH. She estimates that at least a third of the adolescents who commit suicide have untreated or undiagnosed affective disorder.

Like depression, suicide seems to be on the rise among adolescents. Between 1960 and 1980, the suicide rate among 15- to 19-year-olds increased by 136 percent. Among boys, who commit three-fourths of the suicides, the suicide rate soared 154 percent. Suicide attempts, three-fourths of which are made by girls, are estimated to be 100 to 150 times more common than actual suicides. And although most psychiatrists now believe that the majority of people who attempt suicide do not mean to succeed, as many as 20 percent of those who make an attempt later complete the act.

The best lead so far in predicting which young people are at high risk for ending their lives is a low level of the neurotransmitter serotonin. Yet many girls with low serotonin levels don't try to commit suicide. They tend to develop bulimia instead.

Of all the behavioral problems that strike during adolescence, bulimia is perhaps the most bizarre. The disease is

characterized by an uncontrollable urge to binge—as many as 100 calories a minute or 5,000 calories per 30- to 60-minute binge up to five times a day—and then purge by vomiting or taking laxatives. Like suicide, bulimia is on the rise, and recent studies suggest that as many as 15 percent of adolescent girls may have the disorder. While between 70 and 80 percent of these young women maintain their normal weight and thus are difficult to spot, the remainder lose so much weight they become dangerously malnourished. Between 50 and 70 percent of bulimics have a major affective disorder, and there is a strong link with depression that runs in families.

There is good reason to believe that bulimia is associated with reduced serotonin in the hypothalamus, a small structure in the brain that serves as the master control center, regulating such functions as appetite, weight, body temperature, the endocrine system, and response to stress. The hypothalamus also serves as the major link between the limbic system, the part of the brain that controls mood, and the rest of the body. "We think the key may be found by studying the hypothalamus, the focal point for these illnesses, that something messes up serotonin function there, and that may be related to the symptoms of bulimia," says Harry E. Gwirtsman, assistant professor of psychiatry at University of California, Los Angeles, school of medicine. "We don't know for sure that bulimia is a brain disease, but certain hypothalamic tumors produce a disease indistinguishable from it." Furthermore, some antidepressants that raise brain serotonin levels reduce or eliminate bulimic binges.

This does not explain, though, why girls with low serotonin levels develop bulimia while boys take their own lives. "We find that as kids develop, boys tend to act out their depression as aggressive behavior, while girls seem to direct their feelings into themselves," says Elizabeth Susman, a research psychologist at the NIMH Laboratory of Developmental Psychology. "We believe this has something to do with the different cues that are important in forming a boy's self-image versus a girl's self-image during puberty."

Susman is examining how various hor-

"We're now taking monkeys with a high risk of depression and placing them with nurturing mothers."

monal and physical changes of adolescence combine with social events to affect a child's psychological development. Her work may offer new insight into the factors that set off the profound mood changes that accompany depression, eating disorders, and suicide.

Some researchers believe they have already identified one system in the body that ties all these seemingly disparate behaviors together. "The evidence strongly suggests that an abnormal biochemical response to stress, mediated through the hypothalamus, eventually produces the variety of biochemical and behavioral changes we find in people with these disorders," says Fred Goodwin.

It is known that hormones produced by the hypothalamus in response to stress cause mood changes in animals. Recently receptors for some of these chemicals have been found in the mood-controlling limbic system of humans, suggesting that stress hormones can affect our moods as well. In addition, nerves have been found leading from the limbic system into those parts of the hypothalamus that control hormone release. "What we could be seeing is a reduced ability of the body's stress-handling system to adapt, so that the levels of neurotransmitters, such as serotonin, do not change the way they should," says NIMH psychopharmacologist William Potter. He theorizes that genetic defects somehow diminish the system's resilience, so that repeated stresses wear it

down further, eventually provoking depression or manic-depression.

An intriguing study of primates tends to support Potter's theory. For more than 15 years now, psychologist Steven J. Suomi, who recently moved from the University of Wisconsin, Madison, to a joint position at the National Institute of Child Health and Human Development and NIMH, has been looking at the behavior of infant rhesus monkeys. While at Wisconsin, he noticed that certain offspring, when separated from their mothers for brief periods, showed symptoms that if seen in humans would be diagnosed as depressive: agitation, withdrawal, and certain biochemical abnormalities. They also responded well to antidepressants.

But what makes this a good model for human depression, Suomi says, "is that most of the monkeys in our colony eventually start coping with the separation, just as most humans adjust to stressful situations." Monkeys that become despondent when separated from their mothers also seem to get depressed if they are faced with stressful situations as they get older. When there is no stress, however, these monkeys behave in a perfectly normal way.

Suomi points out other parallels with human depression. "In infancy and childhood, the depressed animals are withdrawn and shy. During adolescence they are more agitated and delinquent, and in fact, we think we may be seeing some suicidal behavior. Then in adulthood they become withdrawn again. This

really mirrors the different appearance of human depression throughout development."

From the start Suomi noticed that depressed infants behaved differently from normal babies. They showed greater muscle tension and were more responsive to stimuli. By taking all these factors into consideration, "we were able to develop a profile that was very good at predicting which infants would be susceptible to later behavioral problems."

Suomi also found a hereditary risk factor. "I was very skeptical at first, but one look at the pedigree of the monkey colony, and you have to be convinced there is a genetic component to this disorder." But these unlucky monkeys are not necessarily destined for a sad life. "We're now taking monkeys with a high risk of depression and placing them with extremely nurturing mothers in the colony," says Suomi. "While these experiments have only started recently, the preliminary results seem to indicate that the nurturing environment can ameliorate the genetic susceptibility. The data are very exciting."

Elizabeth Susman thinks so too. "We are finding many of the same signs in children that Steve has seen in his monkeys," she says. "If we can learn enough about the genetics, the biochemistry, and the social factors that combine to produce behavioral problems in kids, we can help reduce the incidence of depression and make life better for them now and in the future."

Resilient Children

Emmy E. Werner

Emmy E. Werner, Ph.D., is Professor of Human Development and Research Child Psychologist, University of California at Davis, Davis, California.

Research has identified numerous risk factors that increase the probability of developmental problems in infants and young children. Among them are biological risks, such as pre- and perinatal complications, congenital defects, and low birth weight; as well as intense stress in the caregiving environment, such as chronic poverty, family discord, or parental mental illness (Honig 1984).

In a 1979 review of the literature of children's responses to such stress and risks, British child psychiatrist Michael Rutter wrote:

> There is a regrettable tendency to focus gloomily on the ills of mankind and on all that can and does go wrong.... The potential for prevention surely lies in increasing our knowledge and understanding of the reasons why some children are *not* damaged by deprivation.... (p. 49)

For even in the most terrible homes, and beset with physical handicaps, some children appear to develop stable, healthy personalities and to display a remarkable degree of resilience, i.e., the ability to recover from or adjust easily to misfortune or sustained life stress. Such children have recently become the focus of attention of a few researchers who have asked *What is right with these children?* and, by implication, *How can we help others to become less vulnerable in the face of life's adversities?*

The search for protective factors

As in any detective story, a number of overlapping sets of observations have begun to yield clues to the roots of resiliency in children. Significant findings have come from the few longitudinal studies which have followed the same groups of children from infancy or the preschool years through adolescence (Block and Block 1980; Block 1981; Murphy and Moriarty 1976; Werner and Smith 1982). Some researchers have studied the lives of minority children who did well in school in spite of chronic poverty and discrimination (Clark 1983; Gandara 1982; Garmezy 1981; 1983; Kellam et al. 1975; Shipman 1976). A few psychiatrists and psychologists have focused their attention on the resilient offspring of psychotic patients (Anthony 1974; Bleuler 1978; Garmezy 1974; Kauffman et al. 1979; Watt et al. 1984; Werner and Smith 1982) and on the coping patterns of children of divorce (Wallerstein and Kelly 1980). Others have uncovered hidden sources of strength and gentleness among the uprooted children of contemporary wars in El Salvador, Ireland, Israel, Lebanon, and Southeast Asia (Ayala-Canales 1984; Fraser 1974; Heskin 1980; Rosenblatt 1983). Perhaps some of the most moving testimonials to the resiliency of children are the life stories of the child survivors of the Holocaust (Moskovitz 1983).

All of these children have demonstrated unusual psychological strengths despite a history of severe and/or prolonged psychological stress. Their personal competencies and some unexpected sources of support in their caregiving environment either compensated for, challenged, or protected them against the adverse effects of stressful life events (Garmezy, Masten, and Tellegren 1984). Some researchers have called these children *invulnerable* (Anthony 1974); others consider them to be *stress resistant* (Garmezy and Tellegren 1984); still others refer to them as *superkids* (Kauffman et al. 1979). In our own longitudinal study on the Hawaiian island of Kauai, we have found them to be *vulnerable, but invincible* (Werner and Smith 1982).

These were children like Michael for whom the odds, on paper, did not seem very promising. The son of teen-age parents, Michael was born prematurely and spent his first three weeks of life in the hospital, separated from his mother. Immediately after his birth, his father was sent with the Army to Southeast Asia for almost two years. By the time Michael was eight, he had three younger siblings and his parents were divorced. His mother left the area and had no further contact with the children.

And there was Mary, born to an overweight, nervous, and erratic mother who had experienced several miscarriages, and a father who was an unskilled farm laborer with only four years of education. Between Mary's fifth and tenth birthdays, her mother had several hospitalizations for repeated bouts with mental illness, after having inflicted both physical and emotional abuse on her daughter.

Yet both Michael and Mary, by age 18, were individuals with high self-esteem and sound values, caring for others and liked by their peers, successful in school, and looking forward to their adult futures.

We have learned that such resilient children have four central characteristics in common:

● an active, evocative approach toward solving life's problems, enabling them to negotiate successfully an abundance of emotionally hazardous experiences;

● a tendency to perceive their experiences constructively, even if they caused pain or suffering;

● the ability, from infancy on, to gain other people's positive attention;

● a strong ability to use faith in order to maintain a positive vision of a meaningful life (O'Connell-Higgins 1983).

Protective factors within the child

Resilient children like Mary and Michael tend to have temperamental characteristics that elicit positive responses from family members as well as strangers (Garmezy 1983; Rutter 1978). They both suffered from birth complications and grew up in homes marred by poverty, family discord, or parental mental illness, but even as babies they were described as active, affectionate, cuddly, good natured, and easy to deal with. These same children already met the world on their own terms by the time they were toddlers (Werner and Smith 1982).

Resilient children tend to have temperamental characteristics that elicit positive responses from family members as well as strangers.

Several investigators have noted *both* a pronounced autonomy and a strong social orientation in resilient preschool children (Block 1981; Murphy and Moriarty 1976). They tend to play vigorously, seek out novel experiences, lack fear, and are quite self-reliant. But they are able to ask for help from adults or peers when they need it.

Sociability coupled with a remarkable sense of independence are characteristics also found among the resilient school-age children of psychotic parents. Anthony (1974) describes his meeting with a nine-year-old girl, whose father was an alcoholic and abused her and whose mother was chronically depressed. The girl suffered from a congenital dislocation of the hip which had produced a permanent limp, yet he was struck by her friendliness and the way she approached him in a comfortable, trustful way.

The same researcher tells of another nine-year-old, the son of a schizophrenic father and an emotionally disturbed mother, who found a refuge from his parents' outbursts in a basement room he had stocked with books, records, and food. There the boy had created an oasis of normalcy in a chaotic household.

Resilient children often find a refuge and a source of self-esteem in hobbies and creative interests. Kauffman et al. (1979) describes the pasttimes of two children who were the offspring of a schizophrenic mother and a depressed father:

> When David (age 8) comes home from school, he and his best friend often go up to the attic to play. This area ... is filled with model towns, railroads, airports and castles.... He knows the detailed history of most of his models, particularly the airplanes.... David's older sister, now 15, is extraordinarily well-read. Her other interests include swimming, her boyfriend, computers and space exploration. She is currently working on a computer program to predict planetary orbits. (pp. 138, 139)

The resilient children on the island of Kauai, whom we studied for nearly two decades, were not unusually talented, but they displayed a healthy androgyny in their interests and engaged in hobbies that were not narrowly sex-typed. Such activities, whether it was fishing, swimming, horseback riding, or hula dancing, gave them a reason to feel proud. Their hobbies, and their lively sense of humor, became a solace when things fell apart in their lives (Masten 1982; Werner and Smith 1982).

In middle childhood and adolescence, resilient children are often engaged in acts of "required helpfulness" (Garmezy, in press). On Kauai, many adolescents took care of their younger siblings. Some managed the household when a parent was ill or hospitalized; others worked part-time after school to support their family. Such acts of caring have also been noted by Anthony (1974) and Bleuler (1978) in their studies of the

---

Most resilient children establish a close bond with at least one caregiver from whom they received lots of attention during the first year of life.

---

resilient offspring of psychotic parents, and by Ayala-Canales (1984) and Moskovitz (1983) among the resilient orphans of wars and concentration camps.

Protective factors within the family

Despite chronic poverty, family discord, or parental mental illness, most resilient children have had the opportunity to establish a close bond with at least one caregiver from whom they received lots of attention during the first year of life. The stress-resistant children in the Kauai Longitudinal Study as well as the resilient offspring of psychotic parents studied by Anthony (1974) had enough good nuturing to establish a basic sense of trust.

Some of this nuturing came from substitute caregivers within the family, such as older siblings, grandparents, aunts, and uncles. Such alternate caregivers play an important role as positive models of identification in the lives of resilient children, whether they are reared in poverty (Kellam et al. 1975), or in a family where a parent is mentally ill (Kauffman et al. 1979), or coping with the aftermath of divorce (Wallerstein and Kelly 1980).

Resilient children seem to be especially adept at actively recruiting surrogate parents. The latter can come from the ranks of babysitters, nannies, or student roomers (Kauffman et al. 1979); they can be parents of friends (Werner and Smith 1982), or even a housemother in an orphanage (Ayala-Canales 1984; Moskovitz 1983).

The example of a mother who is gainfully and steadily employed appears to be an especially powerful model of identification for resilient girls reared in poverty, whether they are Black (Clark 1983), Chicana (Gandara 1982), or Asian-American (Werner and Smith 1982). Maternal employment and the need for sibling caregiving seems to contribute to the pronounced autonomy and sense of responsibility noted among these girls, especially in households where the father is permanently absent.

Structure and rules in the household and assigned chores enabled many resilient children to cope well in spite of poverty and discrimination, whether they lived on the rural island of Kauai, or in the inner cities of the American Midwest, or in a London borough (Clark 1983; Garmezy 1983; Rutter 1979).

Resilient children find a great deal of emotional support outside of their immediate family.

Resilient children also seem to have been imbued by their families with a sense of coherence (Antonovsky 1979). They manage to believe that life makes sense, that they have some control over their fate, and that God helps those who help themselves (Murphy and Moriarty 1976). This sense of meaning persists among resilient children, even if they are uprooted by wars or scattered as refugees to the four corners of the earth. It enables them to love despite hate, and to maintain the ability to behave compassionately toward other people (Ayala-Canales 1984; Moskovitz 1983).

Protective factors outside the family

Resilient children find a great deal of

emotional support outside of their immediate family. They tend to be well-liked by their classmates and have at least one, and usually several, close friends and confidants (Garmezy 1983; Kauffman et al. 1979; Wallerstein and Kelly 1980; Werner and Smith 1982). In addition, they tend to rely on informal networks of neighbors, peers, and elders for counsel and advice in times of crisis and life transitions.

Resilient children are apt to like school and to do well in school, not exclusively in academics, but also in sports, drama, or music. Even if they are not unusually talented, they put whatever abilities they have to good use. Often they make school a home away from home, a refuge from a disordered household. A favorite teacher can become an important model of identification for a resilient child whose own home is beset by family conflict or dissolution (Wallerstein and Kelly 1980).

In their studies of London schools, Rutter and his colleagues (1979) found that good experiences in the classroom could mitigate the effects of considerable stress at home. Among the qualities that characterized the more successful schools were the setting of appropriately high standards, effective feedback by the teacher to the students wtih ample use of praise, the setting of good models of behavior by teachers, and giving students positions of trust and responsibility. Children who attended such schools developed few if any emotional or behavioral problems despite considerable deprivation and discord at home (Pines 1984).

Early childhood programs and a favorite teacher can act as an important buffer against adversity in the lives of resilient young children. Moskovitz (1983), in her follow-up study in adulthood of the childhood survivors of concentration camps, noted the pervasive influence of such a warm, caring teacher.

Participation in extracurricular activities or clubs can be another important informal source of support for resilient children. Many youngsters on Kauai were poor by material standards, but they participated in activities that allowed them to be part of a cooperative enterprise, whether being cheerleader for the home team or raising an animal in the 4-H Club. Some resilient older youth were members of the Big Brothers and Big Sisters Associations

which enabled them to help other children less fortunate than themselves. For still others emotional support came from a church group, a youth leader in the YMCA or YWCA, or from a favorite minister, priest, or rabbi.

There is a shifting balance between stressful life events which heighten children's vulnerability and the protective factors in their lives which enhance their resiliency.

The shifting balance between vulnerability and resiliency

For some children some stress appears to have a steeling rather than a scarring effect (Anthony 1974). But we need to keep in mind that there is a shifting balance between stressful life events which heighten children's vulnerability and the protective factors in their lives which enhance their resiliency. This balance can change with each stage of the life cycle and also with the sex of the child. Most studies in the United States and in Europe, for example, have shown that boys appear to be more vulnerable than girls when exposed to chronic and intense family discord in childhood, but this trend appears to be reversed by the end of adolescence.

As long as the balance between stressful life events and protective factors is manageable for children they can cope. But when the stressful life events outweigh the protective factors, even the most resilient child can develop problems. Those who care for children, whether their own or others, can help restore this balance, either by *decreasing* the child's exposure to intense or chronic life stresses, or by *increasing* the number of protective factors, i.e., competencies and sources of support.

Implications

What then are some of the implications of the still tentative findings from studies of resilient children? Most of all, they provide a more hopeful perspective than can be derived from reading the extensive literature on problem children which predominates in clinical psychology, child psychiatry, special education, and social work. Research on resilient children provides us with a focus on the self-righting tendencies that appear to move some children toward normal development under all but the most persistent adverse circumstances.

Faith that things will work out can be sustained if children encounter people who give meaning to their lives and a reason for commitment and caring.

Those of us who care for young children, who work with or on behalf of them, can help tilt the balance from vulnerability to resiliency if we
- accept children's temperamental idiosyncracies and allow them some experiences that challenge, but do not overwhelm, their coping abilities;
- convey to children a sense of responsibility and caring, and, in turn, reward them for helpfulness and cooperation;
- encourage a child to develop a special interest, hobby, or activity that can serve as a source of gratification and self-esteem;
- model, by example, a conviction that life makes sense despite the inevitable adversities that each of us encounters;
- encourage children to reach out beyond their nuclear family to a beloved relative or friend.

Research on resilient children has taught us a lot about the special importance of surrogate parents in the lives of children exposed to chronic or intense distress. A comprehensive assessment of the impact on siblings, grandparents, foster parents, nannies, and babysitters on the development of high risk children is elaborated upon in Werner (1984).

Outside the family circle there are other powerful role models that give emotional support to a vulnerable child. The three most frequently encountered in studies of resilient children are: a favorite teacher, a good neighbor, or a member of the clergy.

There is a special need to strengthen such informal support for those children and their families in our communities which appear most vulnerable because they lack—temporarily or permanently—some of the essential social bonds that appear to buffer stress: working mothers of young children with no provisions for stable child care; single, divorced, or teen-age parents; hospitalized and handicapped children in need of special care who are separated from their families for extended periods of time; and migrant or refugee children without permanent roots in a community.

Two other findings from the studies of resilient children have implications for the well-being of all children and for those who care for them.

(1) At some point in their young lives, resilient children were required to carry out a socially desirable task to prevent others in their family, neighborhood, or community from experiencing distress or discomfort. Such acts of *required helpfulness* led to enduring and positive changes in the young helpers.

(2) The central component in the lives of the resilient children that contributed to their effective coping appeared to be a feeling of confidence or faith that things *will work out* as well as can be reasonably expected, and that the odds *can* be surmounted.

The stories of resilient children teach us that such a faith can develop and be sustained, even under adverse circumstances, if children encounter people who give meaning to their lives and a reason for commitment and caring. Each of us can impart this gift to a child—in the classroom, on the playground, in the neighborhood, in the family—*if* we care enough.

Bibliography

Anthony, E. J. "The Syndrome of the Psychologically Invulnerable Child." In *The Child in His Family 3: Children at Psychiatric Risk,* ed. E. J. Anthony and C. Koupernik. New York: Wiley, 1974.

Antonovsky, A. *Health, Stress and Coping: New Perspectives on Mental and Physical Well-being.* San Francisco: Jossey-Bass, 1979.

Ayala-Canales, C. E. "The Impact of El Salvador's Civil War on Orphan and Refugee Children." M.S. Thesis in Child Development, University of California at Davis, 1984.

Bleuler, M. *The Schizophrenic Disorders: Long-term Patient and Family Studies.* New Haven: Yale University Press, 1978.

Block, J. H. and Block, J. "The Role of Ego-Control and Ego-Resiliency in the Organization of Behavior." In *The Minnesota Symposia on Child Psychology 13: Development of Cognition, Affect and Social Relations,* ed. W. A. Collins. Hillsdale, N.J.: Erlbaum, 1980.

Block, J. "Growing Up Vulnerable and Growing Up Resistant: Preschool Personality, Pre-Adolescent Personality and Intervening Family Stresses." In *Adolescence and Stress,* ed. C.D. Moore. Washington, D.C.: U.S. Government Printing Office, 1981.

Clark, R. M. *Family Life and School Achievement: Why Poor Black Children Succeed or Fail.* Chicago: University of Chicago Press, 1983.

Fraser, M. *Children in Conflict.* Harmondsworth, England: Penguin Books, 1974.

Gandara, P. "Passing Through the Eye of the Needle: High Achieving Chicanas." *Hispanic Journal of Behavioral Sciences* 4, no. 2 (1982): 167–180.

Garmezy, N. "The Study of Competence in Children at Risk for Severe Psychopathology." In *The Child in His Family 3: Children at Psychiatric Risk,* ed. E. J. Anthony and C. Koupernik. New York: Wiley, 1974.

Garmezy, N. "Children Under Stress: Perspectives on Antecedents and Correlates of Vulnerability and Resistance to Psychopathology." In *Further Explorations in Personality,* ed. A. I. Rabin, J. Aronoff, A.M. Barclay, and R. A. Zucker. New York: Wiley, 1981.

Garmezy, N. "Stressors of Childhood." In *Stress, Coping and Development,* ed. N. Garmezy and M. Rutter. New York: McGraw-Hill, 1983.

Garmezy, N. "Stress Resistant Children: The Search for Protective Factors." In *Aspects of Current Child Psychiatry Research,* ed. J. E. Stevenson. *Journal of Child Psychology and Psychiatry,* Book Supplement 4. Oxford, England: Pergamon, in press.

Garmezy, N.; Masten, A. S.; and Tellegren, A. "The Study of Stress and Competence in Children: Building Blocks for Developmental Psychopathology." *Child Development* 55, no. 1 (1984): 97–111.

Garmezy, N. and Tellegren, A. "Studies of Stress-Resistant Children: Methods, Variables and Preliminary Findings." In *Advances in Applied Developmental Psychology,* ed. F. Morrison, C. Lord, and D. Keating. New York: Academic Press, 1984.

Heskin, K. *Northern Ireland: A Psychological Analysis.* New York: Columbia University Press, 1980.

Honig, A. "Research in Review: Risk Factors in Infants and Young Children." *Young Children* 38, no. 4 (May 1984): 60–73.

Kauffman, C.; Grunebaum, H.; Cohler, B.; and Gamer, E. "Superkids: Competent Children of Psychotic Mothers." *American Journal of Psychiatry* 136, no. 11 (1979): 1398–1402.

Kellam, S. G.; Branch, J. D.; Agrawal; K. C.; and Ensminger, M. E. *Mental Health and Going to School.* Chicago: University of Chicago Press, 1975.

Masten, A. "Humor and Creative Thinking in Stress-Resistant Children." Unpublished Ph.D. dissertation, University of Minnesota, 1982.

Moskovitz, S. *Love Despite Hate: Child Survivors of the Holocaust and Their Adult Lives.* New York: Schocken Books, 1983.

Murphy, L. and Moriarty, A. *Vulnerability, Coping and Growth from Infancy to Adolescence.* New Haven: Yale University Press, 1976.

O'Connell-Higgins, R. "Psychological Resilience and the Capacity for Intimacy." Qualifying paper, Harvard Graduate School of Education, 1983.

Pines, M. "PT Conversation: Michael Rutter: Resilient Children." *Psychology Today* 18, no. 3 (March 1984): 60, 62, 64–65.

Rosenblatt, R. *Children of War.* Garden City, N.Y.: Anchor Press, 1983.

Rutter, M. "Early Sources of Security and Competence." In *Human Growth and Development,* ed. J. Bruner and A. Garton. New York: Oxford University Press, 1978.

Rutter, M. "Protective Factors in Children's Responses to Stress and Disadvantage." In *Primary Prevention of Psychopathology 3: Social Competence in Children,* ed. M. W. Kent and J. E. Rolf. Hanover, N.H.: University Press of New England, 1979.

Rutter, M.; Maughan, B.; Mortimore, P.; and Ouston, J; with Smith, A. *Fifteen Thousand Hours: Secondary Schools and Their Effects on Children.* Cambridge, Mass.: Harvard University Press, 1979.

Shipman, V. C. *Notable Early Characteristics of High and Low Achieving Low SES Children.* Princeton, N.J.: Educational Testing Service, 1976.

Wallerstein, J. S. and Kelly, J. B. *Surviving the Breakup: How Children and Parents Cope with Divorce.* New York: Basic Books, 1980.

Watt, N. S.; Anthony, E. J.; Wynne, L. C.; and Rolf, J. E., eds. *Children at Risk for Schizophrenia: A Longitudinal Perspective.* London and New York: Cambridge University Press, 1984.

Werner, E. E. *Child Care: Kith, Kin and Hired Hands.* Baltimore: University Park Press, 1984.

Werner, E. E. and Smith, R. S. *Vulnerable, but Invincible: A Longitudinal Study of Resilient Children and Youth.* New York: McGraw-Hill, 1982.

This is one of a regular series of Research in Review columns. The column in this issue was edited by Elizabeth H. Brady, M.A., Professor and Chair, Department of Educational Psychology, California State University, Northridge, Northridge, California.

*The way a child wants to play is often very
different from the way his parents want him to. The child, however, knows best*

THE IMPORTANCE OF PLAY

BRUNO BETTELHEIM

*Bruno Bettelheim is the Distinguished Service Professor Emeritus
of Education, Psychology, and Psychiatry at the University of
Chicago, and the Visiting Professor Emeritus of Education, Psychiatry, and the Behavioral Sciences at Stanford University.
He is also a senior research fellow of the Hoover Institution.
Bettelheim, who was born in Austria, studied psychology and
philosophy at the University of Vienna, from which he received
a Ph.D. in 1938. He came to the United States in 1939 and
subsequently became a professor of educational psychology—
and later of psychology and psychiatry, as well—at the University of Chicago. He retired as the head of the university's Sonia
Shankman Orthogenic School, a residential laboratory school
for emotionally disturbed children, in 1973.*

"CHILDREN'S PLAYINGS ARE NOT SPORTS AND should be deemed their most serious actions," Montaigne wrote. If we wish to understand our child, we need to understand his play. Freud regarded play as the means by which the child accomplishes his first great cultural and psychological achievements; through play he expresses himself. This is true even for an infant whose play consists of nothing more than smiling at his mother, as she smiles at him. Freud also noted how much and how well children express their thoughts and feelings through play. These are sometimes feelings that the child himself would remain ignorant of, or overwhelmed by, if he did not deal with them by acting them out in play fantasy.

Child psychoanalysts have enlarged on Freud's insights, which recognized the manifold problems and emotions children express through play; they also have shown how children use play to work through and master quite complex psychological difficulties of the past and present. So valuable is play in this connection that play therapy has become the main avenue for helping young children with their emotional difficulties. Freud said that the dream is the "royal road" to the unconscious, and this is true for adults and children alike. But play is also a "royal road" to the child's conscious and unconscious inner world; if we want to understand his inner world and help him with it, we must learn to walk this road.

From a child's play we can gain understanding of how he sees and construes the world—what he would like it to be, what his concerns and problems are. Through his play he expresses what he would be hard pressed to put into words. A child does not play spontaneously only to while away the time, although he and the adults observing him may think he does. Even when he engages in play partly to fill empty moments, what he chooses to play at is motivated by inner processes, desires, problems, anxieties.

The most normal and competent child encounters what seem like insurmountable problems in living. But by playing them out, in the way he chooses, he may become able to cope with them in a step-by-step process. He often does so in symbolic ways that are hard for even him to understand, as he is reacting to inner processes whose origin may be buried deep in his unconscious. This may result in play that makes little sense to us at the moment or may even seem ill advised, since we do not know the purposes it serves or how it will end. When there is no immediate danger, it is usually best to approve of the child's play without interfering, just because he is so engrossed in it. Efforts to assist him in his struggles, while well intentioned, may divert him from seeking, and eventually finding, the solution that will serve him best.

A four-year-old girl reacted to her mother's pregnancy by regressing. Although she had been well trained, she began to wet again, insisted on being fed only from a baby bottle, and reverted to crawling on the floor. All this greatly distressed her mother, who, anticipating the demands of a new infant, had counted on her daughter's relative maturity. Fortunately, she did not try to prevent her daughter's regressions. After a few months of this behavior, the girl replaced it with much more mature play. She now played "good mother." She became extremely caring for her baby doll, ministering to it much more seriously than ever before. Having in the regressed stage identified with the coming infant, she now identified with her mother. By the time her sibling was born, the girl had done much of the work needed for her to cope with the change in the family and her position in it, and her adjustment to the new baby was easier than her mother had expected.

In retrospect it can be seen that the child, on learning that a new baby was to join the family, must have been afraid that the baby would deprive her of her infantile gratifications, and therefore tried to provide herself with them. She may have thought that if her mother wanted an infant, then she herself would again be an infant. There would be no need for her mother to acquire another, and she might give up on the idea.

Permitted to act on notions like these, the girl must have realized after a while that wetting herself was not as pleasant as she might have imagined; that being able to eat a wide variety of foods had definite advantages when compared with drinking only from the bottle; and that walking and running brought many more satisfactions than did crawling. From this experience she convinced herself that being grown up is preferable to being a baby. So she gave up pretending that she was a baby and instead decided to be like her mother: in play to be like her right now, in imagination to become at some future time a real mother. Play provided the child and her mother with a happy solution to what otherwise might have resulted in an impasse.

B ESIDES BEING A MEANS OF COPING WITH PAST AND present concerns, play is the child's most useful tool for preparing himself for the future and its tasks. Play's function in developing cognitive and motor abilities has been explored by Karl Groos (the first investigator to study it systematically), Jean Piaget (to whom we owe our best understanding of what the child learns intellectually from play), and many others. Play teaches the child, without his being aware of it, the habits most needed for intellectual growth, such as stick-to-itiveness, which is so important in all learning. Perseverance is easily acquired around enjoyable activities such as chosen play. But if it has not become a habit through what is enjoyable, it is not likely to become one through an endeavor like schoolwork. That we rarely succeed at a thing as easily or promptly as we might wish is best learned at an early age, when habits are formed and when the lesson can be assimilated fairly painlessly. A child at play begins to realize that he need not give up in despair if a block doesn't balance neatly on another block the first time around. Fascinated by the challenge of building a tower, he gradually learns that even if he doesn't succeed immediately, success can be his if he perseveres. He learns not to give up at the first sign of failure, or at the fifth or tenth, and 'not to turn in dismay to something less difficult, but to try again and again. But he will not learn this if his parents are interested only in success, if they praise him only for that and not also for tenacious effort. Children are very sensitive to our inner feelings. They are not easily fooled by mere words. Thus our praise won't be effective if, deep down, we are disappointed by the length of time it takes them to achieve their goal or by the awkwardness of their efforts. Further, we must not impose *our* goals on them, either in thought or in action.

Gregory Bateson and others have demonstrated how destructive it is for a child to receive contradictory signals from his parents. Exposed to one message from verbal statements and a contrary one from subliminal signs (which the speaker may be unaware of making), the child is utterly confused, for what he hears is the opposite of what he senses is the truth. This will prevent him from persisting in the face of difficulties as effectively as will criticism for his failure or praise only for his success.

Some parents (usually for reasons of which they are completely unaware) are not satisfied with the way their child plays. So they start telling him how he ought to use a toy, and if he continues to suit his own fancy, they "correct" him, wanting him to use the toy in accordance with its intended purpose or the way they think it ought to be played with. If they insist on such guidance, the child's interest in the toy—and to some extent also in play in general—is apt to wane, because the project has become his parents' and is no longer his own. Such parents are likely to continue to direct and dominate the child's activities in later years, motivated by the same inner tendencies that did not allow them to enjoy his play as *he* developed it. But now everything is happening on a more complex intellectual level. The parents may try to improve the child's homework by suggesting ideas that are much too sophisticated and in any case are not his own. In consequence he may lose interest in developing his own ideas, which pale by comparison with his parents'. What he wanted, in talking with his parents about his homework, was appreciation of his efforts and encouragement that his own ideas were valuable—not a demonstration that his ideas were not good enough. Such parents would be most astonished to learn that their efforts to help were the cause of the child's lack of interest in his homework.

A child, as well as an adult, needs plenty of what in German is called *Spielraum*. Now, *Spielraum* is not primarily "a room to play in." While the word also means that, its primary meaning is "free scope, plenty of room"—to move not only one's elbows but also one's mind, to experiment with things and ideas at one's leisure, or, to put it colloquially, to toy with ideas. The biographies of creative people of the past are full of accounts of long hours they spent sitting by a river as teenagers, thinking their own thoughts, roaming through the woods with their faithful dogs, or dreaming their own dreams. But who today has the leisure and the opportunities for this? If a youngster tries it, as likely as not his parents will fret that he is not using his time constructively, that he is daydreaming when he should be tackling the serious business of life. However, developing an inner life, including fantasies and daydreams, is one of the most constructive things a growing child can do. The days of most middle-class children are filled with scheduled activities—Boy or Girl Scout meetings, music and dance lessons, organized sports—which leave them hardly any time simply to be themselves. In fact, they are continually distracted from the task of self-discovery, forced to develop their talents and personalities as those who are in charge of the various activities think best. Today academic teaching begins in kindergarten, if not in nursery school. Kindergarten, as conceived by Friedrich Froebel in the nineteenth century, was a place where children would play, as if in a garden. During most of the period in which kindergartens have existed, they have been play schools.

A lack of sufficient leisure to develop a rich inner life is a large part of the reason why a child will pressure his parents to entertain him or will turn on the television set. It is not that the bad of such mass-produced entertainment drives out the good of inner richness. It is that, in a vicious circle, the lack of a chance to spend much of his energies on his inner life causes the child to turn to readily available stimuli for filling an inner void, and these stimuli then constitute another obstacle to the child's development of his inner life.

Play, Games, and Rules

MOST ADULTS FIND IT EASIER TO INVOLVE THEM-selves directly in complex, adult games, such as chess or baseball, than in play on simpler levels, such as stacking blocks or riding a hobbyhorse or a toy car. Although the words *play* and *game* may seem synonymous, they in fact refer to broadly distinguishable stages of development, with *play* relating to an earlier stage, *game* to a more mature one. Generally speaking, *play* refers to the young child's activities characterized by freedom from all but personally imposed rules (which are changed at will), by free-wheeling fantasy involvement, and by the absence of any goals outside the activity itself. *Games*, however, are usually competitive and are characterized by agreed-upon, often externally imposed, rules, by a requirement to use the implements of the activity in the manner for which they are intended and not as fancy suggests, and frequently by a goal or purpose outside the activity, such as winning the game. Children recognize early on that play is an opportunity for pure enjoyment, whereas games may involve considerable stress. One four-year-old, when confronted with an unfamiliar play situation, asked, "Is this a fun game or a winning game?" It was clear that his attitude toward the activity depended on the answer he was given.

Piaget stresses how important learning the rules of the game is in the process of socialization; a child must become able to control himself in order to do so, controlling most of all his tendency to act aggressively to reach his goals. Only then can he enjoy the continuous interaction with others that is involved in playing games with partners who are also opponents. But obeying the rules and controlling one's selfish and aggressive tendencies is not something that can be learned overnight; it is the end result of long development. When he begins playing games, a child tries to behave as he could in his earlier play. He changes the rules to suit himself, but then the game breaks down. In a later stage he comes to believe that the rules are unalterable. He treats them as if they were laws handed down from time immemorial, which cannot be transgressed under any circumstances, and he views disobeying the rules as a serious crime. Only at a still later stage—often not until he has become a teenager and sometimes even later than that—can he comprehend that rules are voluntarily

agreed upon for the sake of playing the game and have no other validity, and that they can be freely altered as long as all participants agree to such changes. Democracy, based on a freely negotiated consensus that is binding only after it has been formulated and accepted, is a very late achievement in human development, even in game-playing.

When children are free to do as they like in games not supervised by adults, more often than not arguments over which game they will play and how, and what rules they should follow, take up most of their time, so that little actual playing gets done. Left to their own devices, children may require hours of fruitful deliberation before they agree on the rules and related issues, such as who should begin the game and what role each child is to have in it. And this is how it ought to be, if playing games is to socialize children. Only by pondering at great length the advantages and disadvantages of various possible games, their relative appropriateness to the conditions at hand—such as the size of the group, the available playing area, and so forth—and what rules should apply and why, will children develop their abilities to reason, to judge what is appropriate and what is not, to weigh arguments, to learn how consensus can be reached and how important such consensus is to the launching of an enterprise. Learning all this is infinitely more significant for the child's development as a social human being than is mastering whatever skills are involved in playing the game itself. Yet none of these socializing skills will be learned if adults attempt to control which games are played, or if they prevent experimentation with rules (out of fear that this may lead to chaos), or if they impatiently push for the game to get started without further delay.

When thinking about an organization like Little League, we should keep in mind that the most important function of play and games for the well-being of the child is to offer him a chance to work through unresolved problems of the past, to deal with pressures of the moment, and to experiment with various roles and forms of social interaction in order to determine their suitability for himself.

A freely organized ball game looks very ragged, and it *is* very ragged. The children use the game to serve their individual and group needs, so there are interruptions for displays of temper, digressions for talking things over or to pursue a parallel line of play for a time, surprising acts of compassion ("give the little guys an extra turn")—all acts outside adult game protocols. If adults want to see a polished game of baseball played according to the rule books, they can turn on their television sets. John Locke wrote that "because there can be no *Recreation* without Delight, which depends not always on Reason, but oftener on Fancy, it must be permitted Children not only to divert themselves, but to do it after their own fashion." How wonderful it would be for our children if we adults would heed the advice of this great philosopher!

FOR YEARS THE GROWING CHILD MOVES BACK AND forth among the many demands that playing games imposes on him. When all goes well, a child can do full justice to the game's requirements. But when things become psychologically too bewildering or frustrating for him, he may revert to spontaneous play. Although he may understand the rules governing the game—even insist that others follow them—he himself will be unable to obey them and may assert that they do not apply to him. For example, a young child may know perfectly well how to play checkers. All will go smoothly until he realizes, or believes, that he will lose. Then he may suddenly request, "Let's start over." If the other player agrees and the second game goes more in the child's favor, all is well. But if things look bleak for the child the second time around, he may repeat his request for a fresh start, and he may do so repeatedly. This can be frustrating to an adult, who may decide that the child should learn to finish a game once he has started it, even if he is about to lose. But if the adult is able to be patient and agree to repeated new beginnings, even though the checker game may never be concluded the child will eventually learn to play better.

If the adult insists that the child continue playing when he is likely to lose, he will be asking too much of the child's still weak controls. If the child could articulate his position, he might say, "Obeying the rules when it seems I'm going to lose is just too much for me. If you insist that I go on, I'll just have to give up on games and return to fantasy play, where I can't be defeated." Then the checker, which had been accepted as a marker to be moved only according to established rules, will suddenly be moved as the child's fancy determines, or in a way that seems to assure his winning. If this is not accepted, the marker may become a missile, to be hurled off the board or even at the winning opponent.

The reasons for the child's behavior are not difficult to understand. Feeling himself momentarily defeated by the complex realities of the game—he is losing, and thus his extremely tenuous self-respect is about to be damaged, something to be avoided at all costs—he reverts to a play level at which the rules no longer pertain, in order to rescue his endangered feeling of competence. If the opponent is also a child, he will intuitively understand (although not applaud) his companion's action. The child opponent may say in response, "Come on, now, you're acting like a baby," as if recognizing—probably from his own experience in similar situations—that what has taken place is a regression to an earlier stage of development, because the higher stage has proved too painful to be worth the effort to maintain. Or he may suggest, "Let's play something else," knowing that checkers has become too difficult.

If the opponent is an adult, however, such intuitive understanding may be missing. Some parents, unfortunately, are eager to see their child behave maturely before he is ready to do so. They become unhappy with his behavior when he reverts to simple, unstructured play. But criticism and insistence on mature behavior just when the child feels most threatened merely aggravate his sense of defeat. We ought to recognize that a child may be forced by as-yet-uncontrollable pressures to disregard, or even to pervert, the rules of the game in an instant, and that if he does so, he does it for compelling reasons.

Again we must remember that for a child, a game is not "just a game" that he plays for the fun of it, or a distraction from more serious matters. For him, playing a game can be, and more often than not is, a serious undertaking: on its outcome rest his feelings of self-esteem and competence. To put it in adult terms, playing a game is a child's true reality; this takes it far beyond the boundaries of its meaning for adults. Losing is not just a part of playing the game, as it is for adults (at least most of the time), but something that puts the child's sense of his own competence in question and often undermines it.

What makes it all so confusing is that now and then the child is easily able to finish a game even though he is aware that he is losing. So if he can accept defeat sometimes, why not always? Because he could act mature yesterday, adults expect him to do so today, and they try to hold him to this maturity or are critical if he does not. What they overlook is that they themselves act similarly in real life. They are able to accept defeat with relative equanimity when they feel secure in other important respects; at other times defeat temporarily disintegrates them, makes them depressed and unable to function. Since game-playing is for the child a real-life experience, he behaves accordingly: when feeling relatively strong and secure, he can accept defeat in a game without falling apart, but when insecure, he cannot. Because a child's inability to accept defeat in a game is a sign that at that moment he is quite insecure, it becomes even more important that we do not add to this feeling by criticizing him.

Some children—and most children at some stages in their lives—simply cannot afford to lose. So they correct their fortune in order to win—wanting to move a checker more spaces than they are entitled to, for example, or asking for an extra turn (as opposed to making a move while an opponent is out of the room). It is then wrong to hold them to the rules of the game, because they may give up playing altogether and become utterly dejected, deeply disappointed in themselves. If, instead of objecting to their insisting on changing the rules, we silently accept it and in this way make it possible for them to win, they will enjoy the game and continue playing it. As a child continues to play—and to cheat in this way—he slowly becomes more experienced in playing the game and needs to cheat less often, and less outrageously. This is why it is especially important for parents to play games with their child, because others are not so ready to let him change the rules at will without at least remarking on it. But improving his chances of winning may be necessary if the child is to play often enough to become sufficiently expert to win playing by the rules. Winning makes him more and more secure about his ability to hold his own in the game, and soon he will give up changing the rules altogether, although he will by no means win every time. The ability to win fair and square will provide him with enough security in playing the game that an occasional loss will no longer be experienced as such a severe defeat that he must avoid the game altogether. And the parent's pleasure in playing will increase with the child's.

Toys as Symbols

THERE ARE MANY CONTRIBUTIONS THAT ONLY PARents can make to the play of their children. For example, no teacher, and certainly no age-mate, can be as deeply and personally involved in play that seems to relate to the child's future as are his parents. Play is anchored in the present, but it also takes up and tries to solve problems of the past, and is often future-directed, as well. So a girl's doll play anticipates her possible future motherhood and also helps her to deal with emotional pressures of the moment. If she is jealous of the care a sibling receives from their mother, doll play permits her to act out and master her ambivalent feelings. She deals with their negative aspects by mistreating the doll, who represents her sibling. In this symbolic way a girl is able to punish her sibling for her jealous agonies, of which the sibling is the innocent cause. She can make amends for her negative attitude and satisfy the positive elements of her ambivalence when she takes good care of the doll, just as her mother does of the sibling, and in this way can free herself from guilt and identify with her mother. In taking good care of the doll the girl can also identify with the doll, and thus vicariously receive the care her mother lavishes on the sibling. Thus in many ways doll play is closely connected with a girl's relation to her mother.

It is a misfortune for boys that they are only rarely offered the opportunity to play with dolls and even more rarely encouraged to do so. Many parents feel that doll play is not for boys, and because of this boys are usually prevented from dealing with issues such as sibling rivalry and problems of family constellation (emotional grouping), among many others, in this convenient symbolic way. Perhaps if parents could see how eagerly boys use dolls and doll houses in psychoanalytic treatment—certainly as eagerly and persistently as girls do—to work out family problems and anxieties about themselves, they would be more ready to recognize the value of doll play for both sexes. For example, in doll-house play boys—as eagerly as girls—put a figure representing their sibling out of the house, put a figure representing a parent on the roof or lock it in the basement, place both parents together in bed, seat a figure representing themselves on the toilet or have it mess up the house, and in countless other ways visualize, act out, and thus become better able to deal with pressing family problems.

Some parents, especially fathers, think that doll play is contrary to masculinity. It is not. There is a great deal in a boy's past (just as there is in a girl's)—the way he was fed, held, bathed, and toilet-trained—that he can best master through doll play or through play with doll-house furniture, such as tubs and toilets. There are present-day problems, such as sibling rivalry, for him too. And although child care will probably play a more peripheral role in his

future than in that of a girl, it may be a very important aspect of his life as a father. If parents are worried that doll play may feminize a boy, all they need for reassurance is to watch how boys play with dolls, because it is very different from the way girls play with them. Unless a boy has already embraced femininity by reason of severe neurosis, his approach is quite distinctly masculine, typically much more aggressive and manipulative than that of girls—for example, boys make their dolls have fights much more often than girls do.

True, boys' doll play is usually shorter-lived than girls', and not quite as significant an experience for them; but this is no reason that they should lose out entirely on what doll play can offer them. Actually, toys typically viewed as being for boys (dump trucks, racing cars, railroad sets, and many others), though they may offer a chance to work out problems of the present and anticipate the future, are much less suitable than dolls for mastering difficulties of the past. If parents feel relaxed about their son's playing with dolls, they will provide him with valuable opportunities for enriching his play life. For them to do so, it is not sufficient that they simply refrain from disparaging such play. Because of the still prevalent attitude that doll play is only for girls, both parents need to have a positive feeling about a boy's doll play if he is to be able to take full advantage of it.

It is relatively rare for a parent to become as engrossed in a play activity as his child does, but there are toys that evoke deep feelings in a parent, as they do in a child. Dolls are probably the best example of this. Whether a mother merely watches her daughter play with dolls, encourages her in it, or actively participates, she is often deeply involved on many levels. She may re-experience aspects of her own childhood doll play and her own mother's involvement in such play and in herself. The child as she plays with her doll feels, in some way, strong emotions that reign in her mother's conscious and subconscious mind, and experiences a closeness to her mother based on the deep emotional involvement they both have in the girl's doll play. This closeness gives the play a special significance and depth of meaning for the child which it never could attain without the mother's involvement.

In order for the child's doll play to take on this special significance, the mother need not always be physically present, nor when she is present must she be so personally involved on many levels; it is enough if the child carries a mental image of her mother's involvement. One such experience of involvement with her mother can make an impact so lasting that the child will carry this image within her and reactivate it whenever she plays with her doll—it is that meaningful. She will continue to react to the emotional signals she has received from her mother and to combine them in her doll play with other feelings that originate in her past and present experiences of being mothered and playing at mothering. Important as her feelings are about being mothered and about someday becoming a parent herself, her doll play could not attain the same depth of meaning if her mother had not on occasion been deeply and personally involved because of the recollections it evoked in her.

A Double Standard

CERTAINLY PARENTS ARE HAPPY TO SEE THEIR CHILdren absorbed in play. But are they equally happy to become engrossed in the playing themselves? If a child's play is pleasurable to a parent chiefly because it allows him to pursue his adult activities without feeling bad about neglecting his child, it does not take the child long to realize this. He soon learns that to his parents play itself is not very important, but his being out of their way is; this lesson simultaneously diminishes him and his enjoyment of play and reduces the capacity of play to develop his intelligence and personality.

The true test of a parent's beliefs about play is not what he says but how he behaves. The fact is that parents often behave inconsistently. Sometimes all goes well: The parent is not doing anything of particular importance, and his child asks him to play. The parent obliges. The child wants him to admire what he has built, and the parent again obliges. But if the parent is occupied with something that demands his attention, usually his response to the request is, "Not now—I'm busy." If the parent is in a good mood, he may preface his refusal with an apology or a promise to make up for it later—a promise not always kept. Parents tend to assume that if a child doesn't repeat his plea, he has either lost interest or forgotten about it. But many a child hears "In a few minutes" as a brush-off, and he's not all that eager to receive a second brush-off by repeating his request.

Such parental behavior suggests to children that their activities rarely seem as important to parents as the parents' activities, and hardly ever more important. There is nothing very much wrong with that: if both parties are seriously engaged, why should parents drop what they are doing to join their child? The situation is different, of course, when there is an emergency. In such cases the transfer of our attention is virtually automatic. This is very important for the security of the child, and some bright children test how reliably they can depend on our reaction by claiming that an emergency has arisen. Others, without necessarily wishing to ascertain how dependable their parents will be in a crisis, pretend that an emergency exists in order to bring a parent hurrying to their side when they have a great desire to tell or show the parent something of importance. But this works only a few times. Then the parents cease to respond, and make no bones about their annoyance at being taken advantage of in this way—as in the fable of the child who cried "wolf" once too often. This is understandable. But are parents really being taken advantage of when a child goes to great lengths to signal how important it is that they come to him, emergency or no emergency? Or, to put it differently, is only what parents consider an emergency—such as an actual danger or mishap—truly an emergency? Is not a child's need to reassure himself that he and what he is doing are important also an emergency?

If a parent is just a bit more patient with a child's claim of emergency, even if all the child needs is to convince himself that the parent is ready to drop everything and rush to his side, then the child will feel more secure about his importance to the parent. This improvement in the child's security will be reflected in a parallel improvement in his relationship with his parent. Such a result may well be worth the inconvenience of responding to what we don't regard as real emergencies. As the child grows and matures, he will learn to accept that it is unreasonable to expect that if two people are deeply engaged, one will always be ready to quit what he is doing to join the other.

What happens when a child is engrossed in play and the parents are ready to go out? They call him to come and get dressed. Or perhaps they want him to greet a visitor, or come to the table for lunch. His answer is, as ours would be in an analogous situation, "Not now—I'm busy." Are we prepared to honor our child's statement, as we expect him to honor ours? Or do we insist: "You come here, *right now*"? If we do, then we have once again succeeded in impressing on him that we do not take his activities as seriously as we do our own. Worse, we have demonstrated that we do not take his activities seriously at all when they conflict with our plans. If we truly took our child's play as seriously as we take our own tasks, we would be as loath to interrupt it as we are reluctant to be interfered with when we are working. This is the pattern demanded by consistency and a sense of fairness, and one reward for thus respecting our child's play is that it enhances his own sense of play as an important activity in the whole context of family life.

Despite how important it is that we encourage play, it is never beneficial for parents to play with their children strictly out of a sense of duty. To play because one "should" is simply not the same as playing together with one's child, or even appreciating the importance of his play. This confusion about the parent's intent is precisely what mars so much of the child's play with his parents. Many adults, whether parents or teachers, tend to play with children for purposes outside the play; they may wish to distract, entertain, educate, diagnose, or guide them. But this is not what the child desires. Unless the play itself is the thing, it loses much of its meaning to the child, and adult participation becomes offensive; the child can guess the adult's purpose and becomes annoyed at the pretense of wholehearted participation.

The use of educational toys, so dear to the hearts of many parents, may serve as an illustration. There is really nothing wrong with educational toys *if* the emphasis is entirely on the enjoyment of play and not on the intent of educating. Such toys become problematic, however, when parents emphasize what using the toy supposedly teaches the child over how the child desires to use it. Educational toys become absolutely deadly when the child is expected to learn what they are designed to teach rather than what he wants to learn. A child must be permitted to use a toy the way he wishes to (if the toy is not made of any dangerous materials, of course), not as the parent, teacher, or manufacturer thinks it ought to be used.

It is amazing what an infant can learn just by playing with the cardboard core of a roll of toilet paper, and how constructive, imaginative, and educative a child's play with empty boxes can be. In earlier days, when thread came on wooden spools, young children used the spools as blocks and gained as much pleasure and learning from them as they do now from specially constructed building blocks. Indeed, they probably got something more out of playing with spools than they do with blocks, because they knew that their mothers, too, made use of spools. Thus both child and parent found something important represented in wooden spools, whereas blocks are important only to the child.

Some parents spontaneously realize the value of having a personal investment in their child's play objects, although they are not always conscious that this is what motivates them. They instinctively add a measure of mutuality to their child's pleasure, without setting out to do so. Some of these parents may have the time and inclination to fashion toys for their children, thereby duplicating what their own parents or grandparents did out of necessity. Such parents become emotionally involved in the toys they have created with their own hands. They get enormous enjoyment not only from the task but also from imagining how their child will play with these toys. The meaning the parents have invested in the toys remains active as they play with their child or watch his play.

Other parents make the production of toys a common project. For example, with the child's help they collect scraps of wood. Together parent and child cut and sand the wood; perhaps the child invites some of his friends to help with this labor and with the painting and the shellacking that follows. From then on and ever after these blocks are very special to child and parent. No store-bought blocks can compare in importance to these visible and tangible examples of the child's and the parent's common investment in a toy.

Becoming Civilized

IN PSYCHOLOGICAL TREATMENT A CHILD MIGHT BE ENcouraged to shoot a toy pistol at a figure; this might be done either to free his aggressions or to discover their source and intended target. But this occurs in the presence of an adult acting as a therapist, in an "as if" therapeutic situation. If a parent encourages his child to shoot a toy gun at someone, even at himself, in a normal play setting, it is a mistake—he is not taking the child's play seriously enough. If he were, rather than just pretending to do so without paying close attention to what the play is all about, he could hardly encourage such an unequivocal show of aggression against another person, not to mention against himself.

A common mistake adults make in reacting to a child's play is taking it as "not real." But in more than one sense play is the child's true reality, and we have to respect it as such. This is why we ought not to encourage our child to shoot at anyone. But this caution refers only to our encour-

agement. We may very well give him a toy gun to use as *he* likes or sees fit, be it for his protection or for aggressive play. Whether, when, and how to use such a toy should be entirely the child's own decision. Our giving him the gun implies our permission to use it as he wishes, when and how he feels a desire or need to do so, but no more. More important, it also implies our confidence that he will use it in a way that is appropriate, even wise, as seen from his perspective.

Children have a need to rid themselves of their aggressions, at least through symbolic play, and it is sufficient permission to do so when we give them toys suitable for that purpose. If we encourage a child to play aggressively, we exercise—however subtly—control over the activity, which is likely to increase his frustration or aggression and with it the need for discharge. But if his aggressive play is directed toward us—as it might be, not necessarily because he wishes to hurt us even in play, but because he wishes to discover what our reaction might be—and we do not react appropriately to what he does, then we effectively demonstrate to him that we take neither him nor his aggression very seriously. If we show a contradictory approach to the play by initially intellectualizing ("Let him work off his aggressions") and subsequently attempting to render the activity harmless ("Even though you've just 'shot' me, it means nothing"), our attitudes destroy the serious qualities that play has for the child.

When a child "shoots" his parents, should they shoot back? Certainly not. Counteraggression by an adult—whether in play or in earnest—has never yet proved beneficial to a child. Nevertheless, it is not much help to him to shoot us with his toy gun unless we react appropriately. The reaction, of course, must be not to his action as such but to his intentions. Only our on-the-spot assessment of what motivated the action can tell us whether the best response is admiration of the child's assertiveness—what a powerful warrior he is!—or a playful dramatic collapse to the floor, or a show of anxiety, or a question about how he will manage with us out of the way. A well-placed question such as this one is much more effective in convincing a child that shooting and killing are detrimental to his well-being than any theoretical discussion of the evils of war and violence. This is because the child lives in the immediate present and within the limited confines of his direct experience. Wars, even those he sees on the TV screen, take place in some far-distant place and have no bearing on him that he can understand. And should we succeed in impressing on him the tragic consequences of war, the primary effect will be to infuse him with an overwhelming sense of powerlessness. After all, the youngster is smart enough to figure out that he has no effect on what is going on somewhere far away in the world. But shooting at his parent is something he *can* control. Almost any child realizes that however angry he is at his parent, however much he may want to get rid of him at the moment, he does not want to lose him forever.

Some adults may overreact to shooting play. Parents who fall into this trap are usually concerned more with their own feelings about aggression than with helping a child to master rather than merely repress his aggression through such play. This is also true with respect to bodily and other types of anxiety that many children try to cope with through shooting play, such as with water pistols. So when parents forbid such play, they block the safe and necessary outlet it can provide. At the same time, they rob the child of the valuable lesson that if we try to shoot someone, that person may shoot back, and everybody will lose.

Some parents, out of their abhorrence of war and violence, try to control, or forbid altogether, any play with toy guns, soldiers, tanks, or other toys suggestive of war. Although these feelings toward violence are most understandable, when a parent prohibits or severely criticizes his child's gun play, whatever his conscious reasons for doing so, he is acting not for his child's benefit but solely out of adult concerns or anxieties. Some parents even fear that such play may make a future killer of the child who thoroughly enjoys it, but the pitfalls of such thinking are many and serious.

First, as playing with blocks does not indicate that a child will grow up to be an architect or builder, and playing with cars and trucks does not foretell the future auto mechanic or truck driver, so playing with toy guns tells nothing about what a child will do and be later in life. Second, one may reasonably expect that if through gun play a child feels that he can protect himself, and if he discharges many of his aggressive tendencies, then fewer of these will accumulate and require dangerous ways of discharge in later life. Parental prohibition also leads to additional frustration and anger, because the child is prevented from using an outlet that he sees made available to other children and that is suggested to him by the mass media.

Third, and by far the most important attitude, because whether spoken or implied it is the most pernicious in its consequences, is parental fear that the child may become a violent person. This thought is far more damaging to the child's emotional well-being and his sense of self-worth than any play with guns can possibly be. This is particularly true because of the importance to him of his parents' view of him. After all, a child gains a view of himself primarily from his parents. If they seem to hold such a low opinion of him, it is apt to make him very angry at them and the world, and this increases his propensity to act out his anger, not just in symbolic play but in reality, once he has outgrown parental control.

Girls are as subject to all kinds of frustrations, very much including sibling rivalry and anger at their parents, and so it would serve them equally well to be able to discharge their anger through symbolic play, such as with toy guns. Furthermore, it would prevent their frustration at being denied an important type of symbolic play that is available to boys.

Parents who worry exclusively about shooting play often fail to take into account the duality of our human and animal natures and the distance between them. Certainly there is a great deal of the animal—and with it of violence—in human beings, and sometimes these irrational

forces do appear in children's games, making many parents uncomfortable. But more often it is actually the child's developing sense of humanity that motivates what seems to the uninvolved and uninformed parent to be mere brutality. Since ancient times children have played out war games in which *we* fight *them*, *them* being the enemy of the historical moment.

Children of the Middle Ages surely played at being knights and infidels, just as our own children play at being cops and robbers. Elizabeth I is said to have inquired whether the boys were now playing the war of the English against the Scots. In Europe early in this century much play involved the Foreign Legion against the Arabs. And as soon as the wall went up separating West from East Berlin, German children began shooting at each other across miniature walls. Such battle play invariably features the conflict of good and evil in terms and images that a child can readily grasp.

In a game like cops and robbers a child experiments with moral identities. Such games permit him to realize his fantasies. Acting out the roles of cop and robber permits him to get closer to the reality of these characters and what it might feel like to be them, which reading or watching television cannot provide. A passive, receptive role is no substitute for active encounters with experiential reality.

Psychoanalytically speaking, such conflicts between good and evil represent the battle between tendencies of the asocial id and those of the diametrically opposed superego. Such battles—either dramatized by two groups of children warring against each other or acted out by one or more children manipulating toy soldiers—permit some discharge of aggression either actually or symbolically, through conflict. Only after such a discharge of anger or violence can the forces of the superego gain ascendancy to control or overbalance those of the id; with that, the ego becomes able to function again.

As we watch the progress of aggressive activity in our child, we can gradually discern a developmental move from free play, which permits direct id expression and satisfaction (the unstructured free-for-all shooting match, in which aggression is freely discharged), to a more structured game setting in which not mere discharge of aggression but a higher integration—the ascendancy of good over evil—is the goal. So *we* destroy *them*: the Greeks defeat the knavish Trojan wrongdoers, the Christian knights destroy the infidels, the cops corner the robbers, the cowboys crush the savage Indians.

As objective adults, we may know that the Trojan culture was perhaps superior to that of the Bronze Age Greeks, and that the case of the Indian was at least as strong as that of the cowboy. But such objectivity is the end product of a protracted intellectual and moral struggle, a long process of cleansing, tempering, and refining the emotions. For the child such objectivity is not yet possible, because emotions, not intellect, are in control during the early years. Our children *want* to believe that good wins out, and they *need* to believe it for their own well-being, so that they can turn into good people. It serves their developing humanity to repeat the eternal conflict of good and evil in a primitive form understandable to them, and to see that good triumphs in the end.

When play and games have firmly established the ascendancy of good in the child's mind, so that the outcome of the fight is no longer at issue, he can turn to humanitarian refinements of the original war game—the enemy becomes imaginary, and the child's attention is focused instead on the good feeling of comradeship against a common enemy. Then the issue expands to encompass no longer merely order against chaos and good versus evil but sublimation of violent emotions.

At this point the problem ceases to be whether the knight will win out over the infidel (of course he will) and becomes whether he will be able to do so with elegance, according to the protocols of the ring or of knightly virtue. Thus the game determines not merely which is stronger—id or superego, my primitive I or my socialized I—but also whether the ego can ensure the victory of the superego in ways that enhance self-respect. Good must triumph over evil—and it must do so in a way that demonstrates the value of our higher humanity. When the knight errant slays the monster, he does so to free the captive maiden. Good has prevailed, but it has prevailed for a purpose, gaining erotic (id) satisfaction as part of the bargain. Thus ego and superego combine to promise the id a reward if it does their bidding. Serving the good is reinforced by the motivating force of a higher purpose.

When a child acts out this understanding, he begins to appreciate a lesson that cannot be taught to him in a purely didactic fashion: to fight evil is not enough; one must do so in honor of a higher cause and with knightly valor—that is, according to the rules of the game, the highest of which is to act with virtue. This, in turn, will promote self-esteem, a powerful incentive to further integrate id, ego, and superego—to become more civilized.

Practical Piaget: Helping Children Understand

Sharon L. Pontious

SHARON L. PONTIOUS, RN, PhD, is an assistant professor at the University of Texas at El Paso, College of Nursing. She was a doctoral student at New Mexico State University, Las Cruces, N.M., when she wrote this article.

Three-year-old Jennifer asks her mother over and over for a cookie while her mother is busy cooking. In frustration, her mother finally says, "Hold your horses, Jennifer!" Bewildered, Jennifer goes to her room, picks up three toy horses and holds them. Five minutes later, Jennifer goes back to the kitchen still holding the horses and meekly asks, "Mommy, now can I have a cookie?"

A child's thought process and understanding are not just a miniature version of adult thought and comprehension. Nursing care of children is incomplete, unhelpful, and even detrimental for the child when nurses use adult language to explain or teach. Inappropriate language may be used because adults overestimate the child's level of understanding(1).

According to Jean Piaget, the way the child perceives the world, thinks, reasons, and uses language is qualitatively different in each of the following age groups: sensorimotor (birth to 2 years), preoperational (2 to 7 years), concrete operations or school-age (7 to 12 years), and formal operations, or adolescence through adulthood (12 years on)(2). While it is true that children between five to eight years demonstrate certain characteristics of both preoperational and concrete thinking stages, I have found children up to age seven consistently understand words used for preoperational children. However, they do not understand many of the words appropriate for children in the concrete stage, especially when they are sick.

According to Piaget, children up to age two learn about the world by using their senses—seeing, hearing, touching, smelling, and tasting. Thought and reasoning are just emerging, and language is primarily imitation of what is heard. For this reason, verbal explanations are virtually meaningless.

Preoperational child. The child from 2 to 7 years sees the world strictly from his own viewpoint. Unable to even imagine there is another way of seeing things, he refuses to accept anyone else's view.

In this phase, the child perceives the gross outward appearance of objects, but is capable of seeing only one aspect of that object at a time. For example, if a nurse tells this age child, "Take a red pill," he will be confused because he can focus on either "red" or "pill," but not both. At this stage, thinking is literal, concrete, and in the present(3). Thus, drinking a medicine to take away a headache makes no sense because the medicine goes into the stomach, not the head.

The preoperational child thinks in absolutes; things are either good or bad, right or wrong, white or black, hurt or don't hurt. He reasons transductively: going to the hospital is punishment for something he did.

He also is animistic: all objects have life and intention. For example, Jami, a 4-year-old, says the moon follows him wherever he goes because it is his best friend. The combination of transductive reasoning, animism, and focusing on only one aspect of an object leads a child of this age to say, "The needle (on the syringe) makes people well." He does not understand that there is medicine in the syringe and that it is pushed into the body through the needle(4). In addition, the 2- to 7-year-old judges people by the consequence to himself. Thus, the nurse who gives him a shot is bad because she hurt him.

Since a child of this age believes everyone sees things the way he does and knows what he's thinking, language is used only to mimic adult conversation. Many of the words he uses, he does not yet truly understand.

Concrete operational child. This child views the world more objectively and realistically, since he's capable of understanding others' viewpoints and, later in this stage, of combining them with his own. The 7- to 12-year-old's thinking, although still concrete, now includes the past and some future as well as the present. These children are more flexible and able to see some things as relative. For example, a shot hurts a little or a lot. They are aware of reversibility; objects are the same, even if their form changes: an ice cube is water.

The school-age child can classify objects or experiences by using multiple characteristics—size, color, shape, and mass—in order to conceptualize a whole from parts. Piaget has determined that school-age children learn conservation (aspects of objects remain constant when form changes) in the following sequence: first, size in terms of substance and length, then weight, then volume; then they learn time in terms of past, present, and future, then time independent of perceptual data, such as meals, TV programs, light, or darkness; then one hour is one hour, whether it seems to go quickly or slowly(3,5).

The school-age child's ability to reason is the beginning of deductive logic. He solves problems by trial and error(3). Animism is now only used for natural phenomena beyond his manipulation. For example, an 8-year-old said, "The sun comes up to give people warmth and light."

The 7- to 12-year-old judges actions by their logical effect. He can separate the intent from the behavior and outcome. This child now will say, "The shot hurt, but it will make me feel better."

Words to the school-age child,

according to Piaget, represent reality—objects or concepts. Language is used to communicate his thoughts to others and to learn viewpoints other than his own. However, he does not thoroughly comprehend words that have multiple meanings or those which stand for things he has not yet experienced.

Applying the Theory

I find that it is best when working with children to use easily understood words that have one meaning, do not instill fear or anxiety, and are not too sophisticated for the child. Some words to avoid:

cut (implies pain to preoperational child);

incision (instills fear of bodily harm or mutilation to school-age child or adolescent);

fix (implies to a 2- to 7-year-old that he is broken like a stepped-on toy);

organs (causes confusion between functional parts of body and musical instruments);

take (implies to a 2-to 7-year old removal of something, such as TPR, blood pressure);

test (is not understood by 2-to 7-year-old, denotes pass or fail to school-age child;

dye (implies "die" to both preoperational and concrete operational children).

These words are not helpful because they increase the child's anxiety and decrease the child's understanding and/or confidence.

A nurse, preparing six-year-old Bobby for heart surgery, said, "The doctor is going to fix your heart, so you will feel better." Immediately, Bobby became very upset, crying and screaming, "Don't let him fix me." Bobby's anxiety and fear were increased because his father frequently had tried to fix his toys, but usually Bobby had to throw them away because his dad couldn't fix them. Bobby feared the doctor would be unable to "fix" him, and he, too, would be thrown away. A better explanation for a school-age child would be to say, "The doctor is going to help you so you can run and play with your friends and you won't get as tired as you do now."

Remember, the 2-to 7-year-old doesn't know anything exists inside

of him. He is concerned primarily with the consequence of the action (surgery) to him.

Instead of saying "take" your temperature or blood pressure, say, "I'm going to see how warm you are" to a preoperational child, or, "I'm going to measure how much heat your body's making" to a school-age child. Measurement of components of blood or urine can be described in like fashion.

Whereas most words with multiple meanings are confusing, especially to the preoperational child and young school-age child, truthfulness is equally important for all ages. Honesty is especially important to older school-age children because they can easily see through "white lies." For example, a seven-year-old became very fearful of an upcoming meatotomy when the nurse said, "This shot will not hurt." A painful injection followed, and then she said, "I'm going to take you for a little ride now." She should have said, "It's time to go to the operating room now." From that time on, this same child refused to let that nurse come near him.

Four-year-old Tami began crying uncontrollably immediately after she vomited. When she finally quieted down, she said, "All my stuffing came out!" She had previously seen her torn Raggedy Ann's stuffing come out. Once the problem was realized, it was explained to Tami that throwing up was OK. Her tummy was like a balloon filled with water and when it was squeezed, the water came out, but the water could be replaced. It was then explained that the tummy holds food and water, and when it feels sick, it pushes out the food and water. Tami then looked in the mirror and, as if relieved, came back smiling and saying, "All of my other stuffing is still there!"

One day, five-year-old Mona came running up to a nurse pointing to her chest and saying, "This is bumping, I can feel it. I'm sick!" Although Mona had just listened to her heart with a stethoscope, she didn't understand that the heart was under her chest skin, nor could she associate it with what she had just heard. Thus, when you try to tell a 2- to 7-year-old child about a part of him he can't see, it is necessary to use things he can see, touch, hear, and relate to concretely.

Manuel, a six-and-one-half-year-old, had a bladder infection but refused to take his medicine. To explain why he needed the medicine, he was told to pretend he could unzip the skin from his head to his toes. Pointing to his bladder, the nurse told him he would see a thing like the water-filled plastic bag that he was holding.

This bladder keeps his urine inside of it until it gets full and he urinates. But his bladder was hurting because tiny things called germs had gotten inside it. The germs made his bladder hurt. Other tiny, white things called blood cells, like the white flecks in the water in the plastic bag he held, were trying to kill the other tiny bad germs.

Manuel shook the plastic bag up and said, "Just like the good army men try to kill the bad people?" The nurse answered, "Yes!" He was then told the medicine he drank was helping the white blood cells kill the germs. Three hours later, I overheard Manuel asking the staff, "Is it time to take my medicine now?"

I find that for school-age children lifelike pictures or drawings, three-dimensional models, or the visible man doll with clear plastic skin and realistic veins, arteries, and body parts shown in proper perspective inside the body make what is inside more real. Then I relate the function of these parts to the problem or treatment a specific child is having. For example, nine-year-old Danny was frightened of IVs. He would scream whenever anyone even touched the IV. He did not know what the IV was for, where it went in his body, and thought the doctors were punishing him by taking one needle out and inserting another one. Danny was also sure the IV was going to kill him by sucking out all his blood, because he remembered seeing blood back up into the tubing.

A nurse showed him with a visible man doll that blood ran all over the body, inside little tubes. Danny asked, "Just like water in little rivers?" The nurse answered, "Yes," and using straws, she inserted an IV needle inside a straw

explaining, "This way water and sugar are added to blood."

She explained further that the water would give each of his body parts a drink to keep them from being thirsty. The sugar would give his body energy even faster than if it went into his stomach first, so the IV was a short cut. He was then reassured that the doctors and nurses did not want to hurt or punish him, but that since his stomach needed a rest (previously explained), the IV was the only way to give him water and energy.

Danny was also given an appropriate way to cope with pain. He was told he could cry when the needle was put into his vein, because it hurts, but to hold that arm still so nursing staff didn't have to stick him again. Danny proceeded to start IVs on all of his dolls. The next time Danny needed the IV

changed, he kept his arm still and told the doctors why he needed it.

To explain what a simple fracture of a leg or arm is to the 2- to 7-year-old child, I use a tissue wrapped around a tongue blade. I tell the child, "Let's pretend this tongue blade is your bone and the tissue is your skin. Now let's break the bone just like you did when you fell." Then, in front of the child, the tongue blade is broken. "OK, does this look as if it is broken?" The child says, "No." Then I hand the tissue to the child, and say, "Unwrap it and see if the tongue blade is broken." The child is surprised to find a broken tongue blade. "OK, wrap up the broken bone and let's see how it looks now." This should be repeated until the child shows understanding by asking something like, "Is this why my leg doesn't *look* broken, too?" At this point, for

Characteristics of Children's Thinking		
	Preoperational (2- to 7-year old)	Concrete (7- to 12-year old)
Perception:	views world in terms of self	views world he can manipulate or experience objectively
	refuses others' viewpoints	combines own with others' viewpoints
Thought:	classifies by one characteristic	classifies by multiple dimensions of object
	literal concrete	concrete present and past, some future
	present tense absolute	begins relativism reversibility conservation of length and substance, size, weight, volume, time
Reasoning:	transductive	begins deductive logic (trial and error problem solving)
	animism	animism only of some natural phenomenon
	judges action by consequence to self	judges action by logical effect separates cause and intent from outcome
Language:	words represent one aspect or use of object; vague and global concepts	comprehends common meaning talks to communicate thoughts and to understand others' viewpoints

school-age children, it is appropriate to show them their x-rays, pointing out where the bone is broken and then explaining how new bone grows.

The same technique can be used to explain casts and traction. For example, a nursing student made a bed and traction apparatus so a doll could be placed in traction by an eight-year-old. Bobbins were used for pulleys, dimes for weights, and heavy thread for the rope. The student explained the purpose and how each part of the apparatus worked.

Both age groups of children should be encouraged to apply casts to a doll to help them understand what a cast is made of, know that it is still their leg or arm under the cast, and to help alleviate fears of how the cast is put on.

Research has shown that effective explanation of cast application and removal must include a description of the procedure, as well as a description of what the child will see, hear, feel, and in what position he will be placed(5). Just before the child's cast is to be removed, have the doll's cast removed in the proper way so that the child sees lots of dust, feels vibrations and possible warmth on his arm or leg, and hears loud noises from the cast cutter.

The nurse can use the way a child perceives the world and his reasoning ability to encourage eating or drinking. Knowing that things appear to be larger to the 2- to 7-year-old child when they seem to take up more space, the nurse can make the same amount of food appear to be less. For example, on a small plate, push vegetables close together in a small pile, place only six to eight pieces of meat close

together, and give a whole piece of bread instead of halves or quarters.

Do not try to encourage the preoperational child to eat by saying, "You can eat all of your food just like Tommy." This age child can't classify, so Tommy is not like him at all, and, furthermore, he won't care what Tommy did, since his only concern is for himself. Instead, you might try using animism. For example, tell a five-year-old to give his hungry tummy some beef, or suggest that the peas really want to live in his tummy and he can "help them go home" by putting them in his mouth.

A 2-to 7-year-old can be encouraged to drink by giving him a short, but large-diameter, glass and filling it a third to half full. You can also tape a picture on the bottom of a clear glass because he will want to drink it empty to see the surprise. Other useful techniques are to mark lines on the cup and give him a star for each line he drinks down to or give him a funny-shaped glass or straw for drinking.

These ideas will not work with a school-age child. With the 7- to 12 year-old, you can concretely explain that in order to get strong like his favorite hero, he is going to have to give his body energy. Food is what makes his body grow and run, just as gasoline makes the car run. When his body runs out of food or the car runs out of gas, both slow down and then stop.

Children from 7 to 12 can be enticed to eat foods that appear fun to eat. For example, cut a piece of meat in a circle shape; use one large olive for each eye, a strip of crunchy carrot or bacon for the mouth, and a dot of ketchup or mustard for the nose. For breakfast serve a "Mickey Mouse" face made from three pan-

cakes, strawberries, and bacon. For lunch, serve a spaceship made from a hot dog, cheese, and potato chips. These ideas are also good for younger children.

Encourage the school-age child to participate in planning his meals by giving him two or more choices of each food group. Be sure all of the choices you offer are available. Do not serve either age group mixed food, such as casseroles or stews. The reason for this is that typically children from four to nine years believe that food sits in their stomach as it appears on their plates.

Be sure to encourage all children (any age) to first eat or drink what they need most. Children live for now, not the future, so as soon as they are full, no amount of persuasion will make them eat or drink more. However, serving small amounts of attractive, nourishing food six to eight times per day may meet the child's needs better, since he will be hungry each time.

Nurses can provide helpful total nursing care to children if they apply Piaget's theory of cognitive development in designing their explanations and teaching plans for children. Words used must be appropriate, simple, nonambiguous, and truthful.

References

1. Smith, E. C. Are you really communicating? Am.J.Nurs. 77:1966-1968, Dec. 1977.
2. Richmond, P. G. An Introduction to Piaget. New York, Basic Books, 1971.
3. Maier, H. W. Three Theories of Child Development: The Contributions of Erik H. Erikson, Jean Piaget, and Robert R. Sears and their Applications. rev. ed. New York, Harper & Row 1978.
4. Steward, M., and Regalbuto, G. Do doctors know what children know? Am.J.Orthopsychiatry 45: 146-149, Jan. 1975.
5. Piaget, Jean, and Inhelder, Barbel. Psychology of the Child. New York, Basic Books, 1969.

PROFILE
ROBERT J. STERNBERG

Three Heads are Better than One

*THE TRIARCHIC THEORY SAYS WE
ARE GOVERNED BY THREE ASPECTS OF INTELLIGENCE
AND SUGGESTS WAYS OF MAXIMIZING STRENGTHS
AND MINIMIZING WEAKNESSES.*

ROBERT J. TROTTER

*Robert J. Trotter is a senior editor at
Psychology Today.*

I really stunk on IQ tests. I was just terrible," recalls Robert J. Sternberg. "In elementary school I had severe test anxiety. I'd hear other people starting to turn the page, and I'd still be on the second item. I'd utterly freeze."

Poor performances on IQ tests piqued Sternberg's interest, and from rather inauspicious beginnings he proceeded to build a career on the study of intelligence and intelligence testing. Sternberg, IBM Professor of Psychology and Education at Yale University, did his undergraduate work at Yale and then got his Ph.D. from Stanford University in 1975. Since then he has written hundreds of articles and several books on intelligence, received nu-merous fellowships and awards for his research and proposed a three-part theory of intelligence. He is now developing an intelligence test based on that theory.

Running through Sternberg's work is a core of common-sense practicality not always seen in studies of subjects as intangible as intelligence. This practical bent, which stems from his early attempts to understand his own trouble with IQ tests, is also seen in his current efforts to devise ways of teaching people to better understand and increase their intellectual skills.

Sternberg got over his test anxiety in sixth grade after doing so poorly on an IQ test that he was sent to retake it with the fifth-graders. "When you are in elementary school," he explains, "one year makes a big difference. It's one thing to take a test with sixth-graders, but if you're taking it with a bunch of babies, you don't have to worry." He did well on the test, and by seventh grade he was designing and administering his own test of mental abilities as part of a science project. In 10th grade he studied how distractions affect people taking mental-ability tests.

After graduating from high school, he worked summers as a research assistant, first at the Psychological Corporation in New York, then at the Educational Testing Service in Princeton, New Jersey. These jobs gave him hands-on experience with testing organizations, but he began to suspect that the intelligence field was not going anywhere. Most of the tests being used were pretty old, he says, and there seemed to be little good research going on.

This idea was reinforced when Sternberg took a graduate course at Stanford from Lee. J. Cronbach, a leader in the field of tests and measurements. Intelligence research is dead, Cronbach said; the psychometric approach—IQ testing—has run its course and people are waiting for something new. This left Sternberg at a loss. He knew he wanted to study intelligence, but he didn't know how to go about it.

*P*SYCHOLOGY COMES OUT OF EVERYDAY EXPERIENCES; WORKING WITH GRADUATE STUDENTS SUGGESTED THAT HIS THEORY OF INTELLIGENCE WAS INCOMPLETE.

telligence, but he didn't know how to go about it.

About this time, an educational publishing firm (Barron's) asked Sternberg to write a book on how to prepare for the Miller Analogies Test. Since Sternberg had invented a scheme for classifying the items on the test when he worked for the Psychological Corporation, which publishes the test, he was an obvious choice to write the book. Being an impecunious graduate student, he jumped at the chance, but he had an ulterior motive. He wanted to study intelligence and thought that because analogies are a major part of most IQ tests, working on the book might help. This work eventually led to his dissertation and a book based on it.

At this stage, Sternberg was analyzing the cognitive, or mental, processes people use to solve IQ test items, such as analogies, syllogisms and series. His research gave a good account of what people did in their heads, he says, and also seemed to account for individual differences in IQ test performance. Sternberg extended this work in the 1970s and in 1980 published a paper setting forth what he called his "componential" theory of human intelligence.

"I really thought I had the whole bag here," he says. "I thought I knew

what was going on, but that was just a delusion on my part." Psychology comes out of everyday experiences, Sternberg says. And his own experiences—teaching and working with graduate students at Yale—gave him the idea that there was much more to intelligence than what his componential theory was describing. He brings this idea to life with stories of three idealized graduate students—Alice, Barbara and Celia.

Alice, he says, is someone who looked very smart according to conventional theories of intelligence. She had almost a 4.0 average as an undergraduate, scored extremely high on the Graduate Record Exam (GRE) and was supported by excellent letters of recommendation. She had everything that smart graduate students are supposed to have and was admitted to Yale as a top pick.

"There was no doubt that this was Miss Real Smarto," Sternberg says, and she performed just the way the tests predicted she would. She did extremely well on multiple-choice tests and was great in class, especially at critiquing other people's work and analyzing arguments. "She was just fantastic," Sternberg says. "She was one of our top two students the first year, but it didn't stay that way. She wasn't even in the top half by the time she finished. It just didn't work out. So that made me suspicious, and I wanted to know what went wrong."

The GRE and other tests had accurately predicted Alice's performance for the first year or so but then got progressively less predictive. And what became clear, Sternberg says, is that although the tests did measure her critical thinking ability, they did not measure her ability to come up with good ideas. This is not unusual, he says. A lot of people are very good

analytically, but they just don't have good ideas of their own.

Sternberg thinks he knows why people with high GRE scores don't always do well in graduate school. From elementary school to college, he explains, students are continuously reinforced for high test-smarts. The first year of graduate school is similar—lots of multiple-choice tests and papers that demand critical thinking. Then around the second year there is a transition, with more emphasis on creative, or synthetic, thinking and having good ideas. "That's a different skill," Sternberg says. "It's not that test taking and critical thinking all of a sudden become unimportant, it's just that other things become more important."

When people who have always done well on tests get to this transition point, instead of being continually reinforced, they are only intermittently reinforced. And that is the kind of reinforcement most likely to sustain a particular type of behavior. "Instead of helping people try to improve their performance in other areas, intermittent reinforcement encourages them to overcapitalize on test-smarts, and they try to use that kind of intelligence in situations in which it is not relevant.

"The irony is that people like Alice may have other abilities, but they never look for them," he says. "It's like psychologists who come up with a theory that's interesting and then try to expand it to everything under the sun. They just can't see its limitations. It's the same with mental abilities. Some are good in certain situations but not in others."

The second student, Barbara, had a very different kind of record. Her undergraduate grades were not great, and her GRE scores were really low by Yale standards. She did, however, have absolutely superlative letters of recommendation that said Barbara was extremely creative, had really good ideas and did exceptional research. Sternberg thought Barbara would continue to do creative work and wanted to accept her. When he was outvoted, he hired her as a research associate. "Academic smarts," Sternberg says, "are easy to find, but creativity is a rare and precious commodity."

Sternberg's prediction was correct. In addition to working full time as a research associate she took graduate classes, and her work and ideas proved to be just as good as the letters said they would be. When the transition came, she was ready to go. "Some of the most important work I've done was in collaboration with her," Sternberg says.

Barbaresque talent, Sternberg emphasizes, is not limited to psychology graduate school. "I think the same principle applies to everything. Take business. You can get an MBA based on your academic smarts because graduate programs consist mostly of taking tests and analyzing cases. But when you actually go into business, you have to have creative ideas for products and for marketing. Some MBA's don't make the transition and never do well because they overcapitalize on academic smarts. And it's the same no matter what you do. If you're in writing, you have to have good ideas for stories. If you're in art, you have to have good ideas for artwork. If you're in law.... That's where Barbaresque talent comes in."

The third student was Celia. Her grades, letters of recommendation and GRE scores were good but not great. She was accepted into the program and the first year, Sternberg says, she did all right but not great. Surprisingly, however, she turned out to be the easiest student to place in a good job. And this surprised him. Celia lacked Alice's super analytic ability and Barbara's super synthetic, or creative, ability, yet she could get a good job while others were having trouble.

Celia, it turns out, had learned how to play the game. She made sure she did the kind of work that is valued in psychology. She submitted her papers to the right journals. In other words, Sternberg says, "she was a street-smart psychologist, very street-smart. And that, again, is something that doesn't show up on IQ tests."

Sternberg points out that Alice, Barbara and Celia are not extreme cases. "Extremes are rare," he says, "but not good. You don't want someone who is incredibly analytically brilliant but never has a good idea or who is a total social boor." Like all of us, Alice, Barbara and Celia each had all three of the intellectual abilities he described, but each was especially good in one aspect.

After considering the special quali-

ties of people such as Alice, Barbara and Celia, Sternberg concluded that his componential theory explained only one aspect of intelligence. It could account for Alice, but it was too narrow to explain Barbara and Celia. In an attempt to find out why, Sternberg began to look at prior theories of intelligence and found that they tried to do one of three things:

Some looked at the relation of intelligence to the internal world of the individual, what goes on inside people's heads when they think intelligently. "That's what IQ tests measure, that's what information processing tasks measure, that's the componential theory. It's what I had been doing," Sternberg says. "I'd take an IQ test problem and analyze the mental processes involved in solving it, but it's still the same damned problem. It's sort of like we never got away from the IQ test as a standard. It's not that I thought the componential work was wrong. It told me a lot about what made Alice smart, but there had to be more."

Other theories looked at the relation of intelligence to experience, with experience mediating between what's inside—the internal, mental world—and what's outside—the external world. These theories say you have to look at how experience affects a person's intelligence and how intelligence affects a person's experiences. In other words, more-intelligent people create different experiences. "And that," says Sternberg, "is where Barbara fits in. She is someone who has a certain way of coping with novelty that goes beyond the ordinary. She can see old problems in new ways, or she'll take a new problem and see how some old thing she knows applies to it."

A third kind of theory looks at intelligence in relation to the individual's external world. In other words, what makes people smart in their everyday context? How does the environment interact with being smart? And what you see, as with Celia, is that there are a lot of people who don't do particularly well on tests but who are just extremely practically intelligent. "Take Lee Iacocca," Sternberg says. "Maybe he doesn't have an IQ of 160 (or maybe he does, I don't know), but he is extremely effective. And there are plenty of people who are that way. And there are plenty of people going

*A*CADEMIC SMARTS ARE EASY TO FIND, BUT CREATIVITY IS RARE AND PRECIOUS.

around with high IQ's who don't do a damned thing. This Celiaesque kind of smartness—how you make it in the real world—is not reflected in IQ tests. So I decided to have a look at all three kinds of intelligence."

He did, and the result was the triarchic theory. A triarchy is government by three persons, and in his 1985 book, *Beyond IQ*, Sternberg suggests that we are all governed by three aspects of intelligence: componential, experiential and contextual. In the book, each aspect of intelligence is described in a subtheory. Though based in part on older theories, Sternberg's work differs from those theories in a number of ways. His componential subtheory, which describes Alice, for example, is closest to the views of cognitive psychologists and psychometricians. But Sternberg thinks that the other theories put too much emphasis on measuring speed and accuracy of performance components at the expense of what he calls "metacomponents," or executive processes.

"For example," he explains, "the really interesting part of solving analogies or syllogisms is deciding what to do in the first place. But that isn't isolated by looking at performance components, so I realized you need to look at metacomponents—how you plan it, how you monitor what you are doing, how you evaluate it after you are done. [See "Stalking the IQ Quark," *Psychology Today*, September 1979.]

"A big thing in psychometric theory," he continues, "is mental speed. Almost every group test is timed, so if you're not fast you're in trouble. But I came to the conclusion that we were really misguided on that. Almost ev-

THE TRIARCHIC THEORY

Componential

Alice had high test scores and was a whiz at test-taking and analytical thinking. Her type of intelligence exemplifies the componential subtheory, which explains the mental components involved in analytical thinking.

Experiential

Barbara didn't have the best test scores, but she was a superbly creative thinker who could combine disparate experiences in insightful ways. She is an example of the experiential subtheory.

Contextual

Celia was street-smart. She learned how to play the game and how to manipulate the environment. Her test scores weren't tops, but she could come out on top in almost any context. She is Sternberg's example of contextual intelligence.

ILLUSTRATIONS BY JEAN TUTTLE

A BIG THING IN IQ TESTING IS SPEED, BUT ALMOST EVERYONE REGRETS SOME DECISION THAT WAS MADE TOO FAST.

eryone regrets some decision that was made too fast. Think of the guy who walks around with President Reagan carrying the black box. You don't want this guy to be real fast at pushing the button. So, instead of just testing speed, you want to measure a person's knowing when to be fast and when to be slow—time allocation—it's a metacomponent. And that's what the componential subtheory emphasizes."

The experiential subtheory, which describes Barbaresque talent, emphasizes insight. Sternberg and graduate student Janet E. Davidson, as part of a study of intellectual giftedness, concluded that what gifted people had in common was insight. "If you look at Hemingway in literature, Darwin in science or Rousseau in political theory, you see that they all seemed to be unusually insightful people," Sternberg explains. "But when we looked at the research, we found that nobody

seemed to know what insight is."

Sternberg and Davidson analyzed how several major scientific insights came about and concluded that insight is really three things: selective encoding, selective combination and selective comparison. As an example of selective encoding they cite Sir Alexander Fleming's discovery of penicillin. One of Fleming's experiments was spoiled when mold contaminated and killed the bacteria he was studying. Sternberg says most people would have said, "I screwed up, I've got to throw this out and start over." But Fleming didn't. He realized that the mold that killed the bacteria was more important than the bacteria. This selective encoding insight—the ability to focus on the really critical information—led to the discovery of a substance in the mold that Fleming called "penicillin." "And this is not just something that famous scientists do," Sternberg explains. "Detectives have to decide what are the relevant clues, lawyers have to decide which facts have legal consequences and so on."

The second kind of insight is selective combination, which is putting the facts together to get the big picture, as in Charles Darwin's formulation of the theory of natural selection. The facts he needed to form the theory were already there; other people had them too. But Darwin saw how to put them together. Similarly, doctors have to put the symptoms together to figure out what the disease is. Lawyers have to put the facts together to figure out how to make the case. "My triarchic theory is another example of selective combination. It doesn't have that much in it that's different from what other people have said," Sternberg admits. "It's just putting it together that's a little different."

A third kind of insight is selective comparison. It's relating the old to the new analogically, says Sternberg. It involves being able to see an old thing in a new way or being able to see a new thing in an old way. An example is the discovery of the molecular structure of benzene by German chemist August Kekule, who had been struggling to find the structure for some time. Then one night he had a dream in which a snake was dancing around and biting its own tail. Kekule woke up and realized that he had solved the puzzle of benzene's structure. In essence, Sternberg explains, Kekule could see the relation between two very disparate elements—the circular image of the dancing snake and the hexagonal structure of the benzene molecule.

Sternberg and Davidson tested their theory of insight on fourth-, fifth- and sixth-graders who had been identified through IQ and creativity tests as either gifted or not so gifted. They used problems that require the three different kinds of insights. A selective-encoding problem, for example, is the old one about four brown socks and five blue socks in a drawer. How many do you have to pull out to make sure you'll having a matching pair? It's a selective-encoding problem because the solution depends on selecting and using the relevant information. (The information about the 4-to-5 ratio is irrelevant.)

As expected, the gifted children were better able to solve all three types of problems. The less gifted children, for example, tended to get hung up on the irrelevant ratio information in the socks problem, while the gifted children ignored it. When the researchers gave the less gifted children the information needed to solve the problems (by underlining what was relevant, for example), their performance improved significantly. Giving the gifted children this information had no such effect, Sternberg explains, because they tended to have the insights spontaneously.

Sternberg and Davidson also found that insight skills can be taught. In a five-week training program for both gifted and less gifted children, they greatly improved children's scores on insight problems, compared with chil-

dren who had not received the training. Moreover, says Sternberg, the gains were durable and transferable. The skills were still there when the children were tested a year later and were being applied to kinds of insight problems that had never appeared in the training program.

Sternberg's contextual subtheory emphasizes adaptation. Almost everyone agrees that intelligence is the ability to adapt to the environment, but that doesn't seem to be what IQ tests measure, Sternberg says. So he and Richard K. Wagner, then a graduate student, now at Florida State University, tried to come up with a test of adaptive ability. They studied people in two occupations: academic psychologists, "because we think that's a really important job," and business executives, "because everyone else thinks that's an important job." They began by asking prominent, successful people what one needs to be practically intelligent in their fields. The psychologists and executives agreed on three things:

First, IQ isn't very important for success in these jobs. "And that makes sense because you already have a restricted range. You're talking about people with IQ's of 110 to 150. That's not to say that IQ doesn't count for anything," Sternberg says. "If you were talking about a range from 40 to 150, IQ might make a difference, but we're not. So IQ isn't that important with regard to practical intelligence."

They also agreed that graduate school isn't that important either. "This," says Sternberg, "was a little offensive. After all, here I was teaching and doing the study with one of my own graduate students, and these people were saying graduate training wasn't that helpful." But Sternberg remembered that graduate school had not fully prepared him for his first year on the job as an academic. "I really needed to know how to write a grant proposal; at Yale, if you can't get grants you're in trouble. You have to scrounge for paper clips, you can't get students to work with you, you can't get any research done. Five years later you get fired because you haven't done anything. Now, no one ever says you are being hired to write grants, but if you don't get them you're dead meat around here." Sternberg, who has had more than $5 mil-

lion in grants in the past 10 years, says he'd be five years behind where he is now without great graduate students.

"What you need to know to be practically intelligent, to get on in an environment," Sternberg says, is tacit knowledge, the third area of agreement. "It's implied or indicated but not always expressed, or taught." Sternberg and Wagner constructed a test of such knowledge and gave it to senior and junior business executives and to senior and junior psychology professors. The results suggest that tacit knowledge is a result of learning from experience. It is not related to IQ but is related to success in the real world. Psychologists who scored high on the test, compared with those who had done poorly, had published more research, presented more papers at conventions and tended to be at the better universities. Business executives who scored high had better salaries, more merit raises and better performance ratings than those who scored low.

The tacit-knowledge test is a measure of how well people adapt to their environment, but practical knowledge also means knowing when not to adapt. "Suppose you join a computer software firm because you really want to work on educational software," Sternberg says, "but they put you in the firm's industrial espionage section and ask you to spy on Apple Computer. There are times when you have to select another environment, when you have to say 'It's time to quit. I don't want to adapt, I'm leaving.'"

There are, however, times when you can't quit and must stay put. In such situations, you can try to change the environment. That, says Sternberg, is the final aspect of contextual, or practical, intelligence—shaping the environment to suit your needs.

One way to do this is by capitalizing on your intellectual strengths and compensating for your weaknesses. "I don't think I'm at the top of the heap analytically," Sternberg explains. "I'm good, not the greatest, but I think I know what I'm good at and I try to make the most of it. And there are some things I stink at and I either try to make them unimportant or I find other people to do them. That's part of how I shape my environment. And that's what I think practical intelligence is about—capitalizing on your strengths and minimizing your weaknesses. It's sort of mental self-management.

"So basically what I've said is there are different ways to be smart, but ultimately what you want to do is take the components (Alice intelligence), apply them to your experience (Barbara) and use them to adapt to, select and shape your environment (Celia). That is the triarchic theory of intelligence."

What can you do with a new theory of intelligence? Sternberg, who seems to have a three-part answer for every question (and whose triangular theory

*I*T'S REALLY IMPORTANT TO ME THAT MY WORK HAS AN EFFECT THAT GOES BEYOND THE PSYCHOLOGY JOURNALS, TO BRING INTELLIGENCE INTO THE REAL WORLD AND THE REAL WORLD INTO INTELLIGENCE.

of love will be the subject of a future *Psychology Today* article), says, "I view the situation as a triangle." The most important leg of the triangle, he says, is theory and research. "But it's not enough for me to spend my life coming up with theories," he says. "So I've gone in two further directions, the other two legs of the triangle—testing and training."

He is developing, with the Psychological Corporation, now in San Antonio, Texas, the Sternberg Multidimensional Abilities Test. It is based strictly on the triarchic theory and will measure intelligence in a much broader way than traditional IQ tests do. "Rather than giving you a number that's etched in stone," he says, "this test will be used as a basis for diag-

nosing your intellectual strengths and weaknesses."

Once you understand the kind of intelligence you have, the third leg of the triangle—the training of intellectual skills—comes into play. One of Sternberg's most recent books, *Intelligence Applied*, is a training program based on the theory. It is designed to help people capitalize on their strengths and improve where they are weak. "I'm very committed to all three aspects," Sternberg says. "It's really important to me that my work has an effect that goes beyond the psychology journals. I really think it's important to bring intelligence into the real world and the real world into intelligence."

How the brain really works its wonders

A new model of the brain is beginning to explain how it can do things the most powerful computers cannot—recognize faces, recall distant memories, make intuitive leaps. The key: Intricate networks that link together the brain's billions of nerve cells

'Imagine a block of wax. . . ." So wrote the Greek philosopher Plato more than 2,000 years ago to describe memory. Since then, scholars have invoked clocks, telephone switchboards, computers—and even a cow's stomach—in equally futile attempts to explain the mysterious workings of the brain.

But an explosion of recent findings in brain science—aided by new computer programs that can simulate brain cells in action—is now revealing that the brain is far more intricate than any mechanical device imaginable. For the first time, brain researchers are beginning to explain how the brain can call up distant memories from a vast storehouse of recollections and instantly recognize faces, odors and other complex patterns—tasks that even the most powerful electronic computers stumble over.

"For physicists, the most exciting time was during the birth of quantum mechanics earlier this century," says Christof Koch, a brain researcher at the California Institute of Technology. "We are seeing the same excitement now in neuroscience—we are beginning to get an understanding of how the brain really works."

Scientists are now coming to regard the brain as far from some kind of orderly, computerlike machine that methodically plods through calculations step by step. Instead, the new image of our "engine of thought" is more like a beehive or a busy marketplace, a seething swarm of densely interconnected nerve cells—called neurons—that are continually sending electrochemical signals back and forth to each other and altering their lines of communication with every new experience. It is in this vast network of neurons that our thoughts, memories and perceptions are generated in a cellular version of a New England town meeting.

This new view of the brain has burst into every corner of science where researchers think about thinking. Brain scientists are hoping that a comprehensive new theory of how the mind works will lead to ways to control afflictions such as epilepsy and Alzheimer's disease. Computer researchers are looking at how the brain computes in an attempt to give robots eyesight, hearing and memory and to build brainlike machines that can learn by themselves. The new model of the mind even has philosophers dusting off hoary questions about the nature of rationality and consciousness.

A meeting of minds

The revolution in understanding the brain has come about because of a marriage of two widely different fields—neurobiology and computer science—that would have been impossible a decade ago. For years, computer researchers attempting to create machines with humanlike intelligence all but ignored the complex details of the brain's anatomy. Instead, they tried to understand the mind at the more theoretical level of psychology—that is, in terms of the brain's behavior.

Neuroscientists, meanwhile, were focusing on the brain's biology, using microscopic probes to sample electrical pulses from the 100 billion neurons that make up the brain and trying to unravel the chemistry of how those neurons communicate with one another. Many neuroscientists, however, are now beginning to realize that the brain is far more than the sum of its parts. "Suppose you wanted to know how a computer worked," says Koch. "You could sample the signals at all the transistors, and you could crush some up and see what they're made of, but when you were finished you still wouldn't know how the computer operated. For

that, you need an understanding of how all the components work together."

With the recent development of inexpensive, powerful computers and the expansion of knowledge about the details of the brain's anatomy, researchers are finally teaming up with computer scientists to simulate the way neurons might join together in the vast networks that make up our mind. No one is suggesting this new approach will explain, neuron by neuron, how we fall in love or laugh at the Marx Brothers. Nor is it yet clear whether different types of neural networks are responsible for producing all the remarkable things the brain can do. But researchers are beginning to see the outlines of the brain's remarkable organization, which allows it to learn new skills, remember old events, see and hear and adapt itself to new situations.

Laboratory models of the brain—called neural networks—consist of a dozen to several hundred artificial neurons whose actions are simulated on a conventional digital computer, just as modern computers can simulate the way millions of particles of air flow around a fighter jet's wings. Just as a single neuron in the brain is connected to as many as 10,000 other neurons, each artificial neuron in a neural network is connected to many others, so that all the neurons can send signals to each other. Simple rules that mimic how actual neurons alter their communication pathways in the brain are programed into the simulations as well.

The result is a device that shares some properties with the real thing but is far easier for scientists to take apart, examine and run experiments on. "These things aren't toys," says Richard Granger, a brain researcher at the University of California at Irvine who uses neural networks to model how the brain processes smell. "These are from real brain. We put data from the lab into our model, and then we run our model to get predictions that we go back and test in the lab."

Researchers are creating neural networks that show how the brain makes general categories of odors such as cheese or fruit and distinguishes between specific odors such as Swiss or cheddar. Others are modeling the way a casual mention of a particular place or event can evoke a memory of a long-lost friend, how the brain organizes incoming signals from the eyes to give us vision and how neurons rearrange their connections to restore operations after a damaging stroke or in response to a new task.

The models are also giving researchers new insights into the dynamic process by which the brain does all these things. A neuron takes a million times longer to send a signal than a typical computer switch, yet the brain can recognize a familiar face in less than a second—a feat beyond the ability of the most powerful computers. The brain achieves this speed because, unlike the step-by-step computer, its billions of neurons can all attack the problem simultaneously.

This massive collection of neurons acting all at once makes decisions more in the manner of a New England town meeting than of a highly structured bureaucracy. The brain's freewheeling, collective style of processing information may explain why it has trouble doing mathematical computations that are easily done by a $5 calculator. But it may also be what gives the brain its enormous flexibility and the power to match patterns that are similar but not exact, draw scattered bits of visual data into a cohesive picture and make intuitive leaps.

Consider what the brain must do to recognize a smell, for example. It's unlikely that one barbecued-rib dinner will smell exactly like another or that the strength of the odor will be the same each time it is encountered. But a neural network doesn't simply check if the pattern of nerve signals coming from the ribs exactly matches any of the patterns stored in memory: Comparing patterns one by one would take far too long.

Instead, the network goes through a process analogous to a group of people debating evidence. Neurons that are highly activated by the odor signal strongly to other neurons, which in turn activate—or in some cases deactivate—others in the group, and those neurons will influence still others and feed back to the original senders. As the neurons signal back and forth, varying their levels of activity, the group as a whole evolves toward a pattern that most closely matches one in memory, a pattern that reflects fundamental similarities among the many variations of how barbecued ribs smell.

Completing thoughts

This type of interactive process may be what allows the brain to recognize patterns that are slightly different or incomplete as nonetheless belonging to the same overall group. We are able to recognize all the different kinds of things we sit on as types of chairs, for example, even though we might have a hard time writing down exactly what it is about them that qualifies them as such. Likewise, small bits of a memory can trigger the whole memory, even if some of the incoming information is faulty: If someone asks if you have read the latest issue of *U.S. News & Global Report,* you still know which magazine he is talking about.

This kind of memory is possible because, just as some members of a town meeting outshout others, some neurons in a network have stronger communications pathways to their neighbors. These "rabble-rousing" neurons can have a strong influence on the way other neurons behave, and so even when only a few of them are activated, they can nudge the network in the right direction.

By simulating these processes in the lab, researchers are gaining surprising insights into how neural networks—and thus perhaps the brain itself—can perform these tasks. Granger and his colleague at the University of California at Irvine, neuroscientist Gary Lynch, used data from their lab experiments on neurons in a rat's olfactory system to create a neural-network simulation of smell recognition. The 500-neuron network was presented with groups of simulated odors, each containing variations of a general pattern such as cheese or flowers.

At first, the network responded with a unique pattern of activity for each odor. But as it processed more and more odors that were similar, those neurons that were repeatedly activated became stronger and stronger, eventually dampening the activity of other neurons that were less active. Eventually, these highly activated neurons became representatives of each category of smells: After a half-dozen samplings of the group, says Granger, the artificial brain circuit responded with the same pattern of neurons on the first sniff of any of several smells within one category. On subsequent sniffs, however, the neural network did something totally unexpected. The old pattern disappeared, and new neurons fired, creating a different pattern for each particular smell. "We're thrilled with it," says Granger. "With the first sniff, it recognizes the overall pattern and says: 'It's a cheese.' With the next sniffs, it distinguishes the pattern and says: 'It's Jarlsberg.' "

Studies of actual brain tissue are continually refining the ground rules that scientists program into these models—thus making them more realistic. One recently confirmed rule—that two neurons communicate more strongly if both have been active at the same time—has been incorporated into many neural network simulations. Often, such simple rules are enough to produce the striking result that a network will organize itself to perform a task such as smell recognition when given repeated stimuli.

Biological studies have also given some exciting confirmation that neural network models are on the right track. Recent experiments with neural networks that model vision in monkeys have also shown a surprising match with the actual biology of the brain. They may also explain how the growing brain of a fetus lays down its neural circuitry. Nearly two decades ago, Harvard University brain researchers Torsten Wiesel and David Hubel discovered that a monkey's brain has neurons that respond to very specific types of visual scenes such as spots of light or dark bars set at different angles. Yet these neurons are developed before birth—and before any light signals can influence the way they are organized.

Ralph Linsker, at the IBM Thomas J. Watson Research Center in Yorktown Heights, N.Y., has created a neural-network model of the brain's visual system that shows how the brain might be able to wire itself up spontaneously to do such tasks. Linsker's network consists of several sheets of neurons arranged in layers, with groups of neurons in one sheet connected to various individual neurons in the sheet above it. To make his network evolve, Linsker uses the same neuroscientific rules that govern how synapses in the brain increase their communication strength when the neurons they connect to are active at the same time.

Linsker starts his model off with random connections between neurons and feeds in a random pattern of stimulation to the neurons at the bottom layer. Just as with Granger's smell model, the network's simple reinforcement rules cause the neurons to organize themselves into groups for specific tasks. By the time the input pattern has worked its way up through the network, the neurons in the top layer have formed into specialized clusters that respond the most when bars of light with specific orientations are presented—just like the specialized neurons in the monkey's brain.

The network organizes itself because each neuron in one layer gets information from a committee of neurons in the layer below it. Those neurons that "vote" with the majority get reinforced while lone dissenters lose their influence. "As the

Accounting for emotion

Fear, happiness and love are all part of the mind's machinery

The brain does a lot more than think. At the very moment you're deciding which chess piece to move or whether to invest in stocks or mutual funds, your brain is regulating your body temperature, making sure you're standing upright, telling you if you're hungry or thirsty and reacting to the attractive man or woman in the next room.

And when it comes to fear, anger, love, sadness or any of the complicated mixtures of feeling and physical response we label emotions, a loose network of lower-brain structures and nerve pathways called the limbic system appears to be key. Researchers stimulating various parts of this system with an electrode can produce strong responses of pleasure, pain or aggression. A cat, for example, will hiss, spit and growl when an electrical probe is inserted at a specific spot in the hypothalamus—a part of the limbic system that is also involved in regulating appetite and other bodily functions. An electrode in another region of the hypothalamus triggers pleasure so intense that a rat will press a bar thousands of times to receive it—and die from starvation in the process.

The most recent research, however, indicates that the experience of emotion has less to do with specific locations in the brain and more to do with the complicated circuitry that interconnects them and the patterns of nerve impulses that travel among them. "It's a little like your television set," says neuroscientist Dr. Floyd Bloom of the Scripps Clinic and Research Foundation. "There are individual tubes, and you can say what they do, but if you take even one tube out, the television doesn't work."

A mugger or a cat? Researchers have been able to find out the most about primitive emotions like fear. Seeing a shadow flit across your path in a dimly lit parking lot will trigger a complex series of events. First, sensory receptors in the retina of your eye detect the shadow and instantly translate it into chemical signals that race to your brain. Different parts of the limbic system and higher-brain centers debate the shadow's importance. What is it? Have we encountered something like this before? Is it dangerous? Meanwhile, signals sent by the hypothalamus to the pituitary gland trigger a flood of hormones

alerting various parts of your body to the possibility of danger, and producing the response called "fight or flight": Rapid pulse, rising blood pressure, dilated pupils and other physiological shifts that prepare you for action. Hormone signals are carried through the blood, a much slower route than nerve pathways. So even after the danger is past—when your brain decides that the shadow is a cat's, not a mugger's—it takes a few minutes for everything to return to normal.

Fear is a relatively uncomplicated emotion, however. Sophisticated sentiments—sadness or joy, for example—are much harder to trace. And even primitive feelings such as fear or rage involve complex interactions with the higher parts of the brain— witness our ability to become fearful or angry about an abstract idea. The mechanics of these interactions are still out of reach, but the same computer models scientists are using now to understand thinking may someday shed light on emotions as well.

by Erica E. Goode

---

group develops a consensus," explains Linsker, "the mavericks get kicked out."

New connections

New studies have shown that, even though much of the brain's wiring is laid down in the womb, the connections between neurons can also be rearranged during adulthood. It is likely, in fact, that your brain has made subtle changes in its wiring since you began reading this article. More-substantial rearrangements are believed to occur in stroke victims who lose and then regain control of a limb. Michael Merzenich of the University of California at San Francisco first mapped the specific areas in a monkey's brain that were activated when different fingers on the monkey's hand were touched, then trained the monkey to use one finger predominantly in a task that earned it food. When Merzenich remapped the touch-activated areas of the monkey's brain, he found that the area responding to signals from that finger had expanded by nearly 600 percent. Merzenich found a similar rearrangement of processing areas when he simulated brain damage caused by a stroke.

Researchers Leif Finkel and Gerald M.

Edelman of Rockefeller University were able to duplicate these overall phenomena in a neural network when they applied a simple rule to the behavior of small groups of neurons. Groups of neurons were set up to "compete" for connections to the sensory nerves. The researchers found that when they gave one group an excessive input—analogous to training the monkey to use a particular finger— that patch grew in size. When that input was stopped, the patch grew smaller.

Working in concert

The biggest impact of neural networks may be in helping researchers explore how the brain does sophisticated information processing. Even though scientists can record signals from the individual neurons in the brain that might be involved in such a task as tracking an object with the eyes, they still don't know how the brain puts those millions of signals together to perform the computation. But because a neural network can adapt its connections in response to its experiences, it can be trained to learn sophisticated brainlike tasks—and then researchers can examine the artificial brain in detail to get

clues to how a real brain might be doing it.

In one study, for example, a neural network helped researchers explain how the brain is able to judge the position of an object from signals sent by neurons connected to the eyes. Brain scientists Richard Andersen of the Massachusetts Institute of Technology and David Zipser of the University of California at San Diego trained a neural network to do the task by giving it data recorded from a monkey's neurons as the animal tracked an object moving in front of it. Since the researchers already knew the position of the object that the nerve signals corresponded to, they were able to "train" the network to do the task: They gave the network a series of recorded input signals and let the network adjust itself until it consistently was able to give the right answer. The researchers then examined the network to reveal the complex calculations it uses to forge all the data into the correct answer.

These experiments suggest that some extremely complex feats of perception can, at least in theory, be explained by the interaction of many neurons, each of which performs a seemingly quite simple task. Terrence Sejnowski of Johns Hopkins University, for example, created a neural network that learned to judge how much a spherical object was curved by the way a beam of light cast a shadow on it. Much to his surprise, Sejnowski found that even though the network was trained to compute the object's shape from its shading, individual neurons within the network actually responded with the most activity when he later tested the network not with curved surfaces but with bars of light. In fact, the neurons responded just like the specialized neurons in the monkey's brain discovered years ago by Hubel and Wiesel—neurons that had long been assumed to be involved in helping the brain detect the straight edges of objects, not their curvature. "My network doesn't prove that those cells in the monkey's brain are actually there to compute curvature and not edges," says Sejnowski. "But it does mean that you can't make quick assumptions about what the entire brain is doing simply by sampling what individual neurons are doing. You need to look at the system as a whole." Several neuroscientists, inspired by Sejnowski's study,

plan to investigate whether such curvature-computing cells actually exist in the brain.

The ability of neural networks to learn to simulate these brainlike tasks has also inspired researchers who are interested in creating machines that act more like real brains. While conventional computers can perform powerful feats of number crunching, they are dismal failures at doing more-brainlike operations such as seeing, hearing, and understanding speech—things we usually take for granted but that are extremely complex computationally. "The things that distinguish us from monkeys—playing chess, for example—are easy for computers to do," says Caltech's Koch. "But when it comes to doing things we share with the animal kingdom, computers are awful. In computing vision or movement, for example, no computer comes even close to matching the abilities of a fly." Engineers at the National Aeronautics and Space Administration, the Defense Department and computer companies around the world are all busily scrambling to find

the best ways to implement neural networks on computer chips.

It may be a long time, however, before anybody is able to build a machine that actually works like a brain. After all, nature has had a 7-million-year head start on engineers, and researchers have never encountered anything as complex and ingeniously designed as the 3-pound lump of tissue inside your skull.

Meanwhile, the first steps at understanding how the brain really works have already been taken. Many brain researchers now believe that the bigger mysteries of how we make choices and use language—or why some memories last forever while others fade—will inevitably yield their secrets. Even the nature of the brain's creativity, attention and consciousness may someday be revealed. "Basically, the brain is a neural network —however complicated," says Andersen. "It will take time, but we will solve it."

by William F. Allman

ILLUSTRATIONS FROM JOURNAL OF EXPERIMENTAL PSYCHOLOGY

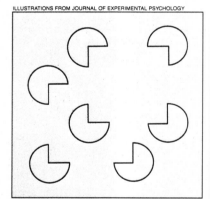 

Perceiving depth: Though the two images above have the same shapes, the right one is perceived by the brain as two overlapping squares—evidence that the brain uses multiple visual clues simultaneously to judge depth

J FRASER © 1979 OXFORD UNIVERSITY PRESS

Spiral or circle? Try tracing it with your finger. Given conflicting cues, the brain chooses one interpretation over another

Dyslexia

Mirror writing and similar problems are usually blamed on defects in visual perception, but in truth dyslexia seems to be a complex linguistic deficiency. The remedy is proper instruction in reading

Frank R. Vellutino

FRANK R. VELLUTINO ("Dyslexia") is professor and research psychologist at the State University of New York at Albany, where he is also director of the Child Research and Study Center. He has been on the Albany faculty since 1966; before that he worked for two years as a clinical psychologist at the Veterans Administration Hospital in Brockton, Mass. He earned both his master's and his Ph.D. in psychology, the latter in 1964, from the Catholic University of America. Most of the research Vellutino cites in his article was funded by the National Institute of Child Health and Human Development.

Dyslexia is a generic term that has come to refer to an extraordinary difficulty experienced by otherwise normal children in learning to identify printed words, presumably as the result of constitutional deficiencies. The condition is commonly believed to originate in the visual-spatial system. Its presence is considered to be signaled by mirror writing and letter reversal. Dyslexics, it is believed, show uncertain hand preference. Children whose first language is based on alphabetic rather than pictographic or ideographic characters are said to be particularly susceptible to the condition. Finally, dyslexia is widely considered to be correctable by means of therapies aimed at "strengthening" the visual-spatial system. Each of these perceptions, contemporary research shows, is seriously flawed.

It was through the work of the U.S. neuropsychiatrist Samuel Torrey Orton in 1925 that the deficiency first came to be perceived as lying in the visual system. Orton suggested that an apparent dysfunction in visual perception and visual memory, characterized by a tendency to perceive letters and words in reverse (*b* for *d* or *was* for *saw*), causes dyslexia. Such a disorder would also explain mirror writing. Orton further suggested that the disorder is caused by a maturational lag: the consequence of a failure of one or the other hemisphere of the brain to dominate the development of language. This last proposal, at least, was and is still a viable hypothesis.

Related hypotheses, not attributable to Orton, are that dyslexia may somehow be caused either by motor and visual defects or by eye-movement defects affecting binocular coordination, eye tracking and directional scanning. Both the concept of dyslexia as a visual problem and its presumed association with uncertain cerebral dominance still underlie many therapeutic approaches to the condition.

Working at the Child Research and Study Center of the State University of New York at Albany, my colleagues and I have begun to examine, and to challenge, common beliefs about dyslexia, including the notion that the condition stems primarily from visual deficits. Along with other researchers in this country and abroad, we have been finding that dyslexia is a subtle language deficiency. The deficiency has its roots in other areas: phonological-coding deficits (inability to represent and access the sound of a word in order to help remember the word); deficient phonemic segmentation (inability to break words into component sounds); poor vocabulary development, and trouble discriminating grammatical and syntactic differences among words and sentences. Far from being a visual problem, dyslexia appears to be the consequence of limited facility in using language to code other types of information.

To understand this definition of the disorder, one can conceive of the mind as an extremely sophisticated reference library. The library model is appropriate because recent studies seem to indicate that dyslexia is as closely aligned with the cross-refer-

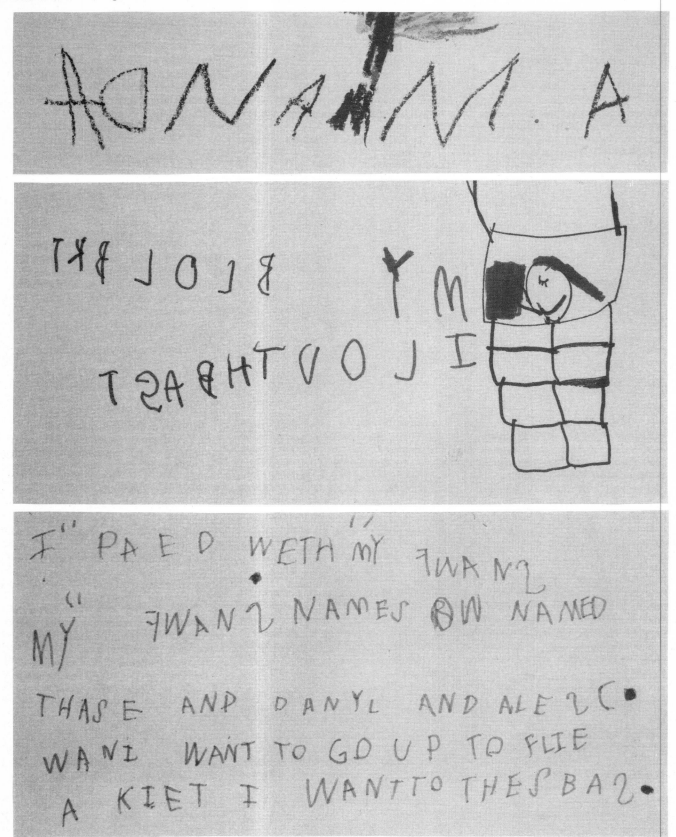

PRESUMED SIGNS OF DYSLEXIA such as mirror writing are often seen in the early stages of normal development of writing skills. At three Amanda mirror-writes her name (*top*). The habit persists when, at four, she writes about "my blanket I love the best" (*middle*). At five she writes about playing with her friends Tracy, Daniel and Alex. Mirror writing has all but disappeared; there is a clear ability to encode a desired sound by means of phonologically appropriate (if not always correct) letters (*bottom*).

encing and retrieval of coded information already stored in memory as it is with the storing and coding of new information.

The library model is based on the assumption that the processing of information to be stored in memory proceeds in stages. The first stage of processing takes place in a sensory storage system, where a replica of a given stimulus is held briefly. The second stage is believed to take place in a short-term "working" memory: a limited-capacity system in which an en-coded (transformed) version of the stimulus is available for no longer than 30 seconds. In this working memory, physical information is transformed into a more abstract symbolic representation for storage in long-term memory, which is thought to have an unlimited capacity. During the final stage of memory processing, the encoded form of the stimulus is either categorized and stored in long-term memory, discarded or inadvertently lost from working memory.

In research based on this model we have found that dyslexia is more a symptom of dysfunction during storage and retrieval of linguistic information than it is a consequence of a defect in the visual system. In one experiment poor readers in the second through sixth grades, who frequently made reversal errors, were asked to copy designs, words, scrambled letters and numerals after brief visual presentation. Afterward they were asked to name the stimuli that were actual words. We found that the poor readers could reproduce the letters in a stimulus word in the correct orientation and sequence even when they could not name the word accurately. For example, they typically copied *was* correctly but then often called it "saw." When asked to read out the letters of the words right after naming the words as wholes, they could name the letters of most words in the correct order even when they named the words incorrectly.

The inference from this experiment was clear. Errors such as calling *was* "saw" are the result of difficulties in storing and retrieving the names of printed words rather than of a dysfunction in visual-spatial processing.

This inference was reinforced by the results of a series of studies of children's ability to reproduce, from visual memory, words from an unfamiliar writing system. Groups of dyslexic and normal readers were asked to print Hebrew words and letters in the proper sequence and orientation after brief exposure. Some children who were already learning to read and write Hebrew were also tested to see how they would compare with the first two groups.

The important finding here was that the dyslexic readers did as well as the normal ones on this task—although, to be sure, neither group did as well as the children who were learning Hebrew. The result seemed to underscore the fact that when complex, wordlike symbols lacked any linguistic associates—had no meaning or sound, in effect—the visual recall of those symbols was no less difficult for the normal readers than it was for the poor readers. This implies that memory for visual symbols representing words is mediated by the linguistic properties of those words, particularly their meanings and sounds.

We found too that both groups of children who were not familiar with Hebrew manifested identical tendencies to process the Hebrew letters from left to right, suggesting that dyslexics are not inherently impaired in their ability to maintain left-to-right directionality. In other words, if dyslexics have difficulty maintaining proper di-

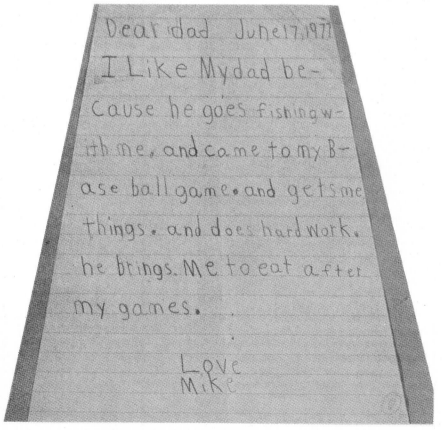

NORMAL DEVELOPMENT is evident in Father's Day cards written by a boy in first grade (*top*) and then in second grade, after he had a year of reading instruction (*bottom*).

rectionality, it is a symptom of a reading disorder rather than the cause of the disorder.

A secondary inference from this study (one that has been documented by other studies directly evaluating sensory storage in dyslexic and normal readers) is that the dyslexic readers were able to hold a memory trace for as long as the normal readers. That is to say, visual traces dissipated no more rapidly in a dyslexic's sensory memory than in a normal reader's: visual form perception seems to be comparable in the two groups.

If dyslexic readers are at least capable of perceiving and reproducing letters at roughly the same level of accuracy as normally developing readers, then the problem is again thrown back on linguistic rather than visual coding systems. Printed words can be identified either through whole-word processing based on their salient visual features, their meanings and the context in which they appear, or through "part-whole" processing based on alphabetic mapping: breaking down words into letter sounds.

Because learning to read is inherently difficult, the beginning reader must be able to adopt both strategies to identify words. If the child leans too heavily on a whole-word strategy—and does not use the sounds associated with alphabetic characters to help decode new words—visual memory is inordinately taxed; errors such as was/saw and lion/loin result. On the other hand, children who rely exclusively on alphabetic mapping—and do not use salient visual features, word meanings and context to facilitate word identification—find it hard to read fluently and have trouble understanding what they read.

The implication is clear that dyslexia may be caused, on the one hand, by highly specific linguistic deficiencies (such as vocabulary weakness or sound-mapping deficits) affecting only certain of the subskills that are necessary components of the ability to read. On the other hand, it is equally plausible that more general language deficiencies are present, which affect all subskills. A hypothesis supporting the first view has been put forward by Isabelle Y. Liberman and Donald P. Shankweiler of the Haskins Laboratories, Inc., in New Haven.

According to Liberman and Shankweiler, poor readers are not explicitly aware that spoken and printed words can be segmented into individual phonemes; this makes it hard for them to learn to identify words through alphabetic mapping and letter-sound syn-

TYPICAL MIRROR WRITING is displayed in a note from a five-year-old boy to "Grandmum." This child's development was not normal: when seen at 11, he was dyslexic, according to the English investigators Macdonald Critchley and Eileen A. Critchley.

thesis, or what is termed phonetic decoding. Poor phoneme segmentation is said to be a manifestation of a more general problem in phonological coding, characterized by the storage in memory of impoverished representations of letter sounds and word names.

Such a dysfunction could theoretically lead to difficulty not only in learning the sounds associated with given letters and combinations of letters but also in learning the names of printed words as whole entities. Words are therefore stored without complete phonological codes—file cards, in the library model. Asked to call up the proper word, the child finds that he or she has not retained enough clues to the name of the word.

Results of other studies done in our laboratory and elsewhere support the concept that deficiencies in alphabetic mapping and phonetic decoding are major factors in reading difficulties. The studies show consistently that severely impaired readers are much less proficient than normal readers in learning to use letter sounds to decode pseudowords (meaningless wordlike letter assemblages used in testing) and words they have never seen before. Such deficiencies seem to be the result of a poor grasp of phoneme values. It has also been found that kindergarten and first-grade children who have some ability to segment spoken words into syllables and phoneme-size units learn to read better than children who have little or no such ability. Perhaps the most impressive support comes from studies showing that children trained to identify phonemes have an increased ability to map alphabetically and therefore an enhanced capacity to identify printed words.

If relative lack of awareness of phonemes and deficiency in phonetic decoding are rooted in more basic difficulties in phonological coding, one might expect to find that poor readers have trouble remembering words they hear. This turns out to be the case. In a large number of studies done in our laboratory, at the Haskins Laboratories and elsewhere, poor readers did less well than normal readers when they were asked to recall lists of words they had just heard.

A number of investigators—notably Martha B. Denckla and the late Rita G. Rudel of the Columbia University College of Physicians and Surgeons—have also found that dyslexics tend to be slower and less accurate than normal readers not only in naming letters and words but also in naming common objects, colors and numerals. The performance of many dyslexics in these studies was often characterized by severe blocking: circumlocutions, long hesitations and such substitution errors as saying "dog" when confronted with a picture of a cat.

Deficiencies in vocabulary development and semantic ability in general also seem to make the identification of words difficult, as several studies have suggested. A deficiency in syntactic competence may be yet another factor. Studies in our laboratory and elsewhere show that poor readers seem to be less proficient than normal readers in comprehending sentences, particularly those that are syntactically complex; in making use of inflectional morphemes (such as -ed and -ing) to specify such things as tense and number; in distinguishing between grammatical and ungrammatical sentences; in using complex sentences in a grammatically correct way, and in making fine distinctions among abstract words, particularly such "noncontent" words as if, but and their.

In fact, considerable evidence suggests that poor readers are more deficient (compared with normally developing readers) in their ability to identify noncontent words than they are in their ability to identify content words such as dog or cat. Poor readers also seem to have difficulty using sentence context to help them identify printed words. It should be pointed out, however, that a causal connection between reading disability and deficiencies in processing the semantic and syntactic attributes of printed words has not yet been established.

The possibility that dyslexics may be impaired by deficiencies in language raises the question of whether or not they are basically impaired in auditory processing. One possibility is that their auditory sensory, or echoic, memory is deficient, which would mean the auditory trace is dissipated faster in poor readers than in normal readers. This possibility has been evaluated and dismissed by Randall W. Engle and his associates at the University of South Carolina. We have obtained similar results.

A second possibility is that poor readers are generally limited in their ability to store acoustic information in permanent memory. Susan Brady of the Haskins Laboratories evaluated this question by comparing the ability of dyslexic readers and of normal readers to remember verbal and nonverbal information (words and environmental sounds). They found that dyslexics performed below the normal readers only on the verbal-memory tasks. In terms of our memory model such results indicate that poor and normal readers are equally capable in initial-stage auditory processing and that the poor readers do not sustain any generalized deficiencies in memory for material presented auditorily. Poor readers do, however, appear to be deficient in their ability to retrieve linguistic representations stored in long-term memory. Such results seem to be consistent with linguistic-coding theories of reading disability.

Several other hypotheses propose that the cause of dyslexia lies in nonlinguistic functions, but none of them is very persuasive. One, the "attention deficit" hypothesis, relates difficulty in reading to a generalized inability to concentrate and pay attention. Some workers have found evidence associating this pattern with physiological abnormalities. Children exhibiting it, however, have difficulty with subjects other than reading and may not be representative of those whose problems are limited to reading.

Another theory relates dyslexia to deficits in "cross-modal transfer," or an inability to relate stimuli perceived through one sensory system to stimuli perceived through another system. The theory suffers both from lack of experimental support and from logical consistency; it would seem improbable that a child whose intelligence is average or above average (as is true of dyslexics) could suffer from crossmodal-transfer deficits, given the degree of cross-modal learning required to score in at least the average range on any test of intelligence.

Some other theories deserve mention. One posits deficiencies in associative learning. Another argues that dys-

lexics have trouble detecting patterns and learning "invariant relationships," such as the rules that govern alphabetic mapping or number concepts. Again, both theories stumble on the unlikelihood that an individual who scores well on intelligence tests could suffer from such pervasive handicaps. Our own studies show that poor readers who do not do well on tests of these abilities were limited by their linguistic-coding ability. When the tests did not rely on linguistic coding, the poor readers were able to improve their performance on association and rulelearning tasks.

For example, in one such study we asked a group of poor and normal readers to learn to associate pairs of novel visual symbols with two-syllable nonsense words; each symbol always represented the same nonsense syllable [see illustration on page 121]. The child's task was to learn to say the twosyllable nonsense word when presented with the pair of symbols representing that word. To help remember the whole word the child was told to try to remember the individual symbol that represented a given syllable. After practice on this task the symbols were rearranged; the child was then shown the rearranged set of symbols to see if he or she could transfer, or generalize, the symbol-syllable units learned initially to associate the rearranged pairs. This is analogous to learning letter sounds in cat, ran and fan and relying on the sounds to decode "new" words such as fat, rat and can.

A second group of poor and normal readers was given a similar transferlearning task, but instead of learning visual-verbal associates these subjects learned to associate and transfer visual pairs; again one of the symbols in a stimulus pair always represented one of the symbols in a response pair. One significant finding from this study was that poor readers did less well than normal readers on the visual-verbal learning tasks. The implication was that they had difficulty with both ini-

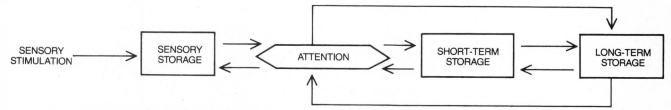

INFORMATION-PROCESSING MODEL of the stages of memory and their interrelations is diagrammed. A literal copy of a visual or auditory stimulus is held in sensory storage briefly. If the subject attends to the stimulus, it enters a short-term memory system, where it is encoded into a representation appropriate for storage in long-term memory; a stimulus that is not attended to is not encoded and is lost from memory. The long arrows suggest some of the interactions among the memory systems. For example, a person is more likely to attend to a familiar stimulus because a representation of it has been filed in long-term memory.

tial-association and transfer learning because they were impaired by inability to remember the nonsense syllables, not by inability to associate or generalize. This conclusion was verified by a second finding, which was that poor readers did as well as normal readers on the visual-visual association and transfer-learning tasks. Poor readers, then, appear to have difficulty with association and rule-learning tasks only when the tasks require them to store and retrieve the auditory representation of words and syllables.

I come, finally, to what might be called the serial-deficit theory. It holds that an inability to remember the sequence of a series of items or events underlies dyslexia. The theory presumes that the brain is equipped with a ubiquitous ordering ability, a supposition that strikes me as highly unlikely. I suspect instead that different cognitive systems have their own rules and algorithms for establishing order and sequence. If this is the case, then such theories are ruled out.

A number of investigators have taken an approach somewhat different from the ones discussed so far. They suggest that reading disability may be associated with a number of different neurologic disorders, each of which underlies one or another of the basic processes involved in learning to read. Such a view implies that there is an array of neurologic disorders characterized respectively by visual deficits, language deficits, deficiencies in cross-modal transfer and so forth. Although reading disability may result from a number of different factors, the cause in an otherwise normal child would seem to be more circumscribed. As I have argued, we believe the problem lies in the linguistic domain. The question remains open, however.

Our research and the work of investigators elsewhere has called into question other perceptions of dyslexia. First, it is important to point out that there are no well-defined reading behaviors that can clearly distinguish a dyslexic from other poor readers whose difficulties stem, for example, from limitations in experience; nor are there distinguishing clinical patterns. All poor readers have difficulty learning to identify and spell printed words, but not all would qualify as dyslexics—if the term dyslexia is used to define a very specific reading disability in an otherwise normal child. Moreover, the reversal errors said to be characteristic of dyslexics account for no more than between 20 and 25 percent of their reading errors, most of which are generalizations promoted by an im-

REAL WORDS		
THREE-LETTER	FOUR-LETTER	FIVE-LETTER
was	**loin**	**blunt**
SCRAMBLED LETTERS		
THREE-LETTER	FOUR-LETTER	FIVE-LETTER
dnv	**jpyc**	**ztbrc**
NUMBERS		
THREE-DIGIT	FOUR-DIGIT	FIVE-DIGIT
382	**4328**	**96842**

VERBAL AND NONVERBAL STIMULI were presented for half a second to poor and normal readers in the second and sixth grades; examples of each are given here. In one phase of the experiment subjects were asked to write down the words, scrambled letters or numbers from memory; in a second phase they were asked instead to name each character of a stimulus in the right order—in the case of the words, after pronouncing them. Poor readers did about as well as normal readers on the copying task but not on the naming; their problems seem to arise from deficiencies in verbal, not visual, processing.

perfect knowledge of linguistic associates (such as *cat* for *fat*, *cat* for *kitty*, *bomber* for *bombardier*).

Moreover, reversal errors can be plausibly explained without invoking spatial confusion. If, for example, a child attempts to remember the words *pot* and *top* only as whole entities and does not know the sounds of the individual letters, there is a stronger inclination to reverse them. We have verified this in experiments in which two groups of dyslexic and normal readers (in the second and sixth grades) learned to identify pseudowords constructed from a novel alphabet, which were designed to prompt reversal errors of the *was/saw* type. Children taught to identify these pseudowords by a whole-word method made many more reversal errors than children who were taught to use alphabetic mapping, but the poor readers made no more reversal errors than the normal readers. It seems clear that the spatial-confusion interpretation of reversal errors is incorrect.

The clinical significance of mirror writing is also commonly misunder-

stood. Some degree of mirror writing can be observed in normally developing readers as well as in poor ones. The tendency is quite likely a vestige of an earlier stage of development, which some poor readers take more time to transcend. It persists in these children, I think, because they find it difficult to remember both the visual-linguistic clues and the clues provided by sentence context that foster accurate judgment about the relative positions of words and letters and the direction in which they run. What causes the habit of mirror writing to persist is a lack of correct practice in writing and spelling that actually results from a child's reading problems. I suggest, in other words, that when mirror writing is observed in poor readers, it is a consequence of their reading difficulty rather than a benchmark of visual-spatial confusion.

Another common misconception is that reading problems are caused by perceptual deficits associated with motor and visual-motor defects or with ocular defects other than loss of visual acuity. If deficiencies in motor

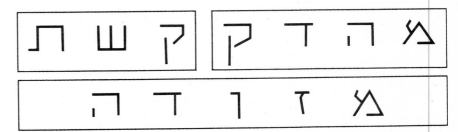

HEBREW WORDS in simplified block form were presented one at a time to three groups (second through sixth grade): poor readers and normal readers unfamiliar with Hebrew and children currently studying it. Three-, four- and five-letter words (sampled here) were shown for three, four and five seconds respectively. Subjects had to reproduce the words on paper from memory. Among those unfamiliar with Hebrew, poor readers did as well as normal readers. (Neither, of course, did as well as children studying Hebrew.)

and visual-motor development or defects in eye movements caused perceptual impairment and reading problems, one would be at a loss to explain how so many children with cerebral palsy and various visual-tracking defects become literate.

Still another misconception is that dyslexia is more prevalent in countries where the writing systems are based on an alphabet than it is in countries where the writing systems either are pictographic or are phonetically less complex. A study done by Harold W. Stevenson and his colleagues at the University of Michigan has yielded some evidence against this idea. The investigators evaluated schoolchildren in comparable cities in the U.S., Japan and Taiwan. The three groups were compared on a large battery of tests measuring school achievement as well as language and cognitive ability. The results yielded no evidence that the

writing systems of Japan and China preclude reading disability. The fact that the Japanese and Chinese languages respectively contain characters representing syllables and entire words seemed to make no difference.

Since dyslexia appears to be commonly associated with brain dysfunction and the brain's ability to store and retrieve information, it is necessary to consider whether constitutional factors—genetic and/or neurologic—contribute to the condition. There are actually no diagnostic criteria that enable one to distinguish clearly between constitutionally and experientially derived origins of reading disability. There is one highly suggestive piece of evidence, however: boys who are impaired in reading outnumber girls who are so impaired by ratios ranging from 4:1 to 10:1.

Taken together with results from de-

velopmental studies showing that boys are in general less capable than girls on language and language-related tasks, such ratios could be taken as support for both constitutional and language-deficit theories of dyslexia. Boys may be either genetically less well endowed with linguistic capabilities than girls are or more vulnerable to neurologic defects affecting language development than girls are. Then, if it is true that reading disability is caused by limitations in language ability, constitutionally derived language deficits would result in a higher incidence of the disorder in boys than in girls.

Support for a genetic basis for dyslexia comes from a small number of familial and twin studies carried out over the years. The early studies in the literature were poorly controlled, but they consistently showed that reading disability occurs more often in near relatives than in the population at

RULE-LEARNING TASKS were given to poor and normal readers in the fourth, fifth and sixth grades. The visual-verbal task *(left)* involved novel visual symbols, each representing a syllable. Subjects were trained *(top)* to match each pair of symbols (stimuli) with the correct response, a two-syllable nonsense word. Then they were shown a transfer series *(bottom)*, in which the symbols and words were reordered, and were asked to say the new words. In the visual-visual task *(right)* visual responses were substituted for the verbal ones. Now each stimulus-response pair was a pair of two-symbol "words." The response dis-

play presented the stimulus *(gray)*; the subject was trained to select the correct response from among a series of five pairs. The asterisks designate the correct responses. Again the symbols had been reordered in the transfer series, but the correspondence between individual stimulus and response symbols was maintained. The poor readers failed to perform as well as the normal readers only on the visual-verbal task, which suggests that they had difficulty learning rules only when the experiment required them to make a verbal response.

large, that reading disability occurs more often in twins than in siblings and that reading disability has a much higher concordance rate in monozygotic (identical) twins than in dizygotic twins. These findings have been verified recently in a more highly controlled study conducted by John C. DeFries and his associates at the Institute for Behavioral Genetics of the University of Colorado at Boulder.

Perhaps an even more exciting finding, from research done by the Boulder group with Shelly D. Smith, is the tentative localization of a particular gene on chromosome 15 in members of families in which there is a history of reading disability. Once a gene that may be responsible for a specific attribute has been localized on a specific chromosome, geneticists are in a position to find the mechanisms whereby the gene gives rise to the attribute. The finding could be a significant breakthrough in the study of dyslexia, but it has not yet been replicated.

A number of investigators of brain function have also begun to study the etiology of dyslexia, and their early findings are promising. A Boulder team headed by David W. Shucard compared dyslexic and normal readers on measures evaluating electrophysiological responses to auditory and visual stimulation. The group's major finding has been that electrical activity in response to reading is characterized in dyslexics by greater amplitude in the left hemisphere than in the right hemisphere, whereas the opposite pattern emerges in normal readers.

A novel technique recently exploited in this area of inquiry is "brain electrical activity mapping" (BEAM), developed by Frank H. Duffy and his colleagues at Children's Hospital in Boston. The BEAM technique produces topographic maps that are based on brain functions, not brain structures. Duffy and his associates have now obtained evidence that left-hemisphere functioning in dyslexics is qualitatively different from that in normally developing readers. The differences were particularly prominent in adjacent regions of the left parietal and temporal lobes, areas of the brain known to support speech, language and related linguistic activities.

Finally, I should mention the neuroanatomical studies conducted by Albert M. Galaburda and his colleagues at the Harvard Medical School. Two types of anatomic anomalies were disclosed by postmortem analysis of the brains of several male dyslexic subjects. First, there was a consistent absence of the standard pattern of brain

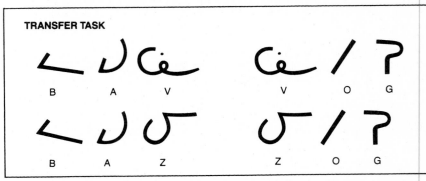

PSEUDOWORDS were shown to two groups of second- and sixth-grade children (both poor and normal readers) who had been given a week's intensive instruction in reading by one of two methods: training in "alphabetic mapping," designed to make them sensitive to the sounds of individual letters, or training in "whole-word meaning," in which pictures of imaginary animals were associated with such nonsense syllables as "zab." For the experiment both groups were shown pseudowords (top), made up of novel alphabetic characters, representing the nonsense syllables presented in the whole-word-meaning training. Children learned to say the right nonsense word when it was presented with its pseudoword—just what a child learning to read new words must do. In essence, the alphabetic mappers had to learn to read the pseudowords without benefit of meaning; the children trained in whole-word meaning had to learn to read them without benefit of alphabetic mapping. After several learning trials, subjects were presented with the same characters in reverse order (bottom) and asked to learn the names of the new pseudowords. Both the poor and the normal readers who had been trained in alphabetic mapping made very few reversal errors, whereas children trained in whole-word meaning made many such errors.

asymmetry in language regions: atypically, the left-hemisphere regions were no more highly developed than the right-hemisphere regions. Second, in the cerebral cortex of the language-related areas there were multiple sites where the microarchitectural arrangement and position of neurons, or nerve cells, was distorted, particularly in the left hemisphere.

Since the absence of asymmetry resulted from the excessive development of right-hemisphere areas that are normally smaller, Galaburda has argued that both the anatomical anomaly and the anomalous neuronal architectures may reflect interference with the normal developmental process whereby undesirable neurons and their connections are eliminated. Presumably the oddity in microanatomical organization of the language areas of the brain

would explain some linguistic deficits seen in the reading disorder.

The late Norman Geschwind hypothesized that the variations from the standard pattern of cerebral lateralization observed by Galaburda were significantly linked to disorders involving an immune dysfunction. He suggested that common developmental mechanisms acting during fetal life might lead to the abnormal development of parts of the immune system and also to anomalous asymmetrical development of the brain; he proposed an important role for the male sex hormone testosterone in these mechanisms. Although testosterone was postulated to interfere with the normal development of language areas, Geschwind also suggested that it may lead to the superior development of certain brain areas involved in spatial visualization and visual-motor coordination. If this suggestion is verified, the failure of investigators to obtain evidence for visual-deficit theories of dyslexia would be at least partly explained.

The nonlinguistic theories of dyslexia discussed here have each given rise to remedial techniques designed to correct the cognitive deficits implicated by the theories. For example, visual-deficit theories have spawned remedial approaches designed to improve visual perception, such as optometric training to facilitate binocular coordination, eye tracking and so on. Similarly, cross-modal and serial-memory theories have given birth to a variety of remedial exercises designed to improve these functions, presumably through the direct stimulation of certain brain centers.

If the logical and empirical arguments against the theories supporting these approaches are valid, then the approaches themselves must be seriously questioned. The fact is that research evaluating the efficacy of these approaches for the remediation of reading disabilities has produced no convincing evidence to support them—and a considerable amount of evidence against them. In any case, not enough is yet known about how the brain works to enable anyone to devise activities that would have a direct and positive effect on neurologic functions responsible for such basic processes as visual perception, cross-modal transfer and serial memory.

More conventional approaches to remedial instruction have had greater success, particularly in educational settings equipped to provide dyslexics with the type and amount of help they need. We have found that early remediation of reading difficulties is indicated. It should be based on intensive one-to-one tutoring and a balanced reading program—one that makes generous use of both the holistic/meaning and the analytic/phonetic approaches. The training in reading should be supplemented with enrichment activities to foster language development. Such a program can help a child to develop functional and independent reading skills and so remove him or her from the disabled list. A consensus is emerging among investigators that there is no substitute for direct remedial instruction in reading.

Family, School, and Cultural Influences on Child Development

Child-rearing advice has changed over the ages. In the early 1900s the view held by behaviorists was that children should be reared strictly in order to correctly shape their behavior. Post-World War II advice emphasized rearing children in a democratic and permissive atmosphere. Today, child-rearing advice seems to have struck a middle road between the strict and permissive approaches. Parents are encouraged to provide their children with ample love, to cuddle their infants, to use reason as the major disciplinary technique, and to encourage verbal interaction—all in an environment where rules are clearly spelled out and enforced. Suggestion, persuasion, and explanation have become the preferred techniques of rule enforcement, rather than spanking or withdrawal of love.

Perhaps our modern-day opinions on child rearing merely reflect the ebb and flow of advice over the ages, as discussed in "The Child Yesterday, Today, and Tomorrow." On the other hand, contemporary child-rearing advice may reflect a growing awareness of effective child rearing based upon knowledge gained from the scientific study of human development, and a growing aversion to the excessive violence, aggression, and alienation in contemporary American society. Weighted against this breath of optimism are daily reports of sexual and physical abuse and extreme neglect of children, sexism, racism, and countless instances of family stress that contribute to teenage pregnancy, suicide, and delinquency.

Parenting is not an easy task. At minimum parents must be flexible and willing to try different approaches, while constantly evaluating these approaches against "expert" opinion and against their own common sense. The marked increase in single-parent families, the decline of intergenerational families, and the increase in teenage mothers have added to the problems of parenting. These factors and others have strained the support services available for families in our society. Families that are stressed, either because of external or internal conditions, may turn to expedient rearing techniques even when they are aware of "expert" opinion. One such expediency is the use of physical punishment in order to "discipline" children. However, physical punishment is an ineffective form of discipline that teaches lack of self-control and the use of aggression in order to control someone else's behavior.

Although it is easy to bemoan the decline of the nuclear family, the fact of its decline remains. Increasing numbers of American women have sole responsibility for child care and child support; that is, most single-parent families are father-absent families. In "The Importance of Fathering," Alice Honig reviews evidence suggesting that both father absence and father presence have important consequences for sex-role development, social adjustment, and cognitive achievement. The role of the father in child development deserves special consideration in discussions of divorce, as examined in the article, "Helping Children Cope with Divorce." Although fathers are awarded custody more frequently than in the past, in most divorces child custody is still awarded to the mother. Regardless of who is awarded custody, children do not have an easy time adjusting to the divorce of their parents. Suggestions for helping children to cope often require a level of parental cooperation that may be unrealistic. In spite of these forces threatening to tear families apart, however, the influence of the family on the life of an individual remains strong and pervasive. Leonard Sagan, in his article "Family Ties," reveals that family relationships and dynamics contribute to the good health, and thus the longevity of people today.

The family is the first major socialization force in a child's life. However, as early as infancy, children become involved with schools, the second major force in socialization. Infants go to day-care centers or supplementary

dren adapt easily to school, preschool-age children may have more difficulty than older children. School expands a child's social network beyond the neighborhood peer group and often presents new social adjustment problems. In "Rumors of Inferiority," differences in performance between blacks and whites are linked to self-doubt, feelings of inferiority, and fear of intellectual abilities—byproducts of cultural racism. In the final article, Urie Bronfenbrenner draws attention to the increase in disorganized families and environments, which contribute to alienation from family, friends, school, and work.

Looking Ahead: Challenge Questions

Do you agree with the premise that sex role attitudes about marriage, parenting, and family relationships are fragile and correlate poorly with actual behavior? If not, what kind of evidence would you require in order to be convinced?

It is well documented that a loving, affectionate relationship between caregiver and infant provides the necessary conditions for the development of a healthy personality. What factors in the individual, the family, and American society interfere with the ability of some parents to foster loving, compliant, self-controlled, competent children?

How realistic are Ralph Ranieri's recommendations for helping children cope with divorce? Do you think that most divorces provide an atmosphere in which parents can set aside their conflicts and animosities sufficiently to act in the best interests of their children?

It is relatively easy to blame teen pregnancy and teen substance abuse on the disorganization of the American family. It is far more difficult to suggest effective solutions to the problems. If you were king of the world for a day or two, what changes in American society would you institute to resolve such problems as divorce, child abuse, teen pregnancy, racism, sexism, and substance abuse?

family-care homes, preschoolers attend nursery school, and older children attend public or private elementary school. In addition, many states are preparing guidelines today for introducing formal school experiences to the pre-kindergarten-age child. In "Project Day-Care," Edward Zigler's proposal for school-based day care is examined in detail. School often represents a child's first extended separation from the home and first experience with significant caregivers other than parents. Although most chil-

FAMILY TIES

The Real Reason People Are Living Longer

LEONARD A. SAGAN

LEONARD A. SAGAN is an epidemiologist at the Electric Power Research Institute, in Palo Alto, California. His book THE HEALTH OF NATIONS: TRUE CAUSES OF SICKNESS AND WELL-BEING *was recently published by Basic Books.*

WHEN MODERN MEDICINE made its debut at the Many Farms Navajo Indian community, in 1956, there was every reason to expect decisive results. The two thousand people who inhabited this impoverished and isolated Arizona settlement were living under extremely primitive conditions. Though nutrition was adequate, hygiene was poor, tuberculosis was widespread, and infant mortality rates were three times the national average. To a group of researchers from the Cornell University Medical College and the U.S. Public Health Service, the situation at Many Farms provided a perfect opportunity to introduce modern health care practices and measure the consequences. If the effort proved successful with this target population, they reasoned, it might become an example for underdeveloped communities worldwide.

Almost overnight, the Navajo settlement acquired an array of modern medical resources. The researchers set up a full-service clinic, staffed with physicians and nurses, as well as with public health consultants, a health teacher, and four Navajo health care workers. For medical emergencies, the community got a fleet of radio-equipped vehicles and a light airplane. Over the next six years, ninety percent of the Many Farms residents took advantage of the clinic. Two-thirds of them were seen at least once a year.

The result was a rapid decline in the transmission of tubercle bacillus (the agent that causes tuberculosis) and in the frequency of otitis media (an inflammation of the middle ear). Yet the population's overall health, as reflected in its mortality statistics, was virtually unchanged. Of the sixty-five deaths that occurred during the six-year study period, more than half involved infants, who made up less than four percent of the population. And, despite expert pediatric care, there was no reduction in the pneumonia–diarrhea complex that was the leading cause of childhood illness and death. In the end, the investigators were unsure whether the improved medical care had, on balance, produced any beneficial effect at all.

This outcome would be less unsettling if it were more unusual. Unfortunately, it is not unusual at all. Consider what happened in 1976, when the state legislature of North Carolina sponsored a study to determine the effects of improved maternal and perinatal health care on the state's poorer communities. Researchers identified a number of counties, similar in racial and socioeconomic characteristics, that had suffered high rates of infant mortality over the preceding decades. For the next five years, residents of some of those counties received state-of-the-art treatment at the medical centers of Duke University and the University of North Carolina while, for the purpose of comparison, similar counties were essentially left alone. As expected, infant mortality declined considerably in the areas that received the additional care, but it also declined in the areas that did not. In fact, the researchers found no significant differences between the two groups.

Similar stories can be told about much larger popula-

This article is reprinted by permission of *The Sciences* and is from the March/April 1988 issue, pp. 21-29. Individual subscriptions are $13.50 per year. Write to The Sciences, 2 East 63rd Street, New York, NY 10020 or call 1-800-THE-NYAS.

tions. When England established its National Health Service, in 1946, the country's lowest social classes had long suffered the poorest health and the shortest lives—presumably because of economic barriers to adequate health care. The new program effectively removed those barriers. Forty-two years later, however, the disparity in mortality rates remains undiminished; the life expectancy of the most affluent is almost twice that of the least affluent. The economists Lee and Alexandra Benham, of Washington University, in Saint Louis, have noted the similar failure of Medicare and Medicaid to affect mortality rates among the disadvantaged in the United States. This country's least educated classes now experience as much hospitalization and surgery as its most educated classes, yet overall health is still strongly associated with educational achievement.

What are we to make of all this? It is well known that life expectancy has risen dramatically in most societies over the past few centuries. As recently as 1900, the typical American lived only forty-nine years, and one in five children died during infancy. Today we live an average of seventy-five years, and infant mortality has declined to just ten deaths for every thousand births, or one percent. Both physicians and the public credit modern medicine for these bold achievements; we assume, almost reflexively, that people who lack expert medical attention die earlier, and that providing more care is the key to longer life.

Americans, therefore, have invested heavily in medicine. Our expenditures now total more than four hundred billion dollars a year, or eleven percent of the gross national product, the highest rate of any nation on Earth. Yet some measures of ill health, such as the rate of disability due to chronic illness among children, are on the rise. And though life expectancy continues to rise in the United States, it is rising more rapidly in countries that are spending at a lower rate. Many of those countries, including Greece, Spain, and Italy, now enjoy life expectancies greater than our own. And Japan, which leads the world in life expectancy, spends only a third of what the United States spends each year—about five hundred dollars per capita compared with fifteen hundred.

Clearly, we need to take a closer look at the relationship between our efforts at health care, on the one hand, and our actual health, on the other. If the United States is spending more on medicine than any other nation, while suffering poorer health than many, there may be something fundamentally wrong with the country's approach. The urgent questions are: What really makes people healthy? Why do we live so much longer than our ancestors and so much longer than the world's remaining premodern peoples? If medicine is not the source of this blessing, we would do well to find out what is—and to direct our medical and public health efforts accordingly.

THERE IS NO DENYING that modern medicine has accomplished much of value. It has done a great deal to alleviate suffering, and many treatments—including surgery for burns, bleeding, abdominal obstructions, and diabetic coma—undoubtedly save lives. Anything that saves lives would presumably contribute to overall life expectancy. But most therapy is not aimed directly at prolonging life. Rare is the patient for whom death would be the price of missing a doctor's appointment. Moreover, any medical procedure involves some risk; there is always a chance that the patient will have an adverse or fatal reaction to a given treatment—be it surgical, pharmaceutical, or even diagnostic. If treatments were administered only when patients stood to benefit, the net effect on mortality rates might be positive. But physicians have a well-documented tendency to overdo a good thing. And because there are no clear guidelines governing the use of most remedies, the cost of such zeal is that the benefits gained by those who require a particular treatment are often outweighed by the adverse effects on those who receive it unnecessarily. Thus, while such major medical advances as antibiotics, immunization, coronary bypass surgery, chemotherapy, and obstetric surgery all have saved lives, it is impossible to demonstrate that any of them has contributed significantly to overall life expectancy.

The introduction of antibiotics into clinical medicine is generally viewed as the turning point in mankind's war against infectious disease. Clearly, such illnesses as typhoid, cholera, measles, smallpox, and tuberculosis no longer claim lives at the rate they did during the nineteenth century. The decline began at different times in different nations, but it was under way in Scandinavia and the English-speaking countries by the mid-nineteenth century, roughly a hundred years before the first antibiotic drug, penicillin, became available, during the Second World War. By the time streptomycin, isoniazid, and other such agents came into wide use, during the 1940s and 1950s, death rates from the eleven most common infectious diseases had dwindled to a mere fraction of their nineteenth-century levels. Antibiotics did, for a time, hold tremendous therapeutic powers, and had they been used in moderation, they might have remained potent weapons against infection. But overuse has largely destroyed their effectiveness.

The indications that we rely too heavily on antibiotics are myriad. In 1973, scientists at the University of Wisconsin at Madison concluded, after reviewing the findings of other researchers, that enough antibiotics are manufactured and dispensed each year in the United States to treat two illnesses of average duration in every man, woman, and child in the country. The evidence suggests, however, that only once in five to ten years does the average individual experience an infection, such as meningitis or tuberculosis, that antibiotics might help control. The drugs are routinely prescribed for colds and flu, even though there is no evidence they have any effect on such viral ailments, and are given out like vitamins in many hospitals. In one recent survey of hospital patients, the internist Theodore C. Eickhoff, of the University of Colorado Medical Center, in Denver, found that thirty percent were receiving antibiotics—though only half of those receiving the drugs showed signs of infection. Other findings suggest that patients who might actually benefit from an antibiotic frequently receive the wrong one, or an incorrect dose.

One outcome of this overreliance on penicillin and the other so-called wonder drugs is that many bacteria have, through natural selection, become resistant to them, and infections that were easily controlled thirty years ago no longer respond well to treatment. Both gonococcus, the pus-producing bacterium that causes the most common venereal disease, and pneumococcus, a bacterium frequently associated with lobar pneumonia, now show resistance to various antibiotics. And in hospitals, overall infection rates are on the rise. A 1985 study, published by Robert W. Haley and his colleagues at the Centers for Disease Control, concluded that hospital-acquired infections occur in almost six out of every one hundred patients, thereby producing a national toll of four million infections a year, and that this rate is increasing by two percent annually.

Immunization is another therapy widely believed to have reduced death rates from infectious disease. But studies indicate that the use of vaccines and their ostensible benefits are largely unrelated. There is no question that the smallpox vaccine, for one, is effective when properly administered. Historical records show, however, that the number of people dying of smallpox was already falling when the vaccine first became available in Europe, during the early nineteenth century. True, smallpox mortality continued to drop as the vaccine became more accessible, but so did the death rates associated with infectious diseases for which vaccines had *not* been developed. The parallel decline in mortality from typhoid and tuberculosis prompted speculation that the smallpox vaccine was somehow protecting people from those infections, too. But there was never any basis for such a conclusion. A more reasonable inference is that deaths from all three illnesses were declining on account of some other factor.

As in the case of smallpox, vaccines for polio, whooping cough, measles, and diphtheria are effective at protecting individuals from these diseases. As a result, they not only save lives but spare many people permanent disabilities. But the question, for our purposes, is whether such vaccines have caused a significant decline in overall mortality, and the evidence indicates they have not. The historical record shows that death rates for childhood diseases started falling before the vaccines became available, and there is no evidence that forgoing such vaccines shortens people's lives. When concern about the risks associated with the diphtheria vaccine led English physicians to stop administering it during the late 1970s, for example, there was a sharp increase in the incidence of the disease, yet diphtheria mortality barely changed.

LIKE INFECTIOUS DISEASE, cardiovascular illness seems to pose a less dire threat to most of us today than it has in the past. Coronary artery disease appears to be waning both in incidence and in deadliness. A twenty-six-year study of Du Pont Company employees found that the number of people afflicted with the disease declined by twenty-eight percent between 1957 and 1983. Other studies indicate that the rate at which Americans are killed by it fell from about three hundred and fifty for every hundred thousand in 1970 to just two hundred and fifty for every hundred thousand in 1985. If such outcomes could be attributed to medical intervention, they would indeed rank as major accomplishments. But here, as with infectious disease, the link between treatment and health is elusive—whether the treatment is directed at preventing the disease or curing it.

Consider the results of the Multiple Risk Factor Intervention Trial, or "Mr. Fit." In this study, a team of investigators from twenty-two health research centers randomly divided a sample population of nearly thirteen thousand men, aged thirty-five to fifty-seven, into two groups. For the next seven years, members of one group continued to receive routine care from their private physicians, while the other group participated in a therapeutic program to reduce the risk of coronary artery disease. Physicians supervised and monitored efforts to have them avoid smoking, reduce the amount of cholesterol in their diets, and control their blood pressure, using medication if necessary. At the end of the treatment period, the subjects who received the extra medical attention had indeed cut back on cigarettes and cholesterol, and they exhibited less hypertension. But they did not end up living any longer than the subjects who simply went about their business. In fact, their death rate from all causes (41.2 deaths for every thousand subjects) was slightly *higher* than that of the control group (40.4 deaths). The reasons for this failure were not readily evident; the researchers speculated that the ill effects of antihypertensive drugs may have outweighed any benefits derived from the program. Whatever the explanation, such results confirm that the recent decline in death from cardiovascular disease probably is not the fruit of preventive medicine.

If efforts at prevention have not caused the decline, might it reflect the advent of better therapeutic techniques, such as coronary bypass surgery? Saving lives was not the original intent of this operation when surgeons began performing it, during the late 1960s; bypassing portions of coronary arteries that had become partially clogged with fatty deposits was viewed as a way of alleviating the chest pain that accompanies such blockage. But when the operation was found to be effective for that purpose, physicians began touting it as a therapeutic measure—and even a preventive treatment for patients without symptoms—despite an utter lack of clinical evidence. Today coronary bypass is one of America's most commonly performed surgical procedures. Roughly two hundred thousand Americans undergo the operation each year, at a total cost of some five billion dollars. Yet only rarely does it contribute to anyone's survival. A study published in 1983 by the National Institutes of Health concluded that bypass surgery prolongs the life of roughly one bypass patient in ten but that it appears to add nothing to the life expectancies of the other nine.

IS CANCER TREATMENT, another major focus of modern medicine, perhaps the secret of our increased life expectancy? One might guess that it has made a contribution; after all, the average interval between the diagnosis of a malignancy and the death of the patient has increased considerably in recent years. Indeed, the percentage of cancer patients surviving at least five years rose from 38.5 percent in 1973 to 40.1 percent in 1978, an improvement of almost one percent a year. Regrettably, it does not necessarily follow that people with cancer are living longer, let alone that chemotherapy or surgical treatments are extending their lives. Many scientists speculate that earlier detection of the disease merely has created an illusion of increased survival.

There is evidence that physicians are diagnosing cancer at earlier stages of development, thanks largely to more frequent checkups and better diagnostic technology. But there is no indication that earlier treatment has improved patients' overall survival rates. In fact, for some forms of cancer, there is hard evidence that it has not. In one study, sponsored by the National Cancer Institute and published in 1984, a population of adult male smokers, all presumably at high risk of developing lung cancer, was divided into two groups. One group received only annual chest X rays; members of the second group underwent frequent X rays and had their sputum examined regularly for cancer cells. Not surprisingly, there were many more diagnoses of lung cancer among the closely monitored subjects. And because their malignancies were usually detected and treated at early stages, their survival rates from the time of diagnosis were impressive. Even so, the numbers of lung cancer deaths in the two groups were nearly identical. In short, the participants in the early-detection program gained no apparent advantage: they were no less likely to suffer recurrences of the disease, or to die of it, than were members of the control group.

If modern treatments were, on balance, helping cancer patients survive, those patients would be dying at a later average age, and this, in turn, would reduce the average person's chances of dying of cancer at any age. But age-adjusted cancer mortality has not declined at all in the United States during the past fifty years. The death rates have changed for particular forms of cancer (lung cancer mortality has increased, whereas deaths from stomach cancer have declined), and it is possible that treatment has played a role in some of the success stories. The relatively rare cancers of childhood, for example, seem to respond well to treatment. But such situations are the exception, not the rule. Most therapies are introduced without ever being thoroughly evaluated for effectiveness, and they are embraced by physicians and patients who are understandably eager to try anything.

Radical mastectomy, the standard treatment for breast cancer throughout most of this century, is a good example. Studies have shown that patients who have this operation —the mutilating removal of the breast and its underlying tissues—do not, as a group, live any longer than patients who undergo the less radical lumpectomy (removal of the cancerous mass only). In one study, published in 1985 in *The New England Journal of Medicine*, nearly two thousand breast cancer patients were randomly assigned to receive one treatment or the other. Those who had the traditional mastectomy died earlier. Still, most U.S. physicians continue to perform the more extensive operation, and some even recommend it as a preventive measure for women whose cystic (lumpy) breasts place them at a theoretical risk of developing the disease.

The point is not that cancer treatment is never justified, only that it has had no discernible effect on the overall survival rates of cancer patients, let alone the life expectancy of the general population. Indeed, no cancer treatment, however successful, could do much to increase life expectancy, for the disease does little to reduce it. Cancer strikes mostly among the aged. It has long been estimated

that even if it were totally preventable or curable, the increase in U.S. life expectancy would be less than two years. Given that life expectancy has increased by twenty-five years during this century, it is impossible that the treatment of cancer has made much of a difference.

If modern medicine cannot be credited with taming infectious disease, cardiovascular illness, or cancer, one might expect to find that it has at least improved the odds that mothers and infants will survive the birth process. Cesarean section has undoubtedly contributed to the rapid decline in maternal mortality during this century. Like so many other medical procedures, however, it is now so grossly overused that it may be costing as many lives as it saves.

The maternal mortality rate had already dwindled to less than one death in ten thousand deliveries by the mid-1970s. Yet, since then, births by cesarean section have increased by three hundred percent in the United States. To confirm that this trend is not making childbirth any safer, one need only consider the survival statistics for societies that rely less heavily on surgical delivery. In 1965, the rate of cesarean births at Ireland's National Maternity Hospital, in Dublin, was equal to that in the United States—about five percent. Since then, the U.S. rate has climbed to twenty percent, but the rate in Dublin has remained stable, and perinatal mortality has fallen faster there than it has in the United States. The Netherlands, meanwhile, which enjoys one of the lowest perinatal and maternal mortality rates in Europe, also has one of the lowest rates of obstetric surgery.

IT SEEMS CLEAR that modern medicine, whatever it has done to save or improve individual lives, has had little effect on the overall health of large populations. Still, the fact is that life expectancy has increased spectacularly during the nineteenth and twentieth centuries. What else might explain such a change? There is no question that sanitation and nutrition, the other factors most often cited, have been beneficial. But neither of these developments accounts fully for the mystery at hand.

It is true that, toward the end of the nineteenth century, improvements in sanitation coincided with a decline in mortality from various infectious diseases in Europe and America. But there is no evidence of a cause-and-effect relationship. Sanitation worsened in many major cities during the Industrial Revolution, as the prospect of work drew hordes of immigrants from rural areas. Rotting meat, fish, and garbage were heaped in the streets of New York and London, and overflowing privies were still far more common than modern toilets in many crowded neighborhoods. Amazingly, though, mortality rates from infectious disease fell steadily over the same period.

Another problem with the sanitation argument is that the *incidence* of infection decreased little during the nineteenth century. What did decline was the frequency with which infections sickened or killed people. As recently as 1940, long after tuberculosis had ceased to be a major health threat, skin tests showed that ninety-five percent of all Americans were still being infected with tuberculosis bacteria by age forty-five. Yet the vast majority managed to fight it off. Even today, most of the microorganisms that caused so much disease and death in pre-modern times, particularly among children, are omnipresent in the environment. No amount of sanitation could eliminate them, for they are passed directly from one person to another. They exist harmlessly, for the most part, both in and on our bodies.

Could it be that improved nutrition has strengthened our resistance? This idea does not withstand scrutiny, either. If eating well were the key to long life, then the most privileged families of old Europe, who enjoyed better nutrition than their contemporaries, should also have enjoyed longer lives. But they died young (as did the first American settlers, for whom the threat of starvation was not a particular problem). Moreover, there is no evidence that the specific dietary changes that are associated with modernization have even been advantageous. Indeed, it is arguable that, on balance, those changes have been harmful.

In the United States—where modernization has been associated with less physical activity, and with increased consumption of white bread, cookies, doughnuts, alcohol, and red meat from fattened animals—an estimated twenty to twenty-five percent of adult men are overweight. Diet and inactivity are not the only factors that contribute to obesity, of course, but they clearly count. In one recent study, the University of Toronto anthropologists Andris Rode and Roy J. Shephard monitored body fat and physical fitness among members of an Eskimo community during a ten-year period of rapid modernization. They found that the community's adoption of a modern diet, along with its increased use of snowmobiles and snow-clearing equipment, accompanied a significant increase in body fat and decreases in several measures of fitness. If these Eskimo follow the usual pattern, modernization will bring about a net increase in life expectancy. But if their overall health improves, it will have improved *despite* the changes in diet and physical activity, not because of them.

I T IS, IN A WORD, impossible to trace the hardiness of modern people directly to improvements in medicine, sanitation, or diet. There is an alternative explanation for our increased life expectancy, however, one that has less to do with these developments than with changes in our psychological environment. We like to imagine that preindustrial peoples endured (and endure) less stress than we do—that, although they may have lacked physical amenities, they spent peaceful days weaving interesting fabrics and singing folk songs. But the psychic stresses of the simple life are, in fact, far greater than those experienced by the most harried modern executive. It is one thing to fret over a tax return or a real estate deal, and quite another to bury one's children, to wonder whether a fall's harvest will last the winter, or to watch one's home wash away in a flood.

To grow up surrounded by scarcity and ignorance and constant loss—whether in an African village or a twentieth-century urban slum—is to learn that misery is usually a consequence of forces beyond one's control and, by extension, that individual effort counts for naught. And there is ample evidence that such a sense of helplessness is often associated with apathy, depression, and death—

whether in laboratory animals or in prisoners of war. The experimental psychologist Martin E. P. Seligman, of the University of Pennsylvania, has designed some remarkable studies to simulate in dogs the experience of helplessness in humans. His classic experiment involved placing dogs in a box in which they could avoid electric shocks by jumping over a barrier upon the dimming of a light. Naïve dogs quickly learned to avoid shocks entirely, leaping gracefully over the barrier whenever the light dimmed. But Seligman found that dogs responded differently if, before being placed in the box, they were confined and subjected to shocks they could not escape. Those dogs, having learned that effort is futile, just lay down and whined.

In many ways, the experiences and reactions of the second group resemble those of people raised in poverty, a shared feature of most premodern societies. Modernization, through such mechanisms as fire departments, building codes, social insurance, and emergency medical care, has cushioned most of us against physical, psychic, and economic disaster. But, more important, it has created circumstances in which few of us feel utterly powerless to control our lives. We now take for granted that we are, in large part, the masters of our own destinies, and that in itself leaves us better equipped to fight off disease.

How did this happen? What are the sources of this sense of personal efficacy and self-esteem? No institution has been so changed by modernization as the family. Until the late eighteenth century, it existed primarily as an economic unit; marriages were arranged for the purpose of preserving property, and children were viewed as a cheap source of labor or a hedge against poverty in old age. Beating and whipping were favored, even among royalty, as tools for teaching conformity and obedience. Then, during the Enlightenment, the standards and goals of child rearing began to change. If children were going to survive in a disorderly and unpredictable world, philosophers began to argue, they could not rely passively on traditional authority; they needed reasoned judgment. And if children were going to develop such judgment, they needed affection and guidance, not brute discipline. It was only gradually, as these ideas took root, that childhood came to be recognized as a special stage of life, and that affection and nurturing replaced obligation and duty as the cohesive forces among family members.

During the nineteenth century, as the upper classes came to view children as having needs of their own rather than serving the needs of the family—and, accordingly, started having fewer of them—their infant and childhood mortality rates began to fall. And as the trend toward smaller families spread to the lower social classes, theirs fell, too. It is unlikely that this was just coincidence, for family size is an excellent predictor of childhood survival even today. Young children of large families continue to suffer more infections, more accidents, and a higher overall mortality rate than the children of small families, regardless of social class. Indeed, as the Columbia University sociologist Joe D. Wray demonstrated in 1971, the effects of family size can outweigh those of social class: an only child in a poor family has about the same chance of surviving the first year of life as a child who is born into a professional-class family but who has four or more siblings.

Why should this be so? One explanation, supported by various lines of evidence, is that the children of small families are strengthened in every way by the extra nurture they receive from their parents. During the past forty years, studies have demonstrated that infants develop poorly, even die, when they are provided food and physical necessities but are denied intimate contact with care givers. In one experiment, orphans placed in an institution at an early age were separated into two groups. Members of one group stayed in the institution while the others were placed with foster parents. At the end of the first year, the children placed in foster homes were better developed, both mentally and physically, than those who received institutional care. And even after the institutionalized children were assigned to foster homes, they remained less developed than their counterparts for a number of years.

Other studies have produced even more arresting evidence. In 1966, Harold M. Skeels, of the National Institute of Mental Health, reported on an experiment that gauged the long-term effect of individual care on retarded institutionalized children. One group of children received routine institutional care, which is often physically adequate but emotionally sterile, while the other children were moved to a special ward to be cared for individually by retarded women. After three years, most of the children in the first group had lost an average of twenty-six IQ points, whereas those in the second group had *gained* an average of twenty-nine points. The differences were even more pronounced thirty years later. None of the children who received routine care had made it past the third grade, and most remained institutionalized. By contrast, many of those cared for by foster mothers had completed the twelfth grade and gone on to become self-supporting.

W E ARE ONLY BEGINNING to understand the mechanisms linking emotional and physical health (the endeavor has of late given rise to a new branch of medicine, known as psychoneuroimmunology). But whatever the connection, the fact stands that the affection and security associated with the modern family are the best available predictors of good health. In the end, it matters little whether sanitation, nutrition, and medical care are crude or sophisticated; children who receive consistent love and attention—who grow up in circumstances that foster self-reliance and optimism rather than submission and hopelessness—are better survivors. They are bigger, brighter, more resistant, and more resilient. And, as a result, they live longer.

It is ironic, in the light of this, that we continue to fret over the quality of our food and the purity of our environment, to spend billions of dollars on medical procedures of no proven value, and to pay so little attention to the recent deterioration of the American family. The divorce rate in the United States, though it appears to have leveled off during the past few years, has increased enormously since the 1950s, from less than ten percent to more than twenty percent today. The number of children being raised by single parents has doubled during the past decade alone, and divorce is not the only reason. Another ominous development is the rise in pregnancy among unwed teenagers. For whites, the rate increased from eight percent in 1940 to twenty percent in 1970, and to thirty percent in 1980. The problem is even worse among blacks, sixty percent of whom are now born out of wedlock. That this, in itself, constitutes a serious health problem is plain when one considers that fetal and infant death rates are twice as high for illegitimate children as for legitimate ones, and that a teenaged mother is at least seven times more likely than an older mother to abuse her child.

All of this suggests that good health is as much a social and psychological achievement as a physical one—and that the preservation of the family is not so much a moral issue as a medical one. Unless we recognize the medical importance of the family and find ways to stop its deterioration, we may continue to watch our health expenditures rise and our life-spans diminish. We will waste precious resources on unnecessary treatments, while ignoring a preventable tragedy.

The Child Yesterday, Today, and Tomorrow

David Elkind

The child is a gift of nature, the image of the child is man's creation. It is the image of the child, rather than nature's gift, that determines educational practice in any historical epoch. And the image of the child, man's creation, is as often wrong as it is correct. Wrong images are more powerful and more easily grasped than true ones. In the present as in the past, our task as educators of young children is not simply to be true to nature's gift, but also to fight against the false images that, in any age, threaten the healthy education of young children.

Images of the past

The image of the child in antiquity was that of young citizen who had to be educated by the laws and culture of society. The children of Babylon went to school at age 6 and even poor children learned to read and write except that their books were bricks and their writing tools

*This article is, with a few minor changes, the address Dr. Elkind gave at NAEYC's November 1986 Annual Conference at the Opening General Session, Washington Hilton Hotel. **David Elkind** is NAEYC's President.*

a reed and damp clay. Children in ancient Greece played with go-carts and dolls, and at the age of 7 boys went to school. In ancient Rome, women had a more equal place and both boys and girls went to school where the discipline was strict and where they learned to write with a stylus and wax tablet.

During the Middle Ages, children fared less well and the prevailing image of the child was that of chattel, or piece of property consistent with the ideology of serfdom. The medieval castle was no place for a child, built as it was for defense rather than for comfort. The children of serfs worked and lived with the animals. Discipline was strict and punishment harsh. In England, there was a brief, golden era for children during the reign of Good Queen Bess. And during this era, the faithful nanny begins to appear in folklore and literature.

Toward the end of the 17th Century, the struggle between Cavaliers and Puritans was reflected in their quite dissimilar images of children. The Cavaliers held a mixed image of the child as part nuisance, part plaything. In contrast, the Puritans constructed an image of the child as one tainted with original sin. "Your child," wrote James Janeway, "is never too young to go to hell."

In this country the images of children changed with our rapidly changing society. In colonial times children were seen as financial

assets who could help work the farm or be apprenticed out of the home at an early age. The children of slaves were an extreme example of this, but they were not the only children who labored from dawn to dark. With the industrial revolution, children, especially the children of immigrants and the poor, came to be seen as cheap factory workers until the cruelty of child labor was made public. The ensuing social reform movement transformed the image of the child from one of cheap factory labor to one of apprentice to factory work. Instead of being sent to the factory, children were sent to school to prepare them to work in factories. School bells, like factory whistles, signaled the beginning and the end of the school day. And children, like their parents, carried lunch pails to be opened at the noon whistle.

As we see, there have been many different images of children, some of which were more beneficial to child health, welfare, and education than others. And there have always been those who, at any given point in history, have been critical of the image of the child current at that time. Often this criticism took the form of an attack on parents and upon parenting, but in fact it was an attack upon the then "accepted" image of the child. A review of these attacks upon the images of the child that were raised in earlier times is instructive. It tells us that the image of the child at any point in history never goes unchallenged and that the challengers in the past, as today, often come from the ranks of early childhood educators.

The criticism of prevailing images of the child has a long history. For his ideal Republic, Plato wanted children to be raised by professional child caretakers, and St. Augustine proclaimed, "Give me other mothers and I will give you other worlds." Rousseau's opening statement in *Emile* to the effect that everything is good as it comes from the hand of the Maker and deteriorates in the hands of man, is an indictment of the image of the child as a young savage who had to be socialized.

Pestalozzi and Froebel did not criticize parents directly, but did believe that parents needed to be given a truer image of the child that would result in more healthy childrearing practices. Parent education was an important component of early childhood education practiced by Pestalozzi and Froebel. Pestalozzi's book, *How Gertrude Teaches Her Children,* which is subtitled *An Attempt To Help Mothers Teach Their Own Children,* reflects this emphasis upon training parents. The same theme was repeated in Froebel's *The Education of Man* and in his *Songs for Mothers and Nursery Songs.*

Their successor Maria Montessori never criticized parents either, but she had less faith in parent education than her predecessors. Like Plato she wanted children reared by professionals, not by parents. For her, childrearing was too important a task to be left to untrained parents whose image of the child gave too little credit to their budding intellecual powers.

In the past, the prevailing image of the child that dictated childrearing and education was determined by a complex of social, economic, and cultural factors that may have had little or nothing to do with the natural child. And since early times, there have been critics of the prevailing conception of the child. These critics fought to replace the false image of the child with a truer one that would provide for a healthier, happier, and more productive child life.

Images of the present

Historically, predominant images of the child were derived from the prevailing political, social, or religious ethos. What is remarkable about modern images of the child is that they are, or are said to be, scientific in origin. Unfortunately, their scientific origin has not rendered them any more valid than those that had social, political, or religious derivations. In some ways, the scientific origin of some of the contemporary images of the child makes them even more difficult to combat than previous images. I want now to usurp the role of critic and review and comment upon three modern images of the child that have contributed to what I call miseducation, namely putting children at risk for no purpose.

The sensual child

The advent of Freudian psychology gave rise to the image of the sensual child. In this view, the child was "polymorphous perverse" in the sense of having the whole gamut of sexual instincts and proclivities that were once reserved to adults. In Freudian terms, children whose sexual instincts were unduly repressed were destined to become neurotic. The childrearing and educational implications of this image of the sensual child were straightforward. Children had to be allowed to express themselves, and play was the natural medium of self-expression. With adequate self-expression at home and at school, children would develop healthy personalities and their intelligence would take care of itself.

Like so many images of the child, this idea contains a partial truth. Freud made it clear that a certain amount of repression was healthy, indeed necessary, for people to live in a society. It was *excessive* repression, not repression, that produced neuroses. But that point was sometimes lost on those who fought for expression at all costs.

The malleable child

Another image of the child that has dominated contemporary thought has come from the anthropologists who were concerned with the conflict between generations. The leading writers of this genre were Kingsley Davis, Ruth Benedict, and Margaret Mead. Although they differed in detail, they were all making the same point, namely, when it comes to adapting to social change, children are plastic and adaptable whereas adults are rigid and unadaptable. Children, they argued, are better suited to social change than are adults.

Davis, for example, argued that adults are locked into the orientation they received as children and this makes it difficult, if not impossible, for them to appreciate the changed circumstances of their offspring, hence the generational conflict. Benedict said that adults are independent and children are dependent, and that it was the adult's inability to deal with the child's growing independence that was the cause of the generational conflict. And Margaret Mead argued that in a rapidly changing culture, children, who are free of ingrained habits of thought, are much better able to adapt to new and changing technologies than adults.

This image of child malleability in contrast to adult rigidity is sometimes misinterpreted. Anthropologists are talking about change in the overall society, *not* about changes within the immediate family. When a family moves, the children have more trouble with the change than adults. And, while divorce may be hard on adults, it is certainly much harder on children. Children thrive on consistency, stability, and security, while it is adults

Courtesy U.S. Department of Labor

who seek new experience and adventure. Children adapt less easily to change within the family than adults do, but the reverse image fostered by a misapplication of social scientists' ideas about change in society persists and contributes to miseducation.

The introduction of computers into early childhood education, and the teaching of programming to young children, is a direct offshoot of this malleability conception. It is simply a fact of technological development that as technology develops it requires more, rather than less, intellectual maturity. A child can use a shovel but not a power shovel; a child can use a hand saw or hand drill, but not a power saw or power drill; a child can ride a horse, but cannot drive an automobile and certainly cannot fly an airplane. The more advanced the technology, the more advanced the intelligence required to use it. Modern warfare is another example. Modern weapons require college

graduates if they are to be used properly. The modern army has no place for a Sergeant York trained with a hunting rifle. And even when a technology is easy to use, such as television, it can still be dangerous to young children.

Yet the idea that children should be programming and running computers persists despite the fact that the complexity and technological sophistication of computers is far beyond what a young child can really comprehend and master. To be used by young children, computers have to be converted into teaching machines presenting programmed learning. And programmed learning is simply boring. Exposing young children to computers in this way runs the risk that they will get turned off to computers before they have a chance to see what they can really do. It is a good example of miseducation, of putting children at risk for no purpose.

In the same way, I am often asked

about programs to inform young children about the threats of nuclear war. Presumably, children have to be exposed to this idea at an early age so they will be better prepared for a nuclear holocaust when it comes. Even if one accepts this shaky premise, it has to be recognized that the concept of nuclear war is completely foreign to young children, who do not even have a conception of biological death, much less of millions of people and the power of nuclear weapons to destroy them. Recent suggestions that young children be taught about AIDS also stem from this wrongheaded image of child malleability.

To be sure, children are fresh learners to the extent that they are not handicapped by previous ideas and concepts. But this does not mean that they are ready to learn everything and anything—far from it. Their openness to learning is limited and we need to recognize these limitations. There is a time and a place for everything and early childhood education is not the time nor the place to teach children computer programming, the threat of nuclear war, or for that matter, the dangers of AIDS.

The competent infant

Perhaps the most pervasive and most pernicious contemporary image of the child is one that has been promoted by psychologists writing in the 60s. Responding to the Civil Rights Movement, to the War on Poverty, and to the inadequacies of the educational system, many writers gave voice to a vision of childhood that would undo these wrongs and undo them at an early age. All these wrongs, it was said, could be righted if we only got to children early enough. The result was a new image of infants and young children as having much more capacity to learn academic skills than children, regardless of background, actually have. It is true that all young children have intellectual abilities and that their thinking should be encouraged but within the context of their psychological stage of development. This 60s image of the child as consumer of skills has come to haunt us in the 80s.

In his book *The Process of Education,* Jerome Bruner voiced his now famous hypothesis that you can "teach any child, any subject matter at any age in an intellectually responsible way." Bruner was really speaking to curriculum writers and probably did not fully appreciate the extent to which his hypothesis would be accepted, not as a hypothesis, but rather as a fact by the public at large. And it has also become the motto of entrepreneurs hawking flash cards to parents with the proclamation that you can teach a young child "anything."

But is it true? It is only true if you either redefine the child or redefine the subject matter. The curriculum writers of the 60s, academicians such as Max Beberman at the University of Illinois or Robert Karplus at Berkeley knew their subject matter but not young children. The curricula they designed in effect redefined the competence of children without recourse to children's actual abilities or limitations. For example, variable base arithmetic was said to be easier for children to learn than base ten arithmetic. But even parents had trouble with variable base arithmetic! It was also claimed that children would learn math better if it were introduced as a language. Instead of answering what is the sum of 2 + 2, children were asked to "Make this sentence true."

The error here came from confusing what is simple to an expert in a subject with what is simple for the novice. Simplicity is the end result of learning a skill or a discipline, not its starting point. Reading is simple once you know how, but is far from simple when you first start out. Understanding multiple base arithmetic may be simple once you know base ten, but not if you don't. Understanding the relation of lan-

guage to mathematics is simple if you have a firm grasp of language and mathematics, but not if you don't. We have to always be aware of the danger of assuming that the end point for us as adults should be the starting point for children.

The other side of Bruner's hypothesis requires redefining the subject matter. When an infant who responds to flash cards is said to be "reading" or doing "math," these subject matters have been drastically redefined. Suppose, for example, that I tell you that I can balance 100 pounds on my finger. You would not believe me. But suppose I take out a 3 × 5 card and write *100 pounds* on it. Now I put the card on my finger, and voilà, I am holding 100 pounds on my finger. Claiming to teach infants to read and do math is the same; it is a sleight-of-hand trick accomplished by redefining what is usually meant by reading and by math.

Yet people are taken in by this trickery and really believe that they are teaching their children these subjects. And this trickery has another negative fallout effect. Redefining the subject matter makes it much easier to acquire. Parents then believe that their child who is "reading" flash cards at age 2 is a budding genius. But they will be disappointed in the end. Unfortunately, making a task easier does not make children brighter.

Another contribution to the image of the competent infant came from educational psychologist Benjamin Bloom who argued from statistical summaries of IQ data that 4-year-olds had attained half of their intellectual ability and that it was incumbent upon us to impose formal learning on young children because otherwise we might lose out on this period of phenomenal mental growth. This idea that you must teach as much as possible to young children because their minds are growing so rapidly has become part of the contemporary folk wisdom and is deeply ingrained in

our contemporary image of the child.

But is it true? Bloom was talking about mental test data, not about mental growth. Because infants and young children are not good test takers, their intelligence test performance is not a good index of their later test performance. By the age of 4, however, the child is sufficiently verbal and has sufficient ability to concentrate attention, and her or his test performance is a better index of true ability. From the test score a child attains at the age of 4 you can predict with some 50% accuracy what that child's test score will be at age 17. And that is all that a child attaining half of her or his mental ability at age 4 means.

It does not mean that at age 4 the child has half of all the knowledge, skills, and values she or he will ever have. It does not mean that if a child attains an IQ of 100 at age 4 she or he will attain an IQ score of 200 at age 17. It does not mean that a child at age 4 is a better learner than she or he will be at age 17. Even if we grant that mental growth is rapid during the early years of life, it does not follow as dawn follows the night, that this calls for formal, teacher-directed learning. During periods of rapid mental growth, children seek out the stimuli to nourish themselves mentally. We serve them best by providing an environment rich in materials to observe, explore, manipulate, talk, write, and think about. You do not prune during the growing season.

Still a third writer who has contributed to the contemporary image of the competent infants is J. McV. Hunt. In his book *Intelligence and Experience* he surveyed a great deal of evidence and concluded that intelligence was malleable and not fixed, the view he attributed to professionals of the time. But no reputable psychologist ever claimed that intelligence was fixed. In 1954, in a chapter of the *Handbook of Child Psychology,* Florence Goodenough made it clear that all

Courtesy U.S. Department of Labor

the evidence supported the view that the environmental factors accounted for between 20 and 40% of an individual's IQ.

Up until the 60s, however, psychologists were mainly concerned with middle-class children who, presumably, had maximized their environmental potential. It was only when attention was turned to low-income children who had less than optimal environmental input that the significance of environmental input became a matter of concern. Consider the following analogy. Suppose you place a group of undernourished children on a full calorie, well-balanced diet. Surely such children will make significant gains in both height and weight, but similar gains will not be made by children who are already on a full calorie, well-balanced diet. The potential benefits of an improved program are always relative to the quality of the previous environment.

This idea of intellectual malleability has become common cur-

rency among parents who are being told that with the proper program of stimulation they can have a "brighter child" or that they can raise their child's IQ. Yet there is no evidence that children growing up in an environment where they are talked to, played with, and read to, and which is rich in things to look at, listen to, and explore, will derive additional benefit from prescribed exercises and symbolic materials. If anything, most middle-class children today are over- rather than understimulated.

The last contributor to the image of the competent child is not a psychologist but a historian. In his book *Centuries of Childhood*, Phillip Aries argues that childhood is a social invention and that there was no such conception in the Middle Ages when children were depicted and treated much as adults. The implication is that for the last couple of hundred years we have been coddling children and infantalizing them and ignoring their true com-

petence and abilities. This thesis fit in neatly with the other ideas about infant competence and gave it a historical dimension.

More recent historians of childhood, like Pollack, have shown Aries was wrong. Even in the times Aries was writing, diaries of parents show quite clearly that adults appreciated that children were different from adults and had to be treated differently. Sir Francis Bacon, writing in the 16th Century, even talked about the value of "allowances" and the negative effects of not giving a child a sufficient allowance, and suggested that "The proof is best when men keep their authority towards their children, but not their purse."

These four ideas, then, that a child can be taught any subject at any age, that children have half their intellectual ability at age 4 when mental growth is more rapid, that the IQ is malleable, and that childhood is an invention, all emerged in the 1960s to form a new image of child competence. Although this new image may have corrected a previous image that played down child competence, it went to the other extreme. Ideas meant to improve the conditions of low-income children have been taken over by middle-class parents and have become the rationale for much of the miseducation of young children today.

As in the past, we have not only to assert the values of child-centered early childhood education, but we must also struggle to reveal the concepts of early childhood malleability and competence for what they are, namely distortions of how young children really grow and learn.

Images of the future

Given the brief history I have just outlined, it seems reasonable to predict that the false images of children today will be replaced by equally false images tomorrow. I have no crystal ball, only a belief that history is prologue and that the image of the child at any point in history always fills the predominant parent needs and defenses of that developmental epoch. We have to ask then what the needs of future parents will be and how these will be reflected in a new image of the child.

Our society is already a service and information society with more than 70% of our population in these occupations. I believe that we will eventually get high quality child care for all those youngsters who need it and that those who care for infants and young children will have positions of respect and will be paid well. We may even have parent professionals to care for and rear other people's children. This will not happen immediately and without a great deal of hard work and pain, but I do believe we will get there.

What then? What new image will emerge when the image of the malleable competent child has run its course? What sort of image of the child will be most in keeping with the needs of tomorrow's parents? If present trends continue, it appears that parents will spend less time than ever parenting. Once parents no longer feel guilty or uncomfort-

able about this, the need for the image of child intellectual competence will diminish. In its place will emerge a new image of child social sophistication and self-sufficiency. In an information and service society these are the requisite skills. We already see hints of this in the current emphasis upon social cognition. Psychologists are eager to point out that Piaget was wrong and that infants and young children are much more socially skilled than we gave them credit for being.

And while it may be true that children are more socially proficient and self-sufficient than we may have recognized, they will not be as socially proficient as the image of social sophistication will have us believe. And the cycle will once again repeat itself, the next generation of early childhood educators will have to challenge the new image of the child as, to use the computer term that may well become the catchword of this new image, *an expert system* with respect to social interaction. The next generation will once again have to reassert the values of sound early childhood education.

Our task as early childhood educators then is never ending. Each

generation presents a new challenge and a new battle. And it is a battle that we can never really win, because each new generation is prone to the same mistakes. Yet if we do not fight it is a battle we can most assuredly lose. For those of us in early childhood education it is a battle well worth fighting and, even if we fall before our time, we can take comfort in the knowledge that there will always be others, sufficiently committed to the well-being of young children, to carry on the fight.

Bibliography

Aries, P. (1962). *Centuries of childhood.* New York: Knopf.

Benedict, R. (1938). Continuities and discontinuities in cultural conditioning. *Psychiatry, 1,* 161–167.

Bloom, B. (1964). *Stability and change in human behavior.* New York: Wiley.

Bruner, J. (1962). *The process of education.* Cambridge, MA: Harvard University Press.

Davis, K. (1940). The sociology of parent-youth conflict. *American Sociological Review, 5,* 523–525.

Freud, S. (1905). *Three essays on sexuality.* New York: Basic.

Hunt, J. McV. (1961). *Intelligence and experience.* New York: Ronald.

Mead, M. (1970). *Culture and commitment.* New York: Natural History Press/Doubleday.

Pollack, L. (1983). *Forgotten children.* Cambridge: Cambridge University Press.

The Importance of Fathering

Alice Sterling Honig

Alice Sterling Honig is Associate Professor, Department of Child and Family Studies, at Syracuse University, Syracuse, New York.

Margaret Mead once observed that fathers are a "biological necessity but a social accident." Little appreciation of the direct role and importance of fathering was expressed scarcely a quarter of a century ago by the pioneer advocate of infant attachment (to mother), John Bowlby:

In the young child's eyes, father plays second fiddle and his value increases only as the child's vulnerability to deprivation decreases. Nevertheless, as the illegitimate child knows, fathers have their uses even in infancy. Not only do they provide for their wives to enable them to devote themselves to the care of the infant and toddler, but, by providing love and companionship, they support her emotionally and help her maintain that harmonious contented mood in the aura of which the infant thrives (1958, p. 363).

Hand-in-hand with this belief in the minimal direct influence of fathers went societal views that gave approval to those fathers who were at their desks by 7:30 a.m. and who often did not return home from business until well past the bedtime of small children. Thus, society's "theory" of what constitutes a good father focused one-dimensionally on the role of provider and excluded the role of nurturer and socializer of the young.

Research has indeed sometimes tended to confirm the peripheral nature of the fathering role. Knox & Gilman (1974) questioned 102 first-time fathers regarding their preparation for, and adaptation to, fatherhood. Most fathers had had little preparation for their role. They participated minimally in the day-to-day care of the new baby. In another study of middle-class families in Boston, 43 percent of the fathers had never changed a diaper. Rebelsky & Hanks (1971) reported that new fathers spent 37 seconds per day in one-to-one verbal interaction with baby. Finally, to complete this picture of father as "second-class" parent, 40 percent of American children surveyed preferred television to their fathers (Dodson, 1979)!

Father Absence

More than six million children live in fatherless families. Much research on fathering has concentrated on the effects of father absence. The main question seems to be "How, and how much, are children harmed by growing up in a fatherless home?" Reviews of research and bibliographic searches suggest that psychosexual and emotional maladaptive functioning occurs more often in conjunction with father absence than presence (Herzog & Sudia, 1973; Honig, 1977; Lynn, 1974). Father absence has been found to be negatively related to sex role development, moral development, cognitive competence and to social adjustment.

The impact of father absence has to be assessed in terms of total family functioning under conditions of that absence. Much of the research seems to imply that when a father is missing, only that variable alone is affecting the child's development. Yet, family climate and clusters of family attributes concomitant with father absence may be far more important for a child's development than the actual number of parents present in the home. For example, in a supportive neighborhood or in a three-generation household, grandfather or older relatives may be available as masculine and fathering role models for young children.

Another aspect that is critical in reviewing the findings on fatherless children has to do with the mother's expressed attitudes toward the child and toward the absent father. A mother who derogates the absent father and/or the child will very likely cause profound self-doubt in a son and distrust of males in a daughter.

Thus, out of the father-absence studies comes a fairly consistent picture. Fathering is important.

Research on Effects of Fathers Present

Recent societal changes have triggered an upsurge of interest in fatherhood. Twenty-five years ago, only 1.5 million mothers worked, compared to 14 million today. More than one-third of all mothers of children under six work outside the home. Almost 1,400,000 children live in single-parent families headed by fathers (Mendes, 1976). A re-examaination of father roles and the importance of fathering seems then to flow from a rapid increase in working wives and mothers; increasing flexibility in divorce custody arrangements; and increasing pressure on fathers to shoulder some of the child-rearing responsibilities traditionally "relegated" to mothers.

Interviews with college-educated young men and women turned up almost unanimous agreement among these young people that the decision to parent in their lives was closely tied to an egalitarian expectation of fathers' role in sharing the care of the infant and in assuming an important role in child-rearing (Honig, 1980).

Fathers and Infants

A wide variety of recent studies focuses on fathering in the infancy period — that period formerly left to maternal ministrations. Father-infant research of the past few years has consistently shown that fathers can be quite competent. They are sensitive to infant cues and responsive to the signals of newborns. Fathers are quite as likely, if given the chance through sympathetic and facilitative hospital birthing procedures, to bond lovingly with their babies.

When low-income fathers who had not participated in the birth experience were observed in the days after delivery, they proved just as nurturant and stimulating with their infants as the mothers — if the fathers were observed alone with their babies (Parke & Sawin, 1980). The moral may be that father needs time alone with baby to build a love relationship. Mothers have traditionally always had such time together. An intimate father-infant relationship may require the same "twosome" quality that love relationships usually need to grow and deepen.

Infants 12-21 months, brought up by both parents, did not register any preference for either parent, when separation protest, vocalizing and smiles were the measures used (Kotelchuck, 1972; Lamb, 1976). In the former study,

Babies with fathering get more variety in life.

those few infants who did not relate well to father (that is, spend at least 15 seconds near him on his arrival), came from families with the lowest amount of father caregiving. Thus, it looks as if baby attachment to fathers follows a general rule that you get what you give. Fathers who spend loving attentive time with infants will have infants who attach well to them and under ordinary conditions prefer paternal company equally. Data indicate, however, that when distressed, infants may still turn to mother as primary comforter.

Strong differences are reported in the ways in which fathers interact with infants compared to mothers. "When they have the chance, fathers are more visually attentive and playful (talking to the baby, imitating the baby) but they are less active in feeding and caretaking activities such as wiping the child's face or changing diapers" (Parke & Sawin, 1980, p. 204).

Lamb reports that "when both parents are present, fathers are more salient persons than mothers. They are more likely to engage in unusual and more enjoyable types of play, and hence, appear to maintain the infant's attention more than the mothers do" (1976, p. 324). In short, "fathers seem to be more fun for babies! Fathers play different kinds of games with infants, more vigorous games" (Honig, 1979a, p. 247).

Split-screen motion picture work by Daniel Stern and by T. Berry Brazelton at Harvard Medical School reveals distinctive patterns of interaction of father and mother with baby. Mothers and infants play more reciprocal vocalization games. Baby limb movements tend to be smooth and more rounded with mothers. Fathers use more bursts of a tapping kind of touch. Babies show more angular and abrupt body movements to father's touch. What turns out, delightfully, is that the contributions of maternal and paternal touch and vocal reciprocity patterns from earliest infancy differ.

Babies with fathering get more variety in life. There is the priceless redundancy that paternal loving, extra stimulation and sensitivity to signals afford for advancing development. There is also the variation on human interaction themes and variation in gender and style that help a child grow up learning more perceptively to deal with differing patterns of adult interaction, expectancies and styles.

When their fathers have also tended to their needs, babies cope better with strangers. Babies seem to learn more skills to put them at ease socially in strange situations. Father-infant interactions may add social resilience to an infant's social repertoire. The fact that fathers, albeit limited to after-work interaction time, are salient figures in interaction with infants is well illustrated in Friendlander, et al.'s analysis of infants' natural language environments in the home (1972). Systematic analysis of tape-recorded utterances of fathers to infants revealed that one infant, raised primarily by an English-speaking mother, but talked to regularly and prompted abundantly by father in Spanish daily, was able to demonstrate good understanding of many Spanish words at one year of age.

Quality of father's interaction time, as so well predicted by Eriksonian theory for mothers, seems to count strongly for the positive special influence that fathers can have not only as attachment figures but as boosters of infant language learning.

Fathers and the Development of Intellective Achievements

Ainsworth's pioneer work in attachment has delineated the ways in which baby's secure attachment to mother by one year is related to the organization of socio-emotional behavior up to at least five years and to early achievement of developmental milestones (1979). Sroufe and his colleagues (1979) have demonstrated that securely attached infants are better problem solvers and tool users when challenged with somewhat difficult tasks as toddlers and preschoolers.

Just how involved are fathers in teaching their children, and how important is fathering for the development of child intellectual competence?

Social class and sex of child have been found as significant confounding variables in studies that attempt to assess father impact on cognitive achievement. Deal & Montgomery (1971) observed the techniques used by fathers, from professional and nonprofessional families, to teach their five-year-old sons two sorting tasks. Professional fathers verbalized more, used more complete sentences and more verbal rewards than non-professional fathers.

An interview study with black fathers from three social classes revealed that middle-class black fathers' scores in the domains of "provision of developmental stimulation" and "qual-

... Paternal nurturance has been found to be positively related to high child achievement.

ity of the language environment" were significantly higher than scores for fathers in the upper-lower or lower-lower social class groups. And the amount of enriching home stimulation was significantly highest for middle-class fathers of daughters (Honig & Main, 1980).

Paternal verbal and cognitive interactions seem to differ as a function of child sex. The research results are often confusing. McAdoo (1979) found that middle- and working-class black fathers interacted more with sons than with daughters. There was a significant positive relationship between fathers' warmth and nurturance and amount of interaction with child. Epstein & Radin (1975) observed social class differences related to cognitive achievement among male children. Among middle-class boys, there was a positive relationship between paternal nurturance and child's Binet I.Q. Among lower-class boys, paternal restrictiveness had a negative impact on the child's cognitive explorations.

The researchers speculate that "it may be that in the working class, where sex role stereotypes are strongest, intellectual and academically-oriented activities are viewed as feminine and hence not appropriate for boys" (p. 838). They report that fathers seem to interfere with daughters' task motivation by restrictiveness and by offering mixed messages that they will both meet and ignore daughters' explicit needs.

Lamb & Frodi (1980) have speculated that, in effect, when a warm father encourages femininity in a daughter and yet believes, traditionally, that femininity and achievement are incompatible, a girl may have grave doubts or conflict about the appropriateness of achievement for women. Alternatively, fathers may not be as strong role models as mothers for girls' intellective development. Research provides too little and conflicting evidence to decide this point. Crandall and colleagues (1964), for example, found that daughters who demonstrated excellence in reading and arithmetic had mothers who often praised and rewarded their intellectual efforts and seldom criticized them. Yet, in Bing's (1963) study, father's strictness was related to verbal achievements of daughters even more than sons'.

More often than not, paternal nurtur-

ance has been found to be positively related to high child achievement. High-achieving college students reported that their fathers were more accepting and somewhat less controlling than fathers described by low-achievers (Cross & Allen, 1969). When parents are too anxiously intrusive in trying to foster intellectual achievement, their efforts may have unfortunate effects.

Teahan (1963) administered a questionnaire to low-achieving college freshmen (who had previously done well in school) and their parents. These sons had fathers who felt that children should make only minor decisions, that they should always believe their parents, that they should be under their parents' complete control and that it was wicked to disobey parents. The picture emerged of a clash between a domineering, punitive, overprotective father — and his underachieving son.

It is possible that the relationship among the variables of paternal nurturance, high academic expectations, sex-role attitudes, quality of cognitive facilitation, and marital congruence with spouse may interact in subtle and complex ways to foster a child's intellectual competence.

Rapidly changing beliefs about the roles of fathers may increase the confusion among research findings as to paternal influence on achievement. Talcott Parsons' theory (Parsons & Bales, 1955), for example, that fathers were "instrumental" and mothers "expressive" with children may be outdated by evolving new beliefs. Parsons hypothesized that father comes to be seen by the child as representative of the outside world. Father is the significant major parent to make strong demands that expand a child's horizons for achievement. Such a simplified role dichotomy is no longer acceptable in the value system of many families.

But there is some research support for this conceptualization. Cox, in a 1962 dissertation with gifted fourth to sixth graders, found that their fathers and mothers showed a predominant pattern of affection, setting firm limits and positive relationships. Fathers, especially, set high expectations for their gifted high-achieving children.

A summary of the main thrust of the research so far would counsel that a father who wishes to have positive

academic influence should have high academic expectations, be a helpful teacher, remain nurturant, but respond with flexibility, perceptivity and sensitivity over time to a child's changing needs for assuming autonomous responsibility for the child's own learning career and social life.

Fathers and Sex Role Development

Research indicates that fathers are more focused on sex role differences. They influence sex stereotyping more than mothers. Fathers have preference for male offspring. By one year, fathers prefer boy babies. Fathers talk more to sons than to daughters.

The father's character and the extent to which he has made a success of his personal life, plus his easy affection with his son and his loving relationship with the mother, have been suggested as the foundation for a son's ready acceptance of being male and, in turn, confident acceptance of the role of husband and father in adulthood (Green, 1976).

A father who is violent, contemptuous toward women and overbearing may impair his daughter's ability to grow up "feminine." She may grow up without a basic sexual understanding that men and women can be equally accepting of and tender with each other. Young girls who have grown up with a good relationship with an admiring, nurturant father have been found to relate easily and well in college relationships with young men (Johnson, 1963).

Two-thirds of the world's 800 million illiterates are females. Dominant males in many third-world societies devalue the potential for intellectual development of their daughters and wives. Yet, even in the United States, subtle, intellectual devaluating of daughters occurs at many levels. Research on conversational management techniques reveals that fathers interrupt their children more than mothers do, and daughters are interrupted more than sons. Fathers engage in simultaneous speech more with daughters than with sons. Boys and girls get different messages about their status and role in society. Girls are more interruptible than their brothers — which suggests in a not-so-subtle way that they are less important (Greif, 1980).

Such prejudices run deep. Perhaps

Fathers are important for adequate sex role development of daughters as well as sons.

only as this nation commits itself to teaching parenting skills and family life courses at all educational levels will prospective parents begin to become aware of the impelling and compelling influences they can have on the growth of emotionally healthy and self-actualized daughters and sons in the future.

Fathers are important for adequate sex role development of daughters as well as sons. Daughters' conceptualizations of the worth of fathers may be persistently distorted by being raised in households bereft of a father. Girls then may in some way expect that only women are strong and can take care of families and men are weak and cannot be expected to do so.

What does such sex role learning presage for tomorrow's families? Green has summarized the important role that fathers play in a daughter's sex role development: "A young girl learns how to be a female from her similarly shaped mother. But she will learn how to be a girl who likes men, or does not trust or feel affection for men by the way she responds to her father" (p. 165).

Recent studies on androgynous roles for both mothers and fathers are beginning to reveal some of the fathering patterns that may emerge as new beliefs about masculinity and the fathering role become more widespread. Fathers classified as androgynous by the Bem Sex Role Inventory have been found to be more involved in day-to-day care, activities and play with their children than those classified as masculine. Fathers classified as masculine, married to women classified as androgynous or masculine provided the next highest level of involvement with children. The least involved were masculine fathers married to feminine women (Russell, 1978).

Fathering and Prosocial Child Behaviors

School vandalism, rising delinquency at younger ages and crime in the streets have all helped to spur interest in the area of prosocial behaviors such as empathy, altruism, generosity and helpfulness. Most of the research has focused on female parents, models and teachers. In a study that inquired about fathering patterns, Rutherford & Mussen (1968) played a game with nursery school boys who then had an opportu-

nity to share some of their winnings with friends. The most generous boys, by action and by teacher rating, much more frequently described their fathers as nurturant and warm parents and as models of generosity, sympathy and compassion.

Hoffman (1975) used sociometric questionnaries to assess fifth-grade pupils' reputations for altruism and consideration of others. Children nominated the three same-sexed classmates who were most likely to "stick up for a kid that the other kids are making fun of or calling names" and "to care about how other kids feel and try not to hurt their feelings." Parents of the children were then asked to rank 18 life values. The fathers of those boys rated as most helpful and considerate (and the mothers of similar girls) ranked altruism high in their own hierarchy of values.

Yarrow & Scott (1972) found a child's consideration of others, as assessed by classmates' nominations, to be related to maternal and not paternal affection among middle-class children. However, lower-class boys' (but not girls') consideration for others was significantly related to father and mother affection.

Middle-class six- and eight-year-old boys were found to be more generous when high paternal affection and high maternal child-centeredness were present (as measured by parental Q Sorts). These relationships did not hold for girls (Feshbach, 1973). The level of altruism modeled by fathers appears to be a factor in the development of sons' prosocial behaviors. In a review of studies of the development of prosocial behaviors in children, Mussen & Eisenberg-Berg (1977) conclude that "nurturance is most effective in strengthening predispositions toward prosocial behavior when it is part of a pattern of child-rearing that prominently features the modeling of prosocial acts" (p. 92).

The converse has been found also. Where fathers are relatively unaffectionate and controlling, authoritarian and rejecting, and not likely to trust their sons, boys have been found high in aggression (Feshbach, 1973; Stevens & Mathews, 1978).

Fathering in Alternate Life Styles

Just beginning to receive the research attention they so critically de-

serve are stepfathers, divorced fathers with custody, divorced fathers with only visitation rights, and single unwed fathers*. In 1974, slightly over six million children were living with a stepparent. Rallings (1976) has summarized some of the sociological findings on the extent of stepparenting and the adjustment problems faced by stepfathers in particular, and Pannor, et al. (1971) have focused on studies of unwed fathers.

Hetherington, Cox & Cox (1976) have documented the extent of the disruption on children's lives where father is the non-custodial parent. Divorced parents are less consistent with children; they are less likely to use reasoning and explanation. There is a steady decline in nurturance expressed by divorced fathers toward their children. Two years after divorce, negative affect and distressful symptoms were diminished in girls. Yet boys from divorced families were still more hostile and less happy than boys from nuclear families.

When the father was emotionally mature, then frequency of father's contact with the child was associated with more positive mother-child interactions. When the father was poorly adjusted or there was disagreement and inconsistent attitudes toward the child, or there was ill will between the former spouses, then frequent visitation was associated with poor mother-child functioning and disruptions in the children's behavior.

Wallerstein and Kelly (1980) followed children of divorced families for five years. Boys particularly reported depressed and difficult feelings when there was not consistent attentive relationship maintained with the father. One-third of this middle-class sample of children were still considerably disturbed in functioning after five years.

The intimate relationship of fathers and children seems to be particularly crucial for positive adjustment of sons after divorce. Yet, parental conflict and immaturity can vitiate positive effects of frequent contact. Clearly more urgent is sensitivity to children's needs by both parents after divorce. The effect of divorce per se on children

*All of the articles in the Special Issue of the October 1979 *Family Coordinator* (Volume 28) are addressed to "Men's Roles in the Family."

When nourished by father love and intimate responsive care, babies become well attached to their fathers.

is not as critical as parental conflict and immaturity.

Gasser & Taylor (1976) inquired into the role adjustments faced by single-parent fathers. The middle-class fathers interviewed assumed major responsibility for all child-care activities, sought outside supports, and curtailed club meetings and educational attainments. These fathers felt that they were able to cope with the responsibilities of home management although they may have formerly assumed little responsibility while married. Single fathering no more seems to guarantee unhappiness than does nuclear family living (Katz, 1979). The ways in which stresses and burdens are handled seems to be more indicative of whether a family functions fairly happily than whether one or two parents are present.

Father Involvement: Intervention Programs

Although the overwhelming number of programmatic efforts to enhance parenting skills in the past decade has focused on mothers, some programs have been involved with fathers as part of a family focus of intervention. Middle-class mothers and fathers of babies under 12 months were trained to increase their social competence with infants (Dickie & Carnahan, 1979). Post-training home observations showed that training affected trained fathers the most. They were superior to trained mothers and to control mothers and fathers in anticipating infants' needs, responding more appropriately to the infants' cues and providing more frequent verbal and non-verbal contingent responses. Infants sought interaction least with untrained fathers and most with trained fathers. An extra benefit accrued to the marital partners: trained mothers and fathers thought their spouses were more competent than did untrained mothers and fathers.

An experimental program in Chicago with a small number of low-income families in an urban housing project found that fathers could be more actively involved in their children's educational experiences when male workers tailored the home visitation program specially for the fathers. Tuck (1969) has described this model for working with black fathers.

The Importance of Fathering

The importance of fathering has become more and more evident as research in this area proliferates. Some comments are appropriately representative of major findings to date and some as suggestions for needed research.

Men who traditionally have rejected expression of tender feelings or a range of emotional responsiveness as unmanly may need to rethink "what is masculinity" in light of the needs of infants for fathers and the delights of intimacy with infants for fathers. As Lamb elegantly expresses this: "It is important not to confuse conformity to traditional sex role prescriptions with the security of gender identity or with mental health." Provided an individual's gender identity is secure, a wide range of gender roles can be assumed (1979, p. 942).

The myth that only mothers can nurture an infant seems just that — a myth. Fathers are just as upset by squalling babies as are mothers. Fathers can be as attentive to infant cues as mothers. When nourished by father love and intimate responsive care, babies become well attached to their fathers.

Paternal nurturance may be related in complex ways to cognitive achievement in children. The family serves as a nurturing matrix that allows a child's natural curiosity and exploration to flourish into developmental learnings. A child filled with anxiety or despair that he or she is neither cared for nor cared about cannot focus well on learning tasks whether a father is absent or present. Future research should focus on the complex interweave of factors in family and community that facilitate intellectual engagement and achievement, rather than on putative effects of father absence or presence conceived of as a single variable of an heuristically critical nature.

Process rather than status variables have proved more relevant to child intellective attainments (Honig, 1979b). Indeed, in a father-present family where book-learning is considered sissyish, it is not difficult to predict that despite father presence, neither cognitive strivings nor academic excellence may be a goal of father or son.

Because the effects of fathering may

be related to marital harmony, economic stress and a host of social and cultural variables, future emphasis in fathering research needs to enquire into the covariation of factors that affect father influence on child development. For example, Park & Sawin's (1977) studies demonstrated that both mother and father show more interest when they are together with the newborn. They count toes, check ears, and smile at the infant more.

Clarke-Stewart (1979) found, in a small sample-size but provocative study, that as mothers rear a contented, interesting baby, fathers, after the first year, are lured by such an attractive infant into increased interactions. Triadic effects of fathers and mothers and infants need to be examined. Pederson's (1975) finding that father's warmth and affection helps support the mother and make her more effective with the baby is relevant. Research must be sensitive both to direct and to indirect effects of fathering.

The relation of fathers to the development of altruism, empathy and the gentler arts of positive relations with others deserves far more research effort. If a parent preaches "love thy neighbor" but father models proudly his "he-man" imperviousness and insensitivity to the feelings and rights of others, particularly wife and children, then present research suggests that children will practice what they live rather than what they are told.

As divorce statistics increase, more and more ways to help children weather parental storms and uncouplings must be found. Divorce findings reveal that the role of a well-adjusted father who can communicate without strong rancor with his ex-spouse can do much through intimate, consistent contact with children, post-divorce, to help heal the distress and anger that divorce entails for children. Otherwise, what "frees" the parents may engender possible long-lasting grief and academic difficulty, particularly for sons.

Single parents may have a harder job rearing children well because of the extra stresses that may ensue when there is lack of a supporting other person. Yet, single fathers may have a very high motivation to parent. Strong positive motivation has been known to overcome "handicaps" far more severe than those involved in single fathering.

Despite the many studies which sug-

... Fathering may still be a profound and deeply satisfying experience in human intimacy and engagement.

gest that diapering and child care in early infancy are not the occupation of choice for many fathers (and some mothers too), fathering may still be a profound and deeply satisfying experience in human intimacy and engagement. It may be well to remember the impressive findings from the long-term study of Terman's gifted children. When gifted boys were reinterviewed decades later at age 62, they agreed that the greatest source of satisfaction in their lives was their families.

Sears (1977) has commented that in spite of autonomy and great average success in their occupations, these men placed greater importance on having achieved satisfaction in their family life than in their work. Furthermore, these men believed that they had found such satisfaction. May it be so for fathers of the future. Such a deep conviction and satisfaction would augur well for the children of tomorrow.

Fathering education is not yet politically an "in" issue for society. Yet, a man needs to learn fathering the way he would learn to play ball or set up a business or cook a gourmet meal — early and with lots of practice, patience and encouragement. Communities must become alert to the ways schools and service organizations can provide opportunities for boys to learn about and to nurture younger children responsively and responsibly.

REFERENCES

Ainsworth, M.D.S. Attachment: Retrospect and prospect. Presidential address presented at the Biennial Meeting of the Society for Research in Child Development, San Francisco, March, 1979.

Bing, E. Effect of childrearing practices on development of differential cognitive abilities. *Child Development*, 1963, *34*, 631-648.

Bowlby, J. The nature of the child's tie to his mother. *International Journal of Psychoanalysis*, 1958, *39*, 350-373.

Clarke-Stewart, A. The father's impact on mother and child. Paper presented at the Biennial Meeting of the Society for Research in Child Development, New Orleans, March, 1979.

Crandall, V. J., Dewey, R., Katkovsky, W. & Preston, A. Parents' attitudes and behaviors and grade-school children's academic achievements. *Journal of Genetic Psychology*, 1964, *104*, 53-66.

Cross, H. J. & Allen, J. Relationship between memories of parental behavior and academic achievement motivation. Proceedings of the 77th Annual Convention. American Psychological Association, Washington, D.C., September, 1969, 285-286.

Deal, T. N. & Montgomery, L. L. Techniques fathers use in teaching their young sons. Paper presented at the Meeting of the Society for Research in Child Development, Minneapolis, April 1971.

Dickie, J. R. & Carnahan, S. Training in social competence: The effect on mothers, fathers and infants. Paper presented at the Biennial Meeting of the Society for Research in Child Development, San Francisco, March, 1979.

Dodson, F. How to make your man a great father. *Harper's Bazaar*, April, 1979, p. 155, 194.

Epstein, A. S. & Radin, N. Motivational components related to father behavior and cognitive functioning in preschoolers. *Child Development*, 975, *46*, (No. 4), 831-839.

Feshback, N. The relationship of child rearing factors to children's aggression, empathy, and related positive and negative social behaviors. Paper presented at the NATO Conference on the Determinants and Origins of Aggressive Behavior, Monte Carlo, Monaco, July, 1973.

Friedlander, B. Z., Jacobs, A. C., Davis, V. B. & Wetstone, H. S. Time-sampling analysis of infants' natural language environments in the home. *Child Development*, 1972, *43*, 730-740.

Gasser, R. D. & Taylor, C. M. Role adjustment of single parent fathers with dependent children. Family Coordinator, 1976, *25*, (No.4), 397-402.

Green, M. *Fathering*. New York: McGraw-Hill, 1976.

Greif, E. Sex differences in parent-child conversations, ERIC, ED 174 337, 1980.

Herzog, E. & Sudia, C. Children in fatherless families. In B. M. Caldwell & H.N. Ricciuti, (Eds.) *Review of Child Development Research Vol. 3*, Chicago: University of Chicago Press, 1973.

Hetherington, E. M., Cox, M. & Cox, R. Divorced fathers. *Family Coordinator*, 1976, *25*, 417-428.

Hoffman, M. L. Altruistic behavior and the parent-child relationship. *Journal of Personality and Social Psychology*, 1975, *31*, 937-943.

Honig, A. S. *Fathering: A bibliography*. Urbana, Illinois: ERIC (Document Reproduction Service No. 142293: (Cat No. 164) 1977.

Honig, A. S. A review of recent infancy research. *The American Montessori Society Bulletin*, 1979, *17* (No. 3 & 4) (a)

Honig, A. S. *Parent involvement in early childhood education*. 2nd Edition. Washington, D.C.: National Association for the Education of Young Children, 1979. (b)

Honig, A. S. Choices: To parent or not to parent. Paper presented at the 6th Annual Symposium on Sex Education, Toulouse, France, July, 1980.

Honig, A. S. & Main, G. Black fathering in three social class groups. Manuscript submitted for publication, 1980.

Johnson, M. M. Sex role learning in the nuclear family. *Child Development*, 1963, *34*, 319-333.

Katz, A. J. Lone fathers: Perspectives and implications for family policy. *The Family Coordinator*, 1979, *28*, 521-528.

Knox, I. D. & Gilman, R. C. The first year of fatherhood. Paper presented at the National Council on Family Relations, Missouri, 1974.

Kotelchuck, M. *The nature of the child's tie to his father*. Unpublished doctoral dissertation, Harvard University, 1972.

Lamb, M. E. The role of the father: An overview. In M. E. Lamb (Ed.) *The role of the father in child development*, New York: Wiley, 1976.

Lamb, M. E. Paternal influences and the father's role: A personal perspective. *American Psychologist*, 1979, *34*, 938-943.

Lamb, M. E. & Frodi, A. M. The role of the father in child development. In R. R. Abidin (Ed.) *Parent education and intervention handbook*. Springfield, Illinois: Charles C. Thomas, 1980.

Lynn, D. B. *The father: His role in child development*. Belmont, California: Brooks Cole, 1974.

McAdoo, J. L. Father-child interaction patterns and self esteem in black preschool children. *Young Children*, 1979, *34*, 46-53.

Mendes, H. A. Single fatherhood. *Social Work*, 1976, *21*, (No. 4), 308-312.

Mussen, P. & Eisenberg-Berg, N. *Roots of caring, sharing, and helping*. San Francisco: W. H. Freeman, 1977.

Pannor, R., Evans, B. W. & Massarik, F. *The unmarried father*. New York: Springer Publishing Co., 1971.

Parke, R. D. & Sawin, D. B. Fathering: It's a major role. *Psychology Today*, 1977, *11*, 108-113.

Parke, R. D. & Sawin, D. B. Fathering: It's a major role. H. E. Fitzgerald (Ed.) *Human Development* 80/81. Guilford, Connecticut: Dushkin, 1980.

Parsons, T. & Bales, R. F. *Family, socialization and interaction process*. Glencoe, Illinois: Free Press, 1955.

Pederson, F. A. Mother, father, and infant as an interactive system. Paper presented at the Symposium Fathers and Infants at the meetings of the American Psychological Association, Chicago, August, 1975.

Rallings, E. M. The special role of stepfather. *Family Coordinator*, 1976, *25*, 445-450.

Rebelsky, F. & Hanks, C. Fathers verbal interactions with infants in the first three months of life. *Child Development*, 1971, *42*, 63-68.

Russell, G. The father role and its relation to masculinity, femininity, and androgyny. *Child Development*, 1978, *49*, 1174-1181.

Rutherford, E. & Mussen, P. Generosity in nursery school boys. *Child Development*, 1968, *39*, 755-765.

Sears, R. R. Sources of life satisfactions of the Terman gifted men. *American Psychologist*, 1977, *32*, 119-128.

Sroufe, L. A. The coherence of individual development: Daily care, attachment, and subsequent developmental issues. *American Psychologist*, 1979, *34*, 834-341.

Stevens, J. H., Jr. & Mathews, M. (Eds.) *Mother/child, father/child relationships*. Washington, D. C.: National Association for the Education of Young Children, 1978.

Teahan, J. E. Parental attitudes and college success. *Journal of Educational Psychology*, 1963, *54*, 104-109.

Tuck, S. A model for working with black fathers. Paper presented at the Annual Meeting of the American Orthopsychiatric Association, San Francisco, 1969.

Wallerstein, J. S. & Kelly, J. B. Divorce counseling: A community service for families in the midst of divorce. In R. R. Abidin (Ed.) *Parent education and intervention handbook*. Springfield, Illinois: Charles C. Thomas, 1980.

Yarrow, M. R. & Scott, R. M. Imitation of nurturant and non-nurturant models. Journal of Personality and Social Psychology, 1972, *23*, 259-270.

Helping Children Cope with Divorce

For the husband and wife, the divorce may be a reprieve from a depressive existence; for the children, it may actually be the beginning of their suffering and longing for the absent parent

Ralph F. Ranieri

Ralph F. Ranieri is Director of Consultation and Education at the West Central Florida Human Resources Center, a community health center. He has had articles published in U.S. Catholic, Today's Parish, Liguorian, *and* Our Family.

Next to the death of a spouse, one of life's major crises is divorce. It demands a major readjustment on the part of those divorced. Divorce touches not only the social but also the emotional, spiritual, and financial aspects of day-to-day living. Adults must grapple with the many practical aspects of living alone after a divorce, as well as coping with their own diminished feelings of self-esteem.

Probably more than at any other time in our history, it is important for Americans to understand the dynamics of divorce. One out of every three marriages today ends in divorce. An estimated 1,000,000 divorces and annulments took place in 1976, more than double the number in 1966, and triple the 1950 figure. If you are not divorced, chances are you have someone in your family or circle of friends who has experienced a divorce.

While divorce demands major readjustment for adults, it also has its emotional impact upon children. They are very often the forgotten individuals in a divorce; yet they too suffer. Sometimes their suffering can be quite significant because they have had nothing to say about the separation. Their wishes and opinions have not been a part of the decision.

Sixty percent of all divorces involve children. Approximately 1,000,000 children each year are affected by divorce. Adults who find themselves capable of coping with their own painful experience during a divorce often feel unable to be emotionally supportive to their children during this period. The difficulty often lies in the fact that they do not know what to do for their children or what their children are going through.

It is not uncommon for a child who has watched his mother and father go through a divorce to feel responsible for the breakup. Many children of divorce feel that they are the guilty parties. So often this happens because prior to the divorce there may have been a great deal of animosity between the husband and wife. Sometimes this animosity has spilled over into the relationship with the children. The children then feel they have done something wrong to anger their parents. When the divorce finally materializes, the children take this as a proof of their culpability. They perceive that they have caused their mother and father to separate because they have not been good, obedient, or whatever else they may fantasize.

Actually, children are innocent of any responsibility for a divorce. Two mature adults must bear the total burden for the success or failure of their relationship. Even when adults admit their responsibility, the child's misconception of the event can cause him or her to feel guilty.

Immediately following a divorce, children may exhibit symptoms of a mild depression in varying degrees. This usually stems from a feeling of loneliness. Their life-style has been drastically changed. A parent, whom in many cases they have loved dearly, has been removed from the home, and they have been helpless to do anything about it. When one parent leaves the family, the balance of the family system is upset. As the child becomes lonelier, his feelings of missing the absent parent become more acute.

This is a normal reaction; however, it is a reaction that the parent in the home may overlook. For the husband and wife, the divorce may actually be a relief or a reprieve from a depressive existence. They may not realize that for the children it may actually be the beginning of their suffering and their longing for the absent parent.

Distractibility, withdrawing, and a preoccupation with the past are common symptoms of depression in children going through the transitional phase of divorce. There may also be evidence of anger which stems from the depression. A child may act disrespectful at home or be involved in fighting at school. In these cases, he is putting his anger into action.

One very sensitive area in which a child may express anger is involving a parent's new boyfriend or girl friend. Children find it very hard to accept the fact that Mom or Dad no longer love each other, but it is even more difficult for them to realize that the relationship has now been changed to such an extent that their parents will be dating other friends.

The angry reaction to the new boyfriend or girl friend is more out of loyalty to their absent parent than out of dislike for their parent's new friend. At first, the new friend may even try to be a parent in the sense that they correct or make demands on the child. This in-

creases the child's anger all the more. They have not accepted this new friend. Their heart still belongs to the absent parent, and to receive a correction or advice from them brings home the fact that their absent parent is no longer there. This is difficult for them to accept.

Emotional ties cannot be broken quickly. Parents have had the opportunity of working through the divorce. The child has not had the same chance. For most children, the divorce comes rather abruptly because they do not have any reason for a divorce. It may take them awhile to let go emotionally.

In the meantime, they engage in a great deal of fantasy to mitigate the reality of the situation. Most children of divorce have the fantasy that their mother and father will someday be reconciled. This is a very consoling fantasy for children, especially since they are powerless to do anything about the course of events. Fantasy may be their only way of fulfilling some of their wishes and dreams.

While children are going through a trying time, they are not going through a completely hopeless experience. Children of divorce can grow and mature as normal, healthy children if they are helped through their crisis.

It is a handicap to grow up without two parents. But like so many other hardships in life, it can be overcome. It is a handicap to go to college when you are forty instead of when you are twenty, but many people do it. And they do it quite successfully. The same is true of divorce. Children can surmount their handicap if adults are aware of what they are going through.

One-sixth of all children in the U.S. today live in single-parent families. Fifty percent of all women with school-age children are employed. Therefore, the child of divorce is not alone. Many children are in similar situations.

In order to help their children through divorce, parents must first of all be aware of their own feelings. Most likely, the parents have feelings of guilt and anxiety, which have been provoked by the divorce. When parents are aware of their feelings and are able to identify them, they can make sure they do not inject these feelings into the relationship with their children.

Dr. Kenneth M. Magid is the Director of the Children Facing Divorce program in Evergreen, Colorado. He has several practical suggestions for par-

ents and counselors to help children through this critical period (*Personnel and Guidance Journal,* May 1977).

Dr. Magid suggests that parents keep the focus of their emotions and conversation on the child and not their ex-spouse. There is no doubt that divorce has uncovered many negative feelings between the husband and wife. However, when dealing with children, feelings should be centered on them. The divorce has legally separated the person from a situation which he or she may consider intolerable. Now the focus needs to be placed on the children. Even though it may take some time for emotions to catch up to what has been declared in a court of law, parents should make a conscious effort to directly concentrate on their children and not on their ex-spouse.

When discussing the situation with a child, it is best to keep the conversation in the present tense. If the child wants to talk about how things were, he should be allowed to do so. But when initiating the conversation, it should be in the present. Taking a child back in time may only serve to provoke guilt and anger.

What a child is experiencing at this time may be particularly intense. It may even stir up guilt in the parent for causing his or her child to go through this. Remember, many of the child's reactions are normal, and he or she is not alone in the experience. Most children do return to their usual selves after they work through this experience.

There is a certain emotional period of mourning after a loss, even if it is a divorce. The parent has most likely gone through this in preparing for the divorce. This child is just beginning.

It is important to be truthful with children. The divorce has a direct bearing upon their lives. They should know what the consequences are for them. They do not need to be overburdened with the details of the divorce. A general explanation in this area will suffice. Nor do children need to know the specifics of financial arrangements. Finances are a crucial part of the divorce for spouses, and this can be a rather heated subject. Spouses should never argue about financial matters in front of the children. Children can feel responsible if they think such issues as child support are further driving a wedge between their parents.

What children do need to know, however, are the details of custody ar-

PAUL CONKLIN

rangements. They should not be in the dark about their own future. They need to know with whom they are going to live. They should feel that this has been mutually agreed upon, so that they will not feel rejected by the parent with whom they are not living. They should also know what opportunities they will have to visit the absent parent, when these opportunities will be, and how long the visits will last. The more they know about what directly concerns them, the more secure they will be. It is when they do not know what to expect that their anxiety is the most intense.

Dr. Magid also suggests that the divorced parents establish a hot-line between themselves. This suggestion is aimed at keeping the child from manipulating both parents. This is for the welfare of both the parents and child.

Here is an example of how children can play off separated parents. Tommy is eleven years of age. His mother would not allow him to swim in a nearby river because she thought it was too dangerous. When Tommy's friends would go swimming, he would argue with his mother to let him go, but she never did.

When Tommy went to visit his father for the weekend, he asked to go swimming in the same river. The father knew nothing of the mother's regulations nor did he realize that it was a dangerous place to swim. So the father, who was completely innocent of

prior arrangements, allowed Tommy to go swimming.

In this situation, a hot-line between the parents would have enabled them to talk about the issue. Such communication allows for continuity in the child's discipline. Continuity of discipline is a very important aspect of child care which demands cooperation between parents. It may be difficult for parents who are divorced to still cooperate with each other. However, even though they are no longer husband and wife, they are still both parents to their children. They may have to muster the maturity to cooperate with each other so that they may continue to be effective parents to their children.

Although it may be difficult to cooperate, there should never be any need for either parent to degrade the other in front of the children. Sometimes a conscious effort will be needed to steer clear of who was right and who was wrong in the marriage. Frequently, there is no one to blame. The divorce can simply be described as an agreement reached by the two parents.

If a spirit of cooperation and a nonjudgmental attitude can be maintained, a parent can actually use his or her former spouse as a valuable aid in the child's development. Let us take a simple example. Janice, 15, is having difficulty in school. She has not been studying and suddenly becomes belligerent at home. If the absent parent is forewarned, he or she may be able to provide Janice with additional help for her problem.

The absent parent can be a resource for the single parent. It also helps the child to see that the absent parent is not merely there for fun and games, but is also concerned about his growth and development by the way he or she assumes the more serious parental responsibilities.

Keeping a good working relationship between divorced parents also makes it easier for a child to go from one home to the other. If a child has access to both parents and feels equally accepted in both homes, this can only work toward her or his own welfare.

Besides each other, parents should also utilize the extended family to complement their child's development. For the child who is missing a father, male relatives can fill the gap. Female relatives can do the same when there is limited contact with the mother.

Some parents find it difficult to admit to others that they are going through a divorce. It is necessary to be honest in this regard. It is very important to let a child's teacher know about the divorce and to encourage the teacher to be supportive of the child during this period. If the teacher knows about the family situation, he or she will be able to understand the child's possible change in behavior and study habits.

The same also holds true for other significant adults in a child's life: a scoutmaster, clergyman, coach, or music teacher. They all can provide support and understanding for the child coping with divorce.

If it is at all possible, it is best not to separate siblings. They have already experienced the loss of a parent in the home. To separate a child from brothers and sisters may compound the feeling of loss. The sibling relationship is very often underestimated. Frequently, adults do not appreciate the strong bonds which exist between siblings. Sibling relationships form a very closely-knit society. While we often see the external demonstration of fighting, bickering, and jealousy, there are also unexpressed feelings of warmth, love, and concern. When siblings are separated, their feelings of loss may be deeper than most adults realize.

Parents usually consider separating siblings in order to make the single-parent family more manageable. There are definite problems in single-parent families. Mothers may have to work now by necessity rather than choice. Fathers are not available to share in the disciplining or chores around the house. Because there are less people to do the same amount of work, there is the danger that children may be treated as small adults.

Many children in single-parent families are given more responsibility than they can comfortably handle. This is especially true of older children who are put in charge of their young siblings. There is no doubt that older children should help with the younger children. But if they are given too much responsibility, they may react to it with hostility. It is important to make sure that children are given no more responsibility than they can comfortably handle for their age.

Furthermore, children should not be held accountable for their parent's feelings. The family problems that arise out of a divorce are still the primary responsibility of the parent. Children should not be made counselors to their parent's problems. Parents should try to deal with the difficulties in single-parent living as best they can on their own. If a divorced parent feels the need to talk about personal feelings of loneliness or frustration, he or she should seek out an adult friend or a professional. Children can be made aware of some of the realities such as less money or increased responsibility, but they cannot be expected to provide emotional support. They are going through a time when they need all the support they can get. They may not be ready at this time to give support. To divulge all the problems experienced makes a child feel even more inadequate or anxious about the family situation.

This also holds true for using children as messengers between ex-spouses. Some parents may feel that there is nothing wrong with giving their children a message about a bill that is due or a support payment that was expected as they go off to visit the absent parent for the weekend. Using children as a go-between puts the children in a precarious position. If there is a collaborative relationship between divorced parents, it will not be necessary for children to bear this burden.

Parents often feel that they must somehow compensate the children for the divorce. These feelings are provoked by guilt. The absent parent especially may overindulge the child with privileges or gifts when he or she visits. Overindulgence is artificial and does nothing to resolve problems.

There is also the tendency on the part of both parents to promise more than they can actually deliver. There is really no need to make promises to the children, and there is certainly no need to make exorbitant promises when the possibility of fulfilling them is tenuous. Children are disappointed by divorce. This is part of the reality of divorce. They can work through this. But making promises which cannot be fulfilled only compounds a child's disappointment and adds to some of the frustrations they are already feeling.

Parents will not do everything right in helping their child through a divorce any more than they would do everything right if they were still living with their spouses. But parents' love for their children will make up for all their deficiencies. If the love is genuine and consistent despite the emotions of the moment, children can grow happily and healthily through a divorce.

PROJECT DAY-CARE

Ed Zigler, a psychologist who helped start
Project Head Start, wants to use the public schools
to help solve the nation's day-care problems.

ROBERT J. TROTTER

Robert J. Trotter is a senior editor at Psychology Today.

IT'S 3:30. DO YOU KNOW where your kids are? A lot of parents do, and they aren't all that happy about it:

■ Doug and Lisa's 13-year-old daughter, who goes home to an empty house after school, often smokes dope with her friends before her parents get home. She recently announced that she is pregnant.

■ Lucy, a single parent, can barely afford to pay $50 a week to an elderly woman who cares for 2-year-old Toby and eight other babies in a one-room apartment.

■ John and Julie have to pay $300 a week for a live-in nanny to take care of little Jack.

It's the crisis in child-care. We've all heard about it. Newspapers, magazines and talk shows have turned it into one of the hot social issues of the '80s. *Doonesbury* has turned it into black humor. But is there really a crisis? Yes, if you are a working parent who can't find or afford adequate day-care. Yes, if like Yale University psychologist Edward Zigler you realize that the physical and mental development of millions of children are being compromised by inadequate and damaging day-care.

"We expect these children to grow up, take their place in society and provide the work that's going to make this country competitive and productive," he says. "But these children may not be ready. And that's a very frightening prospect for our society."

Zigler, one of the architects of Project Head Start and for 30 years a researcher in the field of child development and social policy, says, "It's not enough for social scientists to say, 'These are my subjects.' We have to say, 'These are human beings that we study to try to help.'" Combining his research expertise and his experience with the federal government, he is trying to help by developing what he thinks is a practical, affordable solution to the number one problem of American families.

"We have all the knowledge necessary to provide absolutely first-rate child-care in the United States," he claims. "What's missing is the commitment and the will. First," he says, "we must convince the nation that when a family selects a child-care center, they are not simply buying a service that allows them to work. They are buying an environment that determines, in large part, the development of their children. "Remember," he says, "child-care is a day-in-day-out, year-in-year-out phenomenon for a child."

The job of convincing the nation that child-care should be a top-priority issue has been made easier by the sheer size and continued growth of the problem. Fifteen years ago, only 52 percent of mothers with school-age children were in the out-of-home work force. Today, that figure is up to 72 percent, while 57 percent of those with preschoolers work and 53 percent of those with infants and toddlers work. And all indications are that this trend will continue.

Zigler and other social scientists saw the problem coming as long ago as 1971 and tried to get the government to take steps to start a national child-care program. Working with Rep. John Brademus (D-Indiana) and Sen. Walter Mondale (D-Minnesota), they helped write the Child Development Act of 1971 and even had the backing of the Nixon administration until right-wing evangelicals, outraged that any self-respecting mother would let someone else care for her child, quashed it with a hate-mail campaign. "That was one of the great, great defeats of my life—for me personally and certainly for the country," Zigler says. "Just think, that bill provided the embryonic child-care system that could have grown up and been fully in place by now. Those of us who knew the demographics knew where we would be now, but there was no outcry from the parents who needed child-care. And no country tries to solve a problem until there is a sense that a problem exists. Demographics drive social policy. Now, the demographics are there."

Many mothers choose to work because they have fulfilling jobs or because they just don't want to be cooped up at home all day, but most work because they need the money. "In the current economic situation," Zigler explains, "a young family needs two incomes if they are to have what we consider a decent level of life. Furthermore, there has been a tremendous shift in the nature of jobs in the United States, with only one job in four paying enough to support a family of four comfortably."

The child-care situation begins to look even worse, Zigler says, when you realize that about one in four children and one in two black children are being reared in single-parent homes. This is actually a euphemism for poverty, he says. "The fact is that 90 percent of those homes are headed by women, and women only earn about 70 percent of what men earn for comparable work." Furthermore, only about 30 percent of the nearly 8 million women rearing children alone receive child-support payments, leading to what we now call the "feminization of poverty." "One of every two children who are living in poverty," Zigler explains, "are from single-parent homes headed primarily by women. These women must work if they are to support themselves and their children, and child-care is their greatest problem."

Unfortunately, the day-care problem has grown so large and intimidating that it seems insoluble. Even Washington, Zigler says, seems to be saying, "It's too big a job, especially with today's budget deficits. We'd break the bank if we provided child-care for the nation."

Zigler, who was responsible for all the

The day-care problem has grown so large that it seems insoluble.

federal programs for children in the early 1970s and has spent many years commuting between Yale and Capitol Hill lobbying for various types of child and family legislation, is well aware of the country's economic situation. "Some people are still arguing for the Swedish model—a day-care center on every corner; the government buys it and you go use it. That's unrealistic," he admits. Zigler estimates that day-care for American children costs between $75 billion and $100 billion a year, and there is no way the federal government can pick up that kind of bill. "But," he says, "there is an affordable alternative. We just have to take the problem apart, look at its various pieces, then find a way to solve each one."

The biggest part of the problem, numerically, is school-aged children. There may be as many as 5 million children who go home to empty houses after school. These so-called latch-key children represent more than 50 percent of the child-care problem and, according to Zigler, they are the easiest part of the problem to solve.

"I think we have to build a new school in America. We have to change the school system. We have to open schools earlier in the morning, keep them open later in the afternoon and during summer."

Don't think of school as an institution, Zigler says. Think of it as a building—one that's already paid for, one that is owned by taxpaying mothers and fathers who need day-care for their children. Part of the school building would be for teaching and the rest of it for child-care and supervision. This kind of system, Zigler says, could provide working parents with good developmental child-care services. And it should be available to every child over age 3. Zigler does not think children should start formal schooling at age 3. They would only be in

the schools for day-care (see "School's Out for 4-Year-Olds," this article).

At the age of 5, Zigler suggests, children should start kindergarten—but only for half-days. If the child has a parent at home, the child would spend the rest of the day at home. If the parents are working, the child would spend the second half of the day in the child-care part of the school. For children ages 6 to 12 there would be before-school, after-school and vacation care for those who need it.

Keeping the schools open, says Zigler, would solve the problem of latchkey children, but it would do much more. "People have got to realize that there is a connection between leaving children unsupervised after school and such social problems as teenage pregnancy, juvenile delinquency and the use of drugs," he says. "We are really precipitating these problems if we do not provide adult supervision for children and allow them to socialize themselves and each other. Children should be in the care of adults. They do not have the ability, the cognitive wherewithal or the experience to socialize themselves."

Zigler doesn't want teachers to provide child-care. Teachers are trained as educators, and they are expensive. What we need, he says, is something called a child development associate (CDA), a person trained to work with children—but one we can afford to pay. Someone with a degree would run the system, but CDA's, with on-the-job training, would do the bulk of the work. We already have CDA's for infants and toddlers, he says, and we should develop the same kind of professional for school-age children. The teachers would go home at three o'clock, and the CDA's would take over.

In 1971, Zigler suggested that the nation begin developing a group of certified workers that parents could trust to provide proper care for their children. "I still like that idea," he says, "but what I had in mind was to see 200,000 such people by now. Unfortunately, our country has produced only 23,000. Hardly enough to meet the need."

The new school that Zigler wants to build would do more than provide on-site child-care. "I like the Parents as First Teachers program that they have in Missouri," Zigler says, "where trained specialists actually go into homes and help teach mothers and fathers about parenting." (See "Making the Grade as Parents," *Psychology Today*, September 1986.) This kind of program could be run out of the local schools, which would also provide an information-referral system and resource center for

School's Out for 4-Year-Olds

"Joe, what'd you learn at school today?"

"Nothing, Dad. I told you, my school isn't a learning school. It's a playing school."

"I'd be in seventh heaven if my 4-year-old said that to me," says Ed Zigler. "Playing is the job of 4-year-olds.

"We've made parenting an anxiety-producing phenomenon," he says. "Parents are scared to death that they are going to make a misstep, that they are not going to buy the latest toy or gadget that teaches. They lose sight of the fundamentals: Babies are learning organisms. What you have to do is set the environment in a loving, warm way and have confidence that they are going to come to grips with their universe if you just structure it a little bit. We shouldn't be trying to raise IQ's. We should be raising socially competent individuals, people who are happy with themselves, who will contribute to society and see that their own children are raised properly."

Despite being a longtime proponent of early-childhood intervention programs, like Project Head Start for disadvantaged children, Zigler is worried about the growing trend toward formal education for 4- and 5-year-olds.

Much of the emphasis on early education is, in fact, based on the successful results of Head Start and similar projects. But, Zigler explains, these programs were successful primarily with low-income, handicapped and non-English-speaking youngsters. "In contrast, there is a large body of evidence indicating that there is little if anything to be gained by exposing middle-class children to early education."

In other words, compulsory schooling for 4-year-olds, which has been called for in some states and by some educators,

would be a waste of time and money. "Spreading our limited education funds to cover preschool education for all 4-year-olds would spread the money too thin. Such an extension would not only have little effect on the more advantaged mainstream but would also diminish our capacity to intervene with those who could benefit the most."

Beyond that, he says, early schooling is inappropriate for many 4-year-olds and may be harmful to their development. "We are putting so much pressure on these children that we are producing achievement anxiety. You can't hurry up human development. It has a pace, and you must respect it. And you must respect individual variation. Some 4-year-olds can handle a five- or six-hour school day. Many others cannot and will feel like failures. So if you insist that 4-year-olds be 6-year-olds, you are going to be giving many of them a built-in failure experience that can have lasting negative effects on their attitudes toward school."

Zigler admits that many parents are pressuring local school systems to institute all-day kindergarten. But he thinks that is not what they really want. "What many parents are expressing," he says, "is less a burning desire for infant academics than their desperate need for quality day-care." The answer, he says, is full-service schools that would provide high-quality day-care programs for 3- and 4-year-olds. These programs would include a developmentally appropriate educational component, but they would primarily be places for recreation and socialization—the real business of preschoolers. "Yes," he concludes, "our 4-year-olds do have a place in school, but it is not at a school desk."

---

parents. Families, for example, would go to the school to find out about local day-care homes for infants and toddlers. The schools would also serve as a resource center for all the day-care homes in the neighborhood.

"If we had that kind of system in place," says Zigler, "the only missing pieces would be child-care needs for infants and toddlers up to 3 years of age. The first chunk of this would be handled by pregnancy- and infant-care leave. South Africa and the United States are the only two industrialized nations in the world that do not provide this option," Zigler points out.

Several years ago, Zigler and a panel of child-development specialists agreed that it is best for parents to care for their infants for the first few months of life and called for an infancy-care bill. The ideal would be six months of leave with three months of it paid at 75 percent of the person's regular salary, Zigler says. This would not be overly expensive for industry, he explains, because there would be a payment plan something like Social Security or workman's compensation, with both employee and employer contributing a set amount of money.

One infant-care plan now making its way

through Congress would provide only 18 weeks of unpaid leave. "This is the most minimal bill of any nation," he says. "But even Yale professors know that something is better than nothing. What we have to do first is get the principle in place. Then we can work toward something better."

Once the mother returns to work, the final piece of the day-care puzzle has to be solved—care for infants and toddlers until they are old enough to enter the school-based system. This part, Zigler admits, is very expensive and will probably call for government subsidies or a negative income tax.

At present, there are three types of child-care available for infants and preschoolers. The first is home care—somebody comes to your home or you take your child to a neighbor or relative. About 31 percent of the day-care in this country is home care. The second type, 37 percent, is family day-care—someone, almost always a woman, takes in four, five or six young children. Another 23 percent is center-based care—parents drop the kids off at an organized day-care center.

The problem, as Zigler sees it, is that

what's available is very uneven in quality. "There is some absolutely wonderful, beneficial day-care available. Parents with the wisdom to seek it out and the money to pay for it are finding it. But it's like a cosmic crapshoot," he says. "If you are lucky enough to find that loving, committed day-care mother, it's like adding another person to your family and you can count your blessings. But there's a lot of mediocre child-care out there and some absolutely horrible day-care—children tied to chairs being cared for by women so senile that they can't care for themselves. So what we have developing is a two-tier system, with affluent people being able to afford to buy into the first tier and the rest having to accept mediocre or even dangerous care for their children.

"We cannot have a society," Zigler goes on, "in which some children at 3 weeks of age are sent into a child-care system that helps their development while another group is put into a system that is damaging." And it is the children in the second tier who need the most help. "They are already vulnerable, or at risk, because they come from single-parent homes or from families with little money, a lot of deprivation and poor health care. And they are being placed at even greater risk by being put into very inadequate child-care settings. We are talking about hundreds of thousands, if not millions, of children."

One way to rectify this situation is to have a set of minimum standards for all day-care settings. These would include such things as the size of the group, the training of the staff and the quality of the program. Zigler and other child-development specialists have drawn up national standards on several occasions during the past 20 years but have never been able to get them through Congress. One criterion these experts agree on, for example, is that no adult should be allowed to care for more than three infants.

"That's what it takes to provide proper stimulation," Zigler explains. "But how many states·meet that standard?" Only three: Kansas, Massachusetts and Maryland. "It's not uncommon," Zigler says, "to find ratios of eight to one. I don't have to see any research to know that this isn't good. I have visited centers with eight-to-one ratios. When you go there you find overworked women, sweat pouring from them, spending their day changing diapers, placing babies in cribs. They are horrible environments. Visit them. I can tell you without any research that it is impossible for anyone, trained or untrained, to care properly for eight babies."

Beyond the lack of standards, says Zigler, there is another problem. "We are only getting from caretakers what we pay for. If you want quality child-care, you have to pay for it. But 58 percent of the caretakers in day-care centers are earning poverty-level wages or less. And in home care, where I am most concerned, 90 percent of the women are earning at that level. An absolutely number one item for us must be improving the training, the status and certainly the pay for people who decide to give their lives to the care of other people's children."

Zigler's plan to use local schools includes using them as the hub of a network for family day-care homes. They would be a resource center for parents looking for a good child-care home or facility. They could provide training for day-care workers and help make sure that the day-care homes in the local network meet the standards.

Even if every day-care center and home met minimum standards and had trained caretakers, there would still be a question about the overall effects of placing very young children in day-care. According to some researchers, day-care can have highly positive effects, especially on children's social development. Others see day-care—if it meets minimum standards—as benign. And still others argue that separating mother and infant can have lasting negative effects, even when quality day-care is provided. Zigler says he finds this ongoing debate interesting but irrelevant for several reasons.

For one thing, he says, much of the research cited in these arguments has been conducted in settings unlike those used by most parents. Even reviews that find day-care to be harmless usually include the caveat that there be no more than three infants per caretaker. "But that standard does not yet exist," says Zigler, "so we ought not be arguing among ourselves."

Furthermore, he says, this argument is not going to end tomorrow. "We will not know the ultimate effects of infant day-care until these infants have grown up and become parents. We have a whole generation that we should be watching."

Finally, for the many parents who have to work, the question is moot. They have no choice but to use the day-care available to them. Their question is, How do we pay for it?

Zigler's answer is, "We have to figure out a package of payers." First, he says, we put the public schools into the system and add a little bit to the local taxes to pay for having them open earlier and later. But even that, he concedes, would not cover the entire cost. There would have to be a realistic fee system built into the plan, like the one incorporated in the 1971 Child Development Act. The exact amount is yet to be worked out, but Zigler suggests that the fee be adjusted according to family income. And for very poor people, the fee would have to be zero.

Some people may complain about paying higher taxes to keep other people's children in day-care, but Zigler says we are in the same situation now as when we started universal education. It is something we decided to pay for for the good of the nation. And as more mothers join the work force and more families begin to use the system, it will seem logical that we all pay for it. "Why do we pay school taxes anyway? Why do we educate children?" Zigler asks. "Because we don't want stupid people.

That's why." And for the same reason, he says, we should be willing to pay taxes for day-care. "We don't want to put children into a system that is going to damage them. We don't want them to grow up to be criminals."

Day-care is likely to be an issue in the upcoming Presidential campaign, and several legislators on Capitol Hill who have taken an interest in this problem are consulting with Zigler. What most of them are proposing, however, Zigler calls Band-Aid solutions, dinky solutions that are just make-do for the next two or three years. "I want a solution that's going to last for the next 100 years ʼand provide quality day-care for everyone. I think my plan can do that. I've been working on this problem for 30 years, and I can't come up with a better solution. It's simple, it's pragmatic, it's economically viable. These are not new ideas. They've been tried, and they work. I'm just putting it together in one system."

And, he says, it wouldn't cost the federal government that much. He wants $120 million to start a pilot program involving 60 of these full-service schools around the country, at least one per state. The schools would be run for three or four years to see how they work and to get the bugs out of the system. Then the states would take over, with the federal government only providing subsidies for poor people and children with handicaps.

Once these model schools are in place and working, Zigler thinks the rest of the schools would start following the same model. "Then," Zigler says, "we would finally have a complete child-care system. Parents would get excited about having a good, safe place just down the street to send their children and wouldn't have to worry about sending them to that questionable lady across the street.

"It's a large vision," he says. "We're talking about a structural change in our society, a new face for our school system. But I am optimistic. I expect to see a bill legislating the experimental schools within the next 18 months. Sens. Alan Cranston (D-California), Orrin Hatch (R-Utah), Ernest Hollings (D-South Carolina), Christopher Dodd (D-Connecticut) and Barbara Mikulski (D-Maryland) are all trying to get legislation passed. It still amazes me that between the fall of 1964 and the summer of 1965 we managed to put 560,000 kids into Head Start. We can do the same thing with day-care."

The hidden obstacles to black success.

RUMORS OF INFERIORITY

JEFF HOWARD AND RAY HAMMOND

Jeff Howard is a social psychologist; Ray Hammond is a physician and ordained minister.

TODAY'S black Americans are the beneficiaries of great historical achievements. Our ancestors managed to survive the brutality of slavery and the long history of oppression that followed emancipation. Early in this century they began dismantling the legal structure of segregation that had kept us out of the institutions of American society. In the 1960s they launched the civil rights movement, one of the most effective mass movements for social justice in history. Not all of the battles have been won, but there is no denying the magnitude of our predecessors' achievement.

Nevertheless, black Americans today face deteriorating conditions in sharp contrast to other American groups. The black poverty rate is triple that of whites, and the unemployment rate is double. Black infant mortality not only is double that of whites, but may be rising for the first time in a decade. We have reached the point where more than half of the black children born in this country are born out of wedlock—most to teenage parents. Blacks account for more than 40 percent of the inmates in federal and state prisons, and in 1982 the probability of being murdered was six times greater for blacks than for whites. The officially acknowledged high school dropout rate in many metropolitan areas is more than 30 percent. Some knowledgeable observers say it is over 50 percent in several major cities. These problems not only reflect the current depressed state of black America, but also impose obstacles to future advancement.

The racism, discrimination, and oppression that black people have suffered and continue to suffer are clearly at the root of many of today's problems. Nevertheless, our analysis takes off from a forward-looking, and we believe optimistic, note: we are convinced that black people today, because of the gains in education, economic status, and political leverage that we have won as a result of the civil rights movement, are in a position to substantially improve the conditions of our communities using the resources already at our disposal. Our thesis is simple: the progress of any group is affected not only by public policy and by the racial attitudes of society as a whole, but by that group's capacity to exploit its own strengths. Our concern is about factors that prevent black Americans from using those strengths.

It's important to distinguish between the specific circumstances a group faces and its capacity to marshal its own resources to change those circumstances. Solving the problems of black communities requires a focus on the factors that hinder black people from more effectively managing their own circumstances. What are some of these factors?

Intellectual Development. Intellectual development is the primary focus of this article because it is the key to success in American society. Black people traditionally have understood this. Previous generations decided that segregation had to go because it relegated blacks to the backwater of American society, effectively denying us the opportunities, exposure, and competition that form the basis of intellectual development. Black intellectual development was one of the major benefits expected from newly won

From *The New Republic*, September 9, 1985, pp. 17-21. Reprinted by permission of *The New Republic*, © 1985 The New Republic, Inc.

access to American institutions. That development, in turn, was expected to be a foundation for future advancement.

Y ET NOW, three decades after *Brown v. Board of Education*, there is pervasive evidence of real problems in the intellectual performance of many black people. From astronomical high school dropout rates among the poor to substandard academic and professional performance among those most privileged, there is a disturbing consistency in reports of lagging development. While some black people perform at the highest levels in every field of endeavor, the percentages who do so are small. Deficiencies in the process of intellectual development are one effect of the long-term suppression of a people; they are also, we believe, one of the chief causes of continued social and economic underdevelopment. Intellectual underdevelopment is one of the most pernicious effects of racism, because it limits the people's ability to solve problems over which they are capable of exercising substantial control.

Black Americans are understandably sensitive about discussions of the data on our performance, since this kind of information has been used too often to justify attacks on affirmative action and other government efforts to improve the position of blacks and other minorities. Nevertheless, the importance of this issue demands that black people and all others interested in social justice overcome our sensitivities, analyze the problem, and search for solutions.

The Performance Gap. Measuring intellectual performance requires making a comparison. The comparison may be with the performance of others in the same situation, or with some established standard of excellence, or both. It is typically measured by grades, job performance ratings, and scores on standardized and professional tests. In recent years a flood of articles, scholarly papers, and books have documented an intellectual performance gap between blacks and the population as a whole.

• In 1982 the College Board, for the first time in its history, published data on the performance of various groups on the Scholastic Aptitude Test (SAT). The difference between the combined median scores of blacks and whites on the verbal and math portions of the SAT was slightly more than 200 points. Differences in family income don't explain the gap. Even at incomes over $50,000, there remained a 120-point difference. These differences persisted in the next two years.

• In 1983 the NCAA proposed a requirement that all college athletic recruits have a high school grade-point average of at least 2.0 (out of a maximum of 4.0) and a minimum combined SAT score of 700. This rule, intended to prevent the exploitation of young athletes, was strongly opposed by black college presidents and civil rights leaders. They were painfully aware that in recent years less than half of all black students have achieved a combined score of 700 on the SAT.

• Asian-Americans consistently produce a median SAT score 140 to 150 points higher than blacks with the same family income.

• The pass rate for black police officers on New York City's sergeant's exam is 1.6 percent. For Hispanics, it's 4.4 percent. For whites, it's 10.6 percent. These are the results *after* $500,000 was spent, by court order, to produce a test that was job-related and nondiscriminatory. No one, even those alleging discrimination, could explain how the revised test was biased.

• Florida gives a test to all candidates for teaching positions. The pass rate for whites is more than 80 percent. For blacks, it's 35 percent to 40 percent.

This is just a sampling. All these reports demonstrate a real difference between the performance of blacks and other groups. Many of the results cannot be easily explained by socioeconomic differences or minority status per se.

W HAT IS the explanation? Clear thinking about this is inhibited by the tendency to equate performance with ability. Acknowledging the performance gap is, in many minds, tantamount to inferring that blacks are intellectually inferior. But inferior performance and inferior ability are not the same thing. Rather, the performance gap is largely a behavioral problem. It is the result of a remediable tendency to avoid intellectual engagement and competition. Avoidance is rooted in the fears and self-doubt engendered by a major legacy of American racism: the strong negative stereotypes about black intellectual capabilities. Avoidance of intellectual competition is manifested most obviously in the attitudes of many black youths toward academic work, but it is not limited to children. It affects the intellectual performance of black people of all ages and feeds public doubts about black intellectual ability.

I. INTELLECTUAL DEVELOPMENT

The performance gap damages the self-confidence of many black people. Black students and professional people cannot help but be bothered by poor showings in competitive academic and professional situations. Black leaders too often have tried to explain away these problems by blaming racism or cultural bias in the tests themselves. These factors haven't disappeared. But for many middle-class black Americans who have had access to educational and economic opportunities for nearly 20 years, the traditional protestations of cultural deprivation and educational disadvantage ring hollow. Given the cultural and educational advantages that many black people now enjoy, the claim that all blacks should be exempt from the performance standards applied to others is interpreted as a tacit admission of inferiority. This admission adds further weight to the questions, in our own minds and in the minds of others, about black intelligence.

The traditional explanations—laziness or inferiority on the one hand; racism, discrimination, and biased tests on the other—are inaccurate and unhelpful. What is required is an explanation that accounts for the subtle influences people exert over the behavior and self-confidence of other people.

Developing an explanation that might serve as a basis for corrective action is important. The record of the last 20 years suggests that waiting for grand initiatives from the outside to save the black community is futile. Blacks will have to rely on our own ingenuity and resources. We need local and national political leaders. We need skilled administrators and creative business executives. We need a broad base of well-educated volunteers and successful people in all fields as role models for black youths. In short, we need a large number of sophisticated, intellectually developed people who are confident of their ability to operate on an equal level with anyone. Chronic mediocre intellectual performance is deeply troubling because it suggests that we are not developing enough such people.

The Competitive Process. Intellectual development is not a fixed asset that you either have or don't have. Nor is it based on magic. It is a process of expanding mental strength and reach. The development process is demanding. It requires time, discipline, and intense effort. It almost always involves competition as well. Successful groups place high value on intellectual performance. They encourage the drive to excel and use competition to sharpen skills and stimulate development in each succeeding generation. The developed people that result from this competitive process become the pool from which leadership of all kinds is drawn. Competition, in other words, is an essential spur to development.

Competition is clearly not the whole story. Cooperation and solitary study are valuable, too. But of the various keys to intellectual development, competition seems to fare worst in the estimation of many blacks. Black young people, in particular, seem to place a strong negative value on intellectual competition.

Black people have proved to be very competitive at some activities, particularly sports and entertainment. It is our sense, however, that many blacks consider intellectual competition to be inappropriate. It appears to inspire little interest or respect among many youthful peer groups. Often, in fact, it is labeled "grade grubbing," and gives way to sports and social activity as a basis for peer acceptance. The intellectual performance gap is one result of this retreat from competition.

II. THE PSYCHOLOGY OF PERFORMANCE

Rumors of Inferiority. The need to avoid intellectual competition is a psychological reaction to an image of black intellectual inferiority that has been projected by the larger society, and to a less than conscious process of internalization of that image by black people over the generations.

The rumor of black intellectual inferiority has been around for a long time. It has been based on grounds as diverse as twisted biblical citations, dubious philosophical arguments, and unscientific measurements of skull capacity. The latest emergence of this old theme has been in the controversy over race and IQ. For 15 years newsmagazines and television talk shows have enthusiastically taken up the topic of black intellectual endowment. We have watched authors and critics debate the proposition that blacks are genetically inferior to whites in intellectual capability.

Genetic explanations have a chilling finality. The ignorant can be educated, the lazy can be motivated, but what can be done for the individual thought to have been born without the basic equipment necessary to compete or develop? Of course the allegation of genetic inferiority has been hotly disputed. But the debate has touched the consciousness of most Americans. We are convinced that this spectacle has negatively affected the way both blacks and whites think about the intellectual capabilities of black people. It also has affected the way blacks behave in intellectually competitive situations. The general expectation of black intellectual inferiority, and the fear this expectation generates, cause many black people to avoid intellectual competition.

OUR HYPOTHESIS, in short, is this. (1) Black performance problems are caused in large part by a tendency to avoid intellectual competition. (2) This tendency is a psychological phenomenon that arises when the larger society projects an image of black intellectual inferiority and when that image is internalized by black people. (3) Imputing intellectual inferiority to genetic causes, especially in the face of data confirming poorer performance, intensifies the fears and doubts that surround this issue.

Clearly the image of inferiority continues to be projected. The internalization of this image by black people is harder to prove empirically. But there is abundant evidence in the expressed attitudes of many black youths toward intellectual competition; in the inability of most black communities to inspire the same commitment to intellectual excellence that is routinely accorded athletics and entertainment; and in the fact of the performance gap itself—especially when that gap persists among the children of economically and educationally privileged households.

Expectancies and Performance. The problem of black intellectual performance is rooted in human sensitivity to a particular kind of social interaction known as "expectancy communications." These are expressions of belief—verbal or nonverbal—from one person to another about the kind of performance to be expected. "Mary, you're one of the best workers we have, so I know that you won't have any trouble with this assignment." Or, "Joe, since everyone else is busy with other work, do as much as you can on this. When you run into trouble, call Mary." The first is a positive expectancy; the second, a negative expectancy.

Years of research have clearly demonstrated the powerful impact of expectancies on performance. The expectations of teachers for their students have a large effect on academic achievement. Psychological studies under a variety of circumstances demonstrate that communicated expectations induce people to believe that they will do well or poorly at a task, and that such beliefs very often trigger responses that result in performance consistent with the expectation. There is also evidence that "reference group expectancies"—directed at an entire category of people rather than a particular individual—have a similar impact on the performance of members of the group.

EXPECTANCIES do not always work. If they come from a questionable source or if they predict an outcome that is too inconsistent with previous experience, they won't have much effect. Only credible expectancies—those that come from a source considered reliable and that address a belief or doubt the performer is sensitive to—will have a self-fulfilling impact.

The widespread expectation of black intellectual inferiority—communicated constantly through the projection of stereotyped images, verbal and nonverbal exchanges in daily interaction, and the incessant debate about genetics and intelligence—represents a credible reference-group expectancy. The message of the race/IQ controversy is: "We have scientific evidence that blacks, because of genetic inadequacies, can't be expected to do well at tasks that require great intelligence." As an explanation for past black intellectual performance, the notion of genetic inferiority is absolutely incorrect. As an expectancy communication exerting control over our present intellectual strivings, it has been powerfully effective. These expectancies raise fear and self-doubt in the minds of many blacks, especially when they are young and vulnerable. This has resulted in avoidance of intellectual activity and chronic underperformance by many of our most talented people. Let us explore this process in more detail.

The Expectancy/Performance Model. The powerful effect of expectancies on performance has been proved, but the way the process works is less well understood. Expectancies affect behavior, we think, in two ways. They affect performance behavior: the capacity to marshal the sharpness and intensity required for competitive success. And they influence cognition: the mental processes by which people make sense of everyday life.

Behavior. As anyone who has experienced an "off day" knows, effort is variable; it is subject to biological cycles, emotional states, motivation. Most important for our discussion, it depends on levels of confidence going into a task. Credible expectancies influence performance behavior. They affect the intensity of effort, the level of concentration or distractibility, and the willingness to take reasonable risks—a key factor in the development of self-confidence and new skills.

Cognition. Expectations also influence the way people think about or explain their performance outcomes. These explanations are called "attributions." Research in social psychology has demonstrated that the causes to which people attribute their successes and failures have an important impact on subsequent performance.

All of us encounter failure. But a failure we have been led to expect affects us differently from an unexpected failure. When people who are confident of doing well at a task are confronted with unexpected failure, they tend to attribute the failure to inadequate effort. The likely response to another encounter with the same or a similar task is to work harder. People who come into a task expecting to fail, on the other hand, attribute their failure to lack of ability. Once you admit to yourself, in effect, that "I don't have what it takes," you are not likely to approach that task again with great vigor.

Indeed, those who attribute their failures to inadequate effort are likely to conclude that more effort will produce a better outcome. This triggers an adaptive response to failure. In contrast, those who have been led to expect failure will attribute their failures to lack of ability, and will find it difficult to rationalize the investment of greater effort. They will often hesitate to continue "banging my head against the wall." They often, in fact, feel depressed when they attempt to work, since each attempt represents a confrontation with their own feared inadequacy.

THIS COMBINED EFFECT on behavior and cognition is what makes expectancy so powerful. The negative expectancy first tends to generate failure through its impact on behavior, and then induces the individual to blame the failure on lack of ability, rather than the actual (and correctable) problem of inadequate effort. This misattribution in turn becomes the basis for a new negative expectancy. By this process the individual, in effect, internalizes the low estimation originally held by others. This internalized negative expectancy powerfully affects future competitive behavior and future results.

The process we describe is not limited to black people. It goes on all the time, with individuals from all groups. It helps to explain the superiority of some groups at some areas of endeavor, and the mediocrity of those same groups in other areas. What makes black people unique is that they are singled out for the stigma of genetic intellectual inferiority.

The expectation of intellectual inferiority accompanies a black person into each new intellectual situation. Since each of us enters these tests under the cloud of predicted failure, and since each failure reinforces doubts about our capabilities, all intellectual competition raises the specter of having to admit a lack of intellectual capacity. But this particular expectancy goes beyond simply predicting and inducing failure. The expectancy message explicitly ascribes the expected failure to genes, and amounts to an open suggestion to black people to understand any failure in intellectual activity as confirmation of genetic inferiority. Each engagement in intellectual competition carries the weight of a test of one's own genetic endowment and that of black people as a whole. Facing such a terrible prospect, many black people recoil from any situation where the rumor of inferiority might be proved true.

For many black students this avoidance manifests itself in a concentration on athletics and socializing, at the expense of more challenging (and anxiety-provoking) academic work. For black professionals, it may involve a tendency to shy away from competitive situations or projects, or an inability to muster the intensity—or commit the time—necessary to excel. This sort of thinking and behavior certainly does not characterize all black people in competitive settings. But it is characteristic of enough to be a serious problem. When it happens, it should be understood as a less than conscious reaction to the psychological burden of the terrible rumor.

The Intellectual Inferiority Game. There always have been constraints on the intellectual exposure and development of black people in the United States, from laws prohibiting the education of blacks during slavery to the Jim Crow laws and "separate but equal" educational arrangements that persisted until very recently. In dismantling these legal barriers to development, the civil rights movement fundamentally transformed the possibilities for black people. Now, to realize those possibilities, we must address the mental barriers to competition and performance.

The doctrine of intellectual inferiority acts on many black Americans the way that a "con" or a "hustle" like three-card monte acts on its victim. It is a subtle psychological input that interacts with characteristics of the human cognitive apparatus—in this case, the extreme sensitivity to expectancies—to generate self-defeating behavior and thought processes. It has reduced the intellectual performance of millions of black people.

Intellectual inferiority, like segregation, is a destructive idea whose time has passed. Like segregation, it must be removed as an influence in our lives. Among its other negative effects, fear of the terrible rumor has restricted discussion by all parties, and has limited our capacity to understand and improve our situation. But the intellectual inferiority game withers in the light of discussion and analysis. We must begin now to talk about intellectual performance, work through our expectations and fears of intellectual inferiority, consciously define more adaptive attitudes toward intellectual development, and build our confidence in the capabilities of all people.

THE expectancy/performance process works both ways. Credible positive expectancies can generate self-confidence and result in success. An important part of the solution to black performance problems is converting the negative expectancies that work against black development into positive expectancies that nurture it. We must overcome our fears, encourage competition, and support the kind of performance that will dispel the notion of black intellectual inferiority.

III. THE COMMITMENT TO DEVELOPMENT

In our work with black high school and college students and with black professionals, we have shown that education in the psychology of performance can produce strong performance improvement very quickly. Black America needs a nationwide effort, now, to ensure that all black people—but especially black youths—are free to express their intellectual gifts. That ef-

fort should be built on three basic elements:

• Deliberate control of expectancy communications. We must begin with the way we talk to one another: the messages we give and the expectations we set. This includes the verbal and nonverbal messages we communicate in day-to-day social intercourse, as well as the expectancies communicated through the educational process and media images.

• Definition of an "intellectual work ethic." Black communities must develop strong positive attitudes toward intellectual competition. We must teach our people, young and mature, the efficacy of intense, committed effort in the arena of intellectual activity and the techniques to develop discipline in study and work habits.

• Influencing thought processes. Teachers, parents, and other authority figures must encourage young blacks to attribute their intellectual successes to ability (thereby boosting confidence) and their failures to lack of effort. Failures must no longer destroy black children's confidence in their intelligence or in the efficacy of hard work. Failures should be seen instead as feedback indicating the need for more intense effort or for a different approach to the task.

The task that confronts us is no less challenging than the task that faced those Americans who dismantled segregation. To realize the possibilities presented by their achievement, we must silence, once and for all, the rumors of inferiority.

Who's Responsible? Expectations of black inferiority are communicated, consciously or unconsciously, by many whites, including teachers, managers, and those responsible for the often demeaning representations of blacks in the media. These expectations have sad consequences for many blacks, and those whose actions lead to such consequences may be held accountable for them. If the people who shape policy in the United States, from the White House to the local elementary school, do not address the problems of performance and development of blacks and other minorities, all Americans will face the consequences: instability, disharmony, and a national loss of the potential productivity of more than a quarter of the population.

However, when economic necessity and the demands of social justice compel us toward social change, those who have the most to gain from change—or the most to lose from its absence—should be responsible for pointing the way.

It is time that blacks recognize our own responsibility. When we react to the rumor of inferiority by avoiding intellectual engagement, and when we allow our children to do so, black people forfeit the opportunity for intellectual development that could extinguish the debate about our capacities, and set the stage for group progress. Blacks must hold ourselves accountable for the resulting waste of talent—and valuable time. Black people have everything to gain—in stature, self-esteem, and problem-solving capability—from a more aggressive and confident approach to intellectual competition. We must assume responsibility for our own performance and development.

ALIENATION

AND THE FOUR WORLDS OF CHILDHOOD

The forces that produce youthful alienation are growing in strength and scope, says Mr. Bronfenbrenner. And the best way to counteract alienation is through the creation of connections or links throughout our culture. The schools can build such links.

Urie Bronfenbrenner

Urie Bronfenbrenner is Jacob Gould Shurman Professor of Human Development and Family Studies and of Psychology at Cornell University, Ithaca, N.Y.

To be alienated is to lack a sense of belonging, to feel cut off from family, friends, school, or work—the four worlds of childhood.

At some point in the process of growing up, many of us have probably felt cut off from one or another of these worlds, but usually not for long and not from more than one world at a time. If things weren't going well in school, we usually still had family, friends, or some activity to turn to. But if, over an extended period, a young person feels unwanted or insecure in several of these worlds simultaneously or if the worlds are at war with one another, trouble may lie ahead.

What makes a young person feel that he or she doesn't belong? Individual differences in personality can certainly be one cause, but, especially in recent years, scientists who study human behavior and development have identified an equal (if not even more powerful) factor: the circumstances in which a young person lives.

Many readers may feel that they recognize the families depicted in the vignettes that are to follow. This is so because they reflect the way we tend to look at families today: namely, that we see parents as being good or not-so-good without fully taking into account the circumstances in their lives.

Take Charles and Philip, for example. Both are seventh-graders who live in a middle-class suburb of a large U.S. city. In many ways their surroundings seem similar; yet, in terms of the risk of alienation, they live in rather different worlds. See if you can spot the important differences.

CHARLES

The oldest of three children, Charles is amiable, outgoing, and responsible. Both of his parents have full-time jobs outside the home. They've been able to arrange their working hours, however, so that at least one of them is at home when the children return from school. If for some reason they can't be home, they have an arrangement with a neighbor, an elderly woman who lives alone. They can phone her and ask her to look after the children until they arrive. The children have grown so fond of this woman that she is like another grandparent—a nice situation for them, since their real grandparents live far away.

Homework time is one of the most important parts of the day for Charles and his younger brother and sister. Charles's parents help the children with their homework if they need it, but most of the time they just make sure that the children have a period of peace and quiet—without TV—in which to do their work. The children are allowed to watch television one hour each

night—but only after they have completed their homework. Since Charles is doing well in school, homework isn't much of an issue, however.

Sometimes Charles helps his mother or father prepare dinner, a job that everyone in the family shares and enjoys. Those family members who don't cook on a given evening are responsible for cleaning up.

Charles also shares his butterfly collection with his family. He started the collection when he first began learning about butterflies during a fourth-grade science project. The whole family enjoys picnicking and hunting butterflies together, and Charles occasionally asks his father to help him mount and catalogue his trophies.

Charles is a bit of a loner. He's not a very good athlete, and this makes him somewhat self-conscious. But he does have one very close friend, a boy in his class who lives just down the block. The two boys have been good friends for years.

Charles is a good-looking, warm, happy young man. Now that he's beginning to be interested in girls, he's gratified to find that the interest is returned.

PHILIP

Philip is 12 and lives with his mother, father, and 6-year-old brother. Both of his parents work in the city, commuting more than an hour each way. Pandemonium strikes every weekday morning as

From *Phi Delta Kappan*, February 1986, pp. 430-436. Reprinted by permission of the author and Phi Delta Kappan.

the entire family prepares to leave for school and work.

Philip is on his own from the time school is dismissed until just before dinner, when his parents return after stopping to pick up his little brother at a nearby day-care home. At one time, Philip took care of his little brother after school, but he resented having to do so. That arrangement ended one day when Philip took his brother out to play and the little boy wandered off and got lost. Philip didn't even notice for several hours that his brother was missing. He felt guilty at first about not having done a better job. But not having to mind his brother freed him to hang out with his friends or to watch television, his two major after-school activities.

The pace of their life is so demanding that Philip's parents spend their weekends just trying to relax. Their favorite weekend schedule calls for watching a ball game on television and then having a cookout in the back yard. Philip's mother resigned herself long ago to a messy house; pizza, TV dinners, or fast foods are all she can manage in the way of meals on most nights. Philip's father has made it clear that she can do whatever she wants in managing the house, as long as she doesn't try to involve him in the effort. After a hard day's work, he's too tired to be interested in housekeeping.

Philip knows that getting a good education is important; his parents have stressed that. But he just can't seem to concentrate in school. He'd much rather fool around with his friends. The thing that he and his friends like to do best is to ride the bus downtown and go to a movie, where they can show off, make noise, and make one another laugh.

Sometimes they smoke a little marijuana during the movie. One young man in Philip's social group was arrested once for having marijuana in his jacket pocket. He was trying to sell it on the street so that he could buy food. Philip thinks his friend was stupid to get caught. If you're smart, he believes, you don't let that happen. He's glad that his parents never found out about the incident.

Once, he brought two of his friends home during the weekend. His parents told him later that they didn't like the kind of people he was hanging around with. Now Philip goes out of his way to keep his friends and his parents apart.

THE FAMILY UNDER PRESSURE

In many ways the worlds of both

> **I**nstitutions that play important roles in human development are rapidly being eroded, mainly through benign neglect.

teenagers are similar, even typical. Both live in families that have been significantly affected by one of the most important developments in American family life in the postwar years: the employment of both parents outside the home. Their mothers share this status with 64% of all married women in the U.S. who have school-age children. Fifty percent of mothers of preschool children and 46% of mothers with infants under the age of 3 work outside the home. For single-parent families, the rates are even higher: 53% of all mothers in single-parent households who have infants under age 3 work outside the home, as do 69% of all single mothers who have school-age children.[1]

These statistics have profound implications for families — sometimes for better, sometimes for worse. The determining factor is how well a given family can cope with the "havoc in the home" that two jobs can create. For, unlike most other industrialized nations, the U.S. has yet to introduce the kinds of policies and practices that make work life and family life compatible.

It is all too easy for family life in the U.S. to become hectic and stressful, as both parents try to coordinate the disparate demands of family and jobs in a world in which everyone has to be transported at least twice a day in a variety of directions. Under these circumstances, meal preparation, child care, shopping, and cleaning — the most basic tasks in a family — become major challenges. Dealing with these challenges may sometimes take precedence over the family's equally important child-rearing, educational, and nurturing roles.

But that is not the main danger. What

threatens the well-being of children and young people the most is that the external havoc can become internal, first for parents and then for their children. And that is exactly the sequence in which the psychological havoc of families under stress usually moves.

Recent studies indicate that conditions at work constitute one of the major sources of stress for American families.[2] Stress at work carries over to the home, where it affects first the relationship of parents to each other. Marital conflict then disturbs the parent/child relationship. Indeed, as long as tensions at work do not impair the relationship between the parents, the children are not likely to be affected. In other words, the influence of parental employment on children is indirect, operating through its effect on the parents.

That this influence is indirect does not make it any less potent, however. Once the parent/child relationship is seriously disturbed, children begin to feel insecure — and a door to the world of alienation has been opened. That door can open to children at any age, from preschool to high school and beyond.

My reference to the world of school is not accidental, for it is in that world that the next step toward alienation is likely to be taken. Children who feel rootless or caught in conflict at home find it difficult to pay attention in school. Once they begin to miss out on learning, they feel lost in the classroom, and they begin to seek acceptance elsewhere. Like Philip, they often find acceptance in a group of peers with similar histories who, having no welcoming place to go and nothing challenging to do, look for excitement on the streets.

OTHER INFLUENCES

In contemporary American society the growth of two-wage-earner families is not the only — or even the most serious — social change requiring accommodation through public policy and practice in order to avoid the risks of alienation. Other social changes include lengthy trips to and from work; the loss of the extended family, the close neighborhood, and other support systems previously available to families; and the omnipresent threat of television and other media to the family's traditional role as the primary transmitter of culture and values. Along with most families today, the families of Charles and Philip are experiencing the unraveling and disintegration of social institutions that in the past were central to the health and well-being of children and their parents.

Notice that both Charles and Philip come from two-parent, middle-class families. This is still the norm in the U.S. Thus neither family has to contend with two changes now taking place in U.S. society that have profound implications for the future of American families and the well-being of the next generation. The first of these changes is the increasing number of single-parent families. Although the divorce rate in the U.S. has been leveling off of late, this decrease has been more than compensated for by a rise in the number of unwed mothers, especially teenagers. Studies of the children brought up in single-parent families indicate that they are at greater risk of alienation than their counterparts from two-parent families. However, their vulnerability appears to have its roots not in the single-parent family structure as such, but in the treatment of single parents by U.S. society.[3]

In this nation, single parenthood is almost synonymous with poverty. And the growing gap between poor families and the rest of us is today the most powerful and destructive force producing alienation in the lives of millions of young people in America. In recent years, we have witnessed what the U.S. Census Bureau calls "the largest decline in family income in the post-World War II period." According to the latest Census, 25% of all children under age 6 now live in families whose incomes place them below the poverty line.

COUNTERING THE RISKS

Despite the similar stresses on their families, the risks of alienation for Charles and Philip are not the same. Clearly, Charles's parents have made a deliberate effort to create a variety of arrangements and practices that work against alienation. They have probably not done so as part of a deliberate program of "alienation prevention" — parents don't usually think in those terms. They're just being good parents. They spend time with their children and take an active interest in what their children are thinking, doing, and learning. They control their television set instead of letting it control them. They've found support systems to back them up when they're not available.

Without being aware of it, Charles's parents are employing a principle that the great Russian educator Makarenko employed in his extraordinarily success-ful programs for the reform of wayward adolescents in the 1920s: "The maximum of support with the maximum of challenge."[4] Families that produce effective, competent children often follow this principle, whether they're aware of it or not. They neither maintain strict control nor allow their children total freedom. They're always opening doors — and then giving their children a gentle but firm shove to encourage them to move on and grow. This combination of support and challenge is essential, if children are to avoid alienation and develop into capable young adults.

From a longitudinal study of youthful alienation and delinquency that is now considered a classic, Finnish psychologist Lea Pulkkinen arrived at a conclusion strikingly similar to Makarenko's. She found "guidance" — a combination of love and direction — to be a critical predictor of healthy development in youngsters.[5]

No such pattern is apparent in Philip's family. Unlike Charles's parents, Philip's parents neither recognize nor respond to the challenges they face. They have dispensed with the simple amenities of family self-discipline in favor of whatever is easiest. They may not be indifferent to their children, but the demands of their jobs leave them with little energy to be actively involved in their children's lives. (Note that Charles's parents have work schedules that are flexible enough to allow one of them to be at home most afternoons. In this regard, Philip's family is much more the norm, however. One of the most constructive steps that employers could take to strengthen families would be to enact clear policies making such flexibility possible.)

But perhaps the clearest danger signal in Philip's life is his dependence on his peer group. Pulkkinen found heavy reliance on peers to be one of the strongest predictors of problem behavior in adolescence and young adulthood. From a developmental viewpoint, adolescence is a time of challenge — a period in which young people seek activities that will serve as outlets for their energy, imagination, and longings. If healthy and constructive challenges are not available to them, they will find their challenges in such peer-group-related behaviors as poor school performance, aggressiveness or social withdrawal (sometimes both), school absenteeism or dropping out, smoking, drinking, early and promiscuous sexual activity, teenage parenthood, drugs, and juvenile delinquency.

This pattern has now been identified in a number of modern industrial societies, including the U.S., England, West Germany, Finland, and Australia. The pattern is both predictable from the circumstances of a child's early family life and predictive of life experiences still to come, e.g., difficulties in establishing relationships with the opposite sex, marital discord, divorce, economic failure, criminality.

If the roots of alienation are to be found in disorganized families living in disorganized environments, its bitter fruits are to be seen in these patterns of disrupted development. This is not a harvest that our nation can easily afford. Is it a price that other modern societies are paying, as well?

A CROSS-NATIONAL PERSPECTIVE

The available answers to that question will not make Americans feel better about what is occurring in the U.S. In our society, the forces that produce youthful alienation are growing in strength and scope. Families, schools, and other institutions that play important roles in human development are rapidly being eroded, *mainly through benign neglect.* Unlike the citizens of other modern nations, we Americans have simply not been willing to make the necessary effort to forestall the alienation of our young people.

As part of a new experiment in higher education at Cornell University, I have been teaching a multidisciplinary course for the past few years titled "Human Development in Post-Industrial Societies." One of the things we have done in that course is to gather comparative data from several nations, including France, Canada, Japan, Australia, Germany, England, and the U.S. One student summarized our findings succinctly: "With respect to families, schools, children, and youth, such countries as France, Japan, Canada, and Australia have more in common with each other than the United States has with any of them." For example:

• The U.S. has by far the highest rate of teenage pregnancy of any industrialized nation — twice the rate of its nearest competitor, England.
• The U.S. divorce rate is the highest in the world — nearly double that of its nearest competitor, Sweden.
• The U.S. is the only industrialized society in which nearly one-fourth of all infants and preschool children live in families whose incomes fall below the

poverty line. These children lack such basics as adequate health care.

• The U.S. has fewer support systems for individuals in all age groups, including adolescence. The U.S. also has the highest incidence of alcohol and drug abuse among adolescents of any country in the world.[6]

All these problems are part of the unraveling of the social fabric that has been going on since World War II. These problems are not unique to the U.S., but in many cases they are more pronounced here than elsewhere.

WHAT COMMUNITIES CAN DO

The more we learn about alienation and its effects in contemporary postindustrial societies, the stronger are the imperatives to counteract it. If the essence of alienation is disconnectedness, then the best way to counteract alienation is through the creation of connections or links.

For the well-being of children and adolescents, the most important links must be those between the home, the peer group, and the school. A recent study in West Germany effectively demonstrated how important this basic triangle can be. The study examined student achievement and social behavior in 20 schools. For all the schools, the researchers developed measures of the links between the home, the peer group, and the school. Controlling for social class and other variables, the researchers found that they were able to predict children's behavior from the number of such links they found. Students who had no links were alienated. They were not doing well in school, and they exhibited a variety of behavioral problems. By contrast, students who had such links were doing well and were growing up to be responsible citizens.[7]

In addition to creating links within the basic triangle of home, peer group, and school, we need to consider two other structures in today's society that affect the lives of young people: the world of work (for both parents and children) and the community, which provides an overarching context for all the other worlds of childhood.

Philip's family is one example of how the world of work can contribute to alienation. The U.S. lags far behind other industrialized nations in providing child-care services and other benefits designed to promote the well-being of children and their families. Among the most needed benefits are maternity and paternity leaves, flex-time, job-sharing

> Caring is surely an essential aspect of education in a free society; yet we have almost completely neglected it.

arrangements, and personal leaves for parents when their children are ill. These benefits are a matter of course in many of the nations with which the U.S. is generally compared.

In contemporary American society, however, the parents' world of work is not the only world that both policy and practice ought to be accommodating. There is also the children's world of work. According to the most recent figures available, 50% of all high school students now work part-time — sometimes as much as 40 to 50 hours per week. This fact poses a major problem for the schools. Under such circumstances, how can teachers assign homework with any expectation that it will be completed?

The problem is further complicated by the kind of work that most young people are doing. For many years, a number of social scientists — myself included — advocated more work opportunities for adolescents. We argued that such experiences would provide valuable contact with adult models and thereby further the development of responsibility and general maturity. However, from their studies of U.S. high school students who are employed, Ellen Greenberger and Lawrence Steinberg conclude that most of the jobs held by these youngsters are highly routinized and afford little opportunity for contact with adults. The largest employers of teenagers in the U.S. are fast-food restaurants. Greenberger and Steinberg argue that, instead of providing maturing experiences, such settings give adolescents even greater exposure to the values and lifestyles of their peer group. And the adolescent peer group tends to emphasize immediate gratification and consumerism.[8]

Finally, in order to counteract the

mounting forces of alienation in U.S. society, we must establish a working alliance between the private sector and the public one (at both the local level and the national level) to forge links between the major institutions in U.S. society and to re-create a sense of community. Examples from other countries abound:

• Switzerland has a law that no institution for the care of the elderly can be established unless it is adjacent to and shares facilities with a day-care center, a school, or some other kind of institution serving children.

• In many public places throughout Australia, the Department of Social Security has displayed a poster that states, in 16 languages: "If you need an interpreter, call this number." The department maintains a network of interpreters who are available 16 hours a day, seven days a week. They can help callers get in touch with a doctor, an ambulance, a fire brigade, or the police; they can also help callers with practical or personal problems.

• In the USSR, factories, offices, and places of business customarily "adopt" groups of children, e.g., a day-care center, a class of schoolchildren, or a children's ward in a hospital. The employees visit the children, take them on outings, and invite them to visit their place of work.

We Americans can offer a few good examples of alliances between the public and private sectors, as well. For example, in Flint, Michigan, some years ago, Mildred Smith developed a community program to improve school performance among low-income minority pupils. About a thousand children were involved. The program required no change in the regular school curriculum; its principal focus was on building links between home and school. This was accomplished in a variety of ways.

• A core group of low-income parents went from door to door, telling their neighbors that the school needed their help.

• Parents were asked to keep younger children out of the way so that the older children could complete their homework.

• Schoolchildren were given tags to wear at home that said, "May I read to you?"

• Students in the high school business program typed and duplicated teaching materials, thus freeing teachers to work directly with the children.

• Working parents visited school classrooms to talk about their jobs and about how their own schooling now helped them in their work.

WHAT SCHOOLS CAN DO

As the program in Flint demonstrates, the school is in the best position of all U.S. institutions to initiate and strengthen links that support children and adolescents. This is so for several reasons. First, one of the major — but often unrecognized — responsibilities of the school is to enable young people to move from the secluded and supportive environment of the home into responsible and productive citizenship. Yet, as the studies we conducted at Cornell revealed, most other modern nations are ahead of the U.S. in this area.

In these other nations, schools are not merely — or even primarily — places where the basics are taught. Both in purpose and in practice, they function instead as settings in which young people learn "citizenship": what it means to be a member of the society, how to behave toward others, what one's responsibilities are to the community and to the nation.

I do not mean to imply that such learnings do not occur in American schools. But when they occur, it is mostly by accident and not because of thoughtful planning and careful effort. What form might such an effort take? I will present here some ideas that are too new to have stood the test of time but that may be worth trying.

Creating an American classroom. This is a simple idea. Teachers could encourage their students to learn about schools (and, especially, about individual classrooms) in such modern industrialized societies as France, Japan, Canada, West Germany, the Soviet Union, and Australia. The children could acquire such information in a variety of ways: from reading, from films, from the firsthand reports of children and adults who have attended school abroad, from exchanging letters and materials with students and their teachers in other countries. Through such exposure, American students would become aware of how attending school in other countries is both similar to and different from attending school in the U.S.

But the main learning experience would come from asking students to consider what kinds of things *should* be happening — or not happening — in American classrooms, given our na-

tion's values and ideals. For example, how should children relate to one another and to their teachers, if they are doing things in an *American* way? If a student's idea seems to make sense, the American tradition of pragmatism makes the next step obvious: try the idea to see if it works.

The curriculum for caring. This effort also has roots in our values as a nation. Its goal is to make caring an essential part of the school curriculum. However, students would not simply learn about caring; they would actually engage in it. Children would be asked to spend time with and to care for younger children, the elderly, the sick, and the lonely. Caring institutions, such as daycare centers, could be located adjacent to or even within the schools. But it would be important for young caregivers to learn about the environment in which their charges live and the other people with whom their charges interact each day. For example, older children who took responsibility for younger ones would become acquainted with the younger children's parents and living arrangements by escorting them home from school.

Just as many schools now train superb drum corps, they could also train "caring corps" — groups of young men and women who would be on call to handle a variety of emergencies. If a parent fell suddenly ill, these students could come into the home to care for the children, prepare meals, run errands, and serve as an effective source of support for their fellow human beings. Caring is surely an essential aspect of education in a free society; yet we have almost completely neglected it.

Mentors for the young. A mentor is someone with a skill that he or she wishes to teach to a younger person. To be a true mentor, the older person must be willing to take the time and to make the commitment that such teaching requires.

We don't make much use of mentors in U.S. society, and we don't give much recognition or encouragement to individuals who play this important role. As a result, many U.S. children have few significant and committed adults in their lives. Most often, their mentors are their own parents, perhaps a teacher or two, a coach, or — more rarely — a relative, a neighbor, or an older classmate. However, in a diverse society such as ours, with its strong tradition of volunteerism, potential mentors

abound. The schools need to seek them out and match them with young people who will respond positively to their particular knowledge and skills.

The school is the institution best suited to take the initiative in this task, because the school is the only place in which all children gather every day. It is also the only institution that has the right (and the responsibility) to turn to the community for help in an activity that represents the noblest kind of education: the building of character in the young.

There is yet another reason why schools should take a leading role in rebuilding links among the four worlds of childhood: schools have the most to gain. In the recent reports bemoaning the state of American education, a recurring theme has been the anomie and chaos that pervade many U.S. schools, to the detriment of effective teaching and learning. Clearly, we are in danger of allowing our schools to become academies of alienation.

In taking the initiative to rebuild links among the four worlds of childhood, U.S. schools will be taking necessary action to combat the destructive forces of alienation — first, within their own walls, and thereafter, in the life experience and future development of new generations of Americans.

1. Urie Bronfenbrenner, "New Worlds for Families," paper presented at the Boston Children's Museum, 4 May 1984.

2. Urie Bronfenbrenner, "The Ecology of the Family as a Context for Human Development," *Developmental Psychology*, in press.

3. Mavis Heatherington, "Children of Divorce," in R. Henderson, ed., *Parent-Child Interaction* (New York: Academic Press, 1981).

4. A.S. Makarenko, *The Collective Family: A Handbook for Russian Parents* (New York: Doubleday, 1967).

5. Lea Pulkkinen, "Self-Control and Continuity from Childhood to Adolescence," in Paul Baltes and Orville G. Brim, eds., *Life-Span Development and Behavior*, Vol. 4 (New York: Academic Press, 1982), pp. 64-102.

6. S.B. Kamerman, *Parenting in an Unresponsive Society* (New York: Free Press, 1980); S.B. Kamerman and A.J. Kahn, *Social Services in International Perspective* (Washington, D.C.: U.S. Department of Health, Education, and Welfare, n.d.); and Lloyd Johnston, Jerald Bachman, and Patrick O'Malley, *Use of Licit and Illicit Drugs by America's High School Students — 1975-84* (Washington, D.C.: U.S. Government Printing Office, 1985).

7. Kurt Aurin, personal communication, 1985.

8. Ellen Greenberger and Lawrence Steinberg, *The Work of Growing Up* (New York: Basic Books, forthcoming).

Development During Adolescence and Early Adulthood

The onset of adolescence is demarcated by the emergence of secondary sex characteristics and the achievement of reproductive maturity. However, the transition to adolescence is not only marked by physical changes; substantive shifts also occur in memory and problem-solving skills, preferred activities, and emotional behavior, as discussed in "The Magic of Childhood."

The onset of adulthood is more difficult to distinguish, particularly in modern industrial societies. In some cultures, a ritualistic ceremony marks the transition to adulthood—a transition that occurs quickly, smoothly, and with relatively few problems. In American culture the transition is vague. Does someone become an adult when he or she achieves the right to vote, the privilege of obtaining a driver's license, the ability to legally order an alcoholic drink in a bar, or the right to volunteer for the armed forces?

Some researchers argue that much of the storm and stress attributed to adolescence is a myth, created from an overemphasis on adolescent fads and rebelliousness and an underemphasis on obedience, conformity, and cooperation. Focusing on the negative aspects of adolescent behavior may create a set of expectations which the adolescent then strives to fulfill. However, for some adolescents the transition to adulthood *is* fraught with despair, loneliness, and interpersonal conflict. The pressures of peer group, school, and family may produce conformity or may lead to rebellion against or withdrawal from friends, parents, or society at large. These pressures may peak as the adolescent prepares to separate from the family and assume the independence and responsibilities of adulthood. Bruce Baldwin's analysis in "Puberty and Parents" suggests that adolescence spans a 20-year period, roughly between the ages of 10 and 30, which can be divided into three periods: early adolescence (adhering to tribal loyalties), middle adolescence (testing adult realities), and late adolescence (joining up). Understanding adolescent attitudes and confronting emotional reactions to adolescents can promote parental growth and development as much as it promotes the development of the adolescent.

Although much attention has been given to the problems of adolescence and the transition to adulthood, developmentalists have shown far less interest in the early years of adulthood. Yet during early adulthood many individuals experience significant changes in their lives. Marriage, parenthood, divorce, single parenting, employment, and the effects of sexism may be powerful influences on ego development, self-concept, and personality. Negative emotions, such as jealousy and envy, subvert efforts to establish effective interpersonal relationships and often lead to hostility or isolation. In "Jealousy and Envy," Jon Queijo describes three strategies—self-reliance, selective ignoring, and self-bolstering—that individuals can use in their attempts to cope with negative emotions. Some individuals seem to grow stronger when confronted by the stresses of daily living, whereas others have great difficulty coping. Several developmentalists have elaborated stage theories to describe major periods of crisis. The article "The Prime of Our Lives" reviews several theories, but suggests that changes during adulthood may better be explained by changing perceptions of self than by universally experienced life stages. Irrespective of the cause of changes in adulthood, it is clear that alarming increases in such problems as suicide (analyzed in the article "Suicide") suggest that too many adolescents and young adults are losing the battle against feelings of hopelessness, despair, alienation, and lack of self-control. The challenge for developmentalists is to discover the factors that contribute to one's ability to cope with stress and the natural crises of life with minimum disruption to the integrity of one's personality. One step in this direction has been taken by neuropsychologists, who attempt to connect behavioral patterns with brain function. The article "The Emotional Brain" describes efforts to link the mood states of depression and euphoria to injuries to the left and right hemispheres of the brain, respectively.

Looking Ahead: Challenge Questions

Describe the adolescent's ability to make use of moral and political principles in organizing his or her thinking about social issues. What can be done within our educational system to promote a broader and more complete sense of the adolescent's social environment?

What techniques do you use to deal with your emotional ups and downs? Which do you think are effective and which are ineffective? Do you see any signs of growth or

change in yourself? Why do you think so many adolescents and young adults find it easier to lose themselves in drugs or cults than to confront their problems and take steps to develop self-control and self-reliance? What kinds of parenting techniques might have given such individuals sufficient self-esteem and coping skills to combat their self-doubts and loneliness?

Most stage theories of adult development are based on studies of individuals who were born during the Depression and, for the major portion of their childhood, adolescent, and young adult years, were being reared when the United States was at war. How much do generational (cohort) effects shape one's development? Will this generation's problem of midlife crisis be the focus of research in 2015, or will future researchers adapt their data to fit stage theories of the twentieth century?

Linking behavioral characteristics to structural areas of the brain helps to identify mechanisms that might explain the source of an individual's dysfunction. But does location of brain mechanisms offer any further insight into effective treatments? What kinds of treatment do you think would be most effective in assisting in the reorganization of neural networks associated with such specific emotional dysfunctions as depression?

The Magic of Childhood

IT TAKES MORE THAN A WAVE OF A WAND TO TURN TODDLERS INTO GROWN-UPS. A GROWING MIND IS A TRICKY THING, WITH MANY DIMENSIONS.

Paul Chance and Joshua Fischman

Paul Chance is a contributing editor to Psychology Today *and Joshua Fischman is an assistant editor at the magazine.*

There is something magical about childhood. Like a magician pulling a rabbit out of a hat, childhood transforms small toddlers into full-grown adults. Though growing up takes a lot longer than a hat trick, the feat is really quite remarkable.

As with any magic trick, however, there is a lot more to growing up than meets the eye. Some of the physical changes are easy to spot, but some of the changes in children's mental abilities and social skills can be very subtle. Psychologists, paying very close attention to the magician's movements, have drastically changed their ideas about how this trick of growing up is done.

Twenty years ago, there was very convincing evidence that as far as mental development was concerned, growing up was a matter of advancing through a series of separate, age-related stages. Much of this evidence came from the innovative work of Swiss psychologist Jean Piaget. Piaget argued that biological growth combines with children's interaction with their surroundings to take them, step by step, up a developmental staircase, with no skipping allowed. A child at step two can do things that were impossible at step one, but cannot yet accomplish tasks performed by a child at step three.

For instance, Piaget found that, around the age of 7 or 8, children reach the step of "concrete operations." This is marked by the appearance of an ability that Piaget called "conservation." Before then, children believe that the amount of water in a short, wide glass changes when it is poured into a tall, thin beaker—they think there is more water in the beaker because it is higher, or less because it is thinner. An 8-year-old is able to "conserve" the amount of water, and understands it remains constant despite the change in shape.

Piaget's staircase went up to adulthood in this step-by-step fashion. Each step signaled an increase in the complexity of a child's thinking, and Piaget argued that changes in memory, perceptual skills, learning ability and other aspects of mental development were all based on these orderly steps. His ideas were borne out by dozens of studies, in many different societies, and the notion of age-related stages of growth became very popular.

But exceptions and inconsistencies in this scheme have cropped up in more recent studies. Some children show the ability to conserve liquids or objects long before others do; moreover, children who at first can't do this learn how by discussing their ideas with classmates who do conserve. Stanford University psychologist John Flavell points out another major problem with Piaget's ideas: Stages imply long periods of stability, followed by abrupt change. "But it doesn't happen that way. Most important changes happen gradually, over months or years."

Psychologists have begun to move away from Piaget's concept of broad developmental stages to a more complex view of growth. Instead of searching for major overall transformations, they are examining separate domains such as memory, problem solving, creativity and social interaction. They are discovering that each of these areas has a flexible schedule for change and growth, and that not only age but the quality of the environment affects this schedule.

In the area of memory, psychologists have linked improved recall to a number of memory strategies that children learn, rather than to a set series of stages. They've found that memory ability catapults upward from birth through age 5, and then advances more slowly through middle childhood and adolescence. This improvement may be due to what psychologist Ann L. Brown of the University of Illinois calls "learning how and when to remember."

One strategy children learn is to repeat the items they want to remember. Children begin to rehearse around the age of 3, according to psychologist Scott G. Paris of the University of Michigan and researcher Jill A. Weissberg-Benchell of Case Western Reserve University, and the improvement that follows is due largely to the rehearsal technique. Weissberg-Benchell and Paris watched as a group of children, ages 3 through 6, pretended to go to a

Intelligence Curve

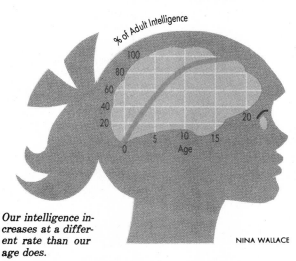

Our intelligence increases at a different rate than our age does.

NINA WALLACE

SOURCE: *STABILITY AND CHANGE IN HUMAN CHARACTERISTICS.* JOHN WILEY & SONS, 1964.

A
CHILD'S MEMORY,
PROBLEM-SOLVING AND
CREATIVE ABILITIES
GROW AT DIFFERENT RATES, NOT IN
OVERALL TRANSFORMATIONS.

grocery store and tried to remember a list of six items to buy. The researchers noticed that the children who rehearsed the list remembered more items. And while 6-year-olds were more likely to rehearse than 3-year-olds, the younger children who rehearsed remembered just as many items as their older playmates did.

As they get older, children replace basic memory strategies like this with more sophisticated methods. Read a list of items, one every five seconds, to 6-year-olds and they are likely to repeat each item over and over until you tell them the next one. Older children usually repeat the last two or three items heard. Adolescents adopt even more sophisticated strategies: Given a list of nouns to recall, they may group them into logical categories such as animals, plants and minerals, or people, places and things.

Piaget might have argued that such sophistication increases naturally as children grow. But children don't grow into these memory strategies automatically; they learn them through experience, often by imitating older children or adults. In school, where they are frequently called upon to memorize, they are taught or encouraged to devise strategies for doing the job. Teachers regularly instruct elementary school students in the use of memory aids such as "Thirty days hath September" and "I before e except after c." Later on, it usually takes teenagers just a few minutes to learn new mnemonic strategies that improve their recall sharply.

Problem solving also improves with age, but here, too, much of the improvement is the result of learning. One of the most important things children learn about problem solving is the value of being systematic. Donald H. Meichenbaum, a psychologist at the University of Waterloo in Canada, has found that children who do poorly at problem solving are often impulsive: They act almost without thinking, rely heavily on trial and error and seldom check for mistakes. Meichenbaum has found that young children can become much better at solving problems if they learn to be a bit more methodical.

Meichenbaum and psychologist Joseph Goodman, who has a private practice in Toronto, had elementary school children watch as an adult worked on a simple problem, copying patterns of lines onto a piece of paper. The adult thought aloud to demonstrate the kinds of mental things good problem-solvers do when at work:

"Okay, what is it I have to do? You want me to copy the picture with the different lines. I have to go slow and careful. Okay, draw the line down, down, good; then to the right, that's it; now down some more and to the left. Good, I'm doing fine so far. Remember, go slow. Now back up again. No, I was supposed to go down. That's okay. Just erase the line carefully . . . . Good. Even if I make an error I can go on slowly and carefully. Okay, I have to go down now. Finished. I did it."

The children then took turns while the adult looked on, giving them constant instruction and encouragement. Even the most impulsive children showed marked improvement.

Surprisingly, young children can also learn to do the kind of systematic hypothesis testing needed to solve scientific problems. In the pendulum problem, devised by Piaget and psychologist Barbel Inhelder, children try to figure out why pendulums of different lengths and weights swing back and forth at different speeds: Is it the length of the string, the weight used, the force with which the weight is released, the height from which it is released or some combination of these variables? (In fact, only the length of the string is important.)

Piaget and Inhelder found that only children 14 and older could solve the pendulum problem on their own. Piaget concluded that children don't develop the capacity for systematically testing their ideas until adolescence. But psychologist Robert S. Siegler, now at Carnegie-Mellon University, and his colleagues found that 10-year-olds could learn to solve the pendulum problem if they were taught some simple skills of logic, and then used those skills to find the right answer. "There is a great deal of difference between what children know how to do and what they can learn to do," Siegler says.

The pendulum problem illustrates a problem with Piaget's work: His stages were centered on children's growing ability to use logical thought and deductive reasoning, a limitation that ignores "nonscientific" areas such as creativity. Researchers have recently started to make up for this neglect, and their exploration has convinced them that creativity is yet another domain that follows its own path of growth and change.

Harvard psychologist Howard E. Gardner has studied children's stories, drawings and play, and notes that creativity develops rapidly in early childhood, often reaching a peak at around 7. At this age, children are willing "to create new figures of speech, to combine forms and colors in innovative ways, to juxtapose elements which are normally kept asunder," Gardner writes. But he finds that as children begin school they become more reluctant to make creative leaps.

Many psychologists believe this creative decline during middle childhood has a lot to do with the behavior of adults. The 5-year-old who colors outside of the lines and makes clouds blue and grass red is praised, but the 7-year-old is told that coloring outside the lines is sloppy work and that the colors must be "true to life." School, with its emphasis on right answers and right ways of doing things, apparently sup-

CHILDREN'S TOP TEN ACTIVITIES
AVERAGE HOURS AND MINUTES SPENT EACH DAY

Activities	WEEKDAY age 3-5	6-8	9-11	12-14	15-17	WEEKEND age 3-5	6-8	9-11	12-14	15-17
SLEEPING	10:30	9:55	9:08	7:53	8:19	10:34	10:41	9:56	10:04	9:22
SCHOOL	2:17	4:52	5:15	5:44	5:14	-	-	-	-	-
TELEVISION	1:51	1:39	2:26	2:22	1:48	2:02	2:16	3:05	2:49	2:37
PLAYING	3:38	1:51	1:05	0:31	0:14	4:27	3:00	1:32	0:35	0:21
EATING	1:22	1:21	1:13	1:09	1:07	1:21	1:20	1:18	1:08	1:05
PERSONAL CARE	0:41	0:49	0:40	0:56	1:00	0:47	0:45	0:44	1:00	0:51
HOUSEHOLD WORK	0:14	0:15	0:18	0:27	0:34	0:17	0:27	0:51	1:12	1:00
SPORTS	0:05	0:24	0:21	0:40	0:46	0:03	0:30	0:42	0:51	0:37
CHURCH	0:04	0:09	0:09	0:09	0:03	0:55	0:56	0:53	0:32	0:37
VISITING SOMEONE ELSE	0:14	0:15	0:10	0:21	0:20	0:10	0:08	0:13	0:22	0:56

SOURCE: "HOW CHILDREN USE TIME" IN *TIME, GOODS, AND WELL-BEING*. © INSTITUTE FOR SOCIAL RESEARCH, UNIVERSITY OF MICHIGAN, 1985.

THE ONE WHO HAS THE MOST TOYS WHEN HE DIES, WINS

That bumper sticker could easily be the motto of today's college freshmen. It represents values that are a far cry from those of students two decades ago, when conservative social critics lambasted college students for being naïve, idealistic intellectuals. What business did students have picketing, campaigning for candidates, burning draft cards, protesting injustice? If they would spend less time filling their heads with literature, philosophy and the social sciences, the critics complained, and more time on practical matters such as making a living, they and the country would be better off.

Those critics stand silent today, in awe perhaps at the extent to which college students are heeding their advice. Today's freshmen seem concerned with little other than the practical matter of making a living. They are pragmatic and materialistic to a fault, or so it seems from the latest survey by the Cooperative Institutional Research Program (CIRP), an affiliate of the University of California, Los Angeles. Every fall since 1966, CIRP has surveyed a representative sample of 300,000 of the nation's college freshmen. A comparison of answers given in 1966 with those given in 1986 reveals some striking—some will say disturbing—trends.

Contrary to popular opinion, students have not become markedly more conservative over the years: They still want arms reduction, less military spending, a clean environment, a fair shake for minorities and women. But they think government should take care of these matters while they busy themselves with the task of getting on in life.

Getting on means, according to CIRP pollsters Alexander W. Astin and Kenneth C. Green, accumulating money, power and status. Money seems to be especially important. In 1986, more than 70 percent of college freshmen said that a major reason for attending college was "to be able to make more money," and a like number of students said that to "be very well-off financially" was an essential or very important goal. These figures are up nearly a third in the past 10 years.

Today's materialism is reflected in the choice of majors and in career plans. Psychology has held its own as a major, but other traditional liberal-arts majors such as foreign languages, history and philosophy have lost so much ground that Astin and Green worry that they may disappear from some campuses entirely. The number of freshmen planning to major in business,

however, has nearly doubled over the past two decades, from about 14 percent to slightly less than 27 percent.

It seems clear that the appeal of business is the promise of high starting salaries and the prospect of even greater wealth later on. And students want to get rich quickly, with a minimum of academic preparation. Any career that requires schooling beyond the bachelor's degree is shunned. This is true

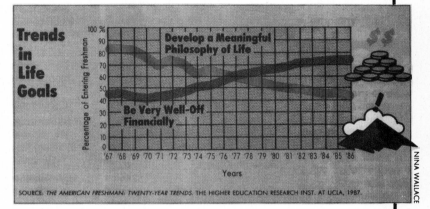

SOURCE: *THE AMERICAN FRESHMAN: TWENTY-YEAR TRENDS.* THE HIGHER EDUCATION RESEARCH INST. AT UCLA, 1987.

Today's college freshman may view making a lot of money as a kind of "philosophy of life" in itself.

not only of professions that are not very lucrative, such as research and college teaching, but of well-paid professions, such as medicine and law. Students have evidently decided that the potential income of these fields does not warrant the extra years of training. In the business jargon these students like so well, becoming a physician, lawyer, social worker, teacher or scientist is not "cost-effective." The fact that these and other fields offer nonmaterial rewards seems not to be part of the formula for career choice.

Astin and Green also suggest that students are steering clear of the arts and sciences because these majors and career choices require high-level verbal and critical-thinking skills, skills that have declined markedly since 1966. Student choices therefore reflect insight into their weaknesses. But Astin and Green believe that a radical change in values has also taken place. Twenty years ago more than 80 percent of freshmen said that developing a meaningful philosophy of life was an important or essential goal. Today only about 41 percent of freshmen consider that goal worthy. Astin and Green note that it could be that making a lot of money precludes the need for a philosophy of life: "It may be that some students view making a lot of money as a kind of 'philosophy of life' in itself."

—*Paul Chance*

THE UPS AND DOWNS OF TEENAGE LIFE

The world of the teenager often appears to be a flat landscape of unchanging, unexciting routine. "Nothing much" is the usual teenage reply when adults ask what they've been doing. "You know, school, homework, hanging out with some friends—the usual."

But these ordinary activities inspire a host of teenage thoughts and feelings that are anything but run-of-the-mill. Inside, daily teenage life consists of a multitude of emotional peaks and valleys, ranging from exhilarating highs to depressing lows. Much of the conflict and growth during the teenage years comes as teenagers try to find their way through this constantly shifting internal landscape, en route to adulthood.

The emotional terrain of teenagers has been charted, in great detail, by University of Chicago psychologist Mihaly Csikszentmihalyi and University of Illinois psychologist Reed Larson. They asked 75 Chicago-area high school students, from a variety of backgrounds, to wear electronic pagers for one week. When they were signaled, every two hours, the teens filled out a report of what they were doing, how they felt while doing it, and answered other questions about their experiences. At the end of the week, the teenagers filled out various questionnaires about their moods, both in general and during specific activities.

The results, published in Csikszentmihalyi and Larson's book, *Being Adolescent: Conflict and Growth in the Teenage Years* (Basic Books), show that teenagers spend about 40 percent of their waking time in leisure activities: socializing with friends, watching TV, playing sports or pursuing hobbies, or simply "thinking." The remainder of their time is split fairly evenly between maintenance activities, such as chores, errands, eating and transportation, and the productive activities of school, studying and jobs.

These broad categories hide a remarkable diversity. School, for example, involves encounters with friends and teachers, and opportunities for sports, music and arts, as well as study. Classes change every hour, and students feel quite differently about each one. And school takes up only part of their lives. During a single day, a teenager can be a student, a worker, a child, a teammate, an artist, an outcast, a friend, a rebel and more.

Three days in the life of Kathy, a very directed student and accomplished violinist, and in the life of Greg, a disaffected, rebellious youth, reveal these rapid shifts (see chart). These changing roles and surround-

ings cause teenagers' moods to vary much more often than the moods of working adults. Csikszentmihalyi, who has studied adults using the same beeper technique, says the typical teenager comes down from extreme happiness or up from deep sadness within 45 minutes; in contrast, adult happy or sad moods usually extend for several hours. Teenagers are unhappy with the drudgery of schoolwork, jobs and chores, and their moods follow suit; they would much rather be with friends, away from the control of adults and adult-oriented activities.

This doesn't mean that teenagers simply like to play and hate to work. They get the most satisfaction from meeting challenges that fit their developing skills and provide them with meaningful rewards. Csikszentmihalyi describes this match between abilities and demand as "flow." Teenagers who

---

ORDINARY ACTIVITIES INSPIRE A HOST OF TEENAGE THOUGHTS AND FEELINGS THAT ARE ANYTHING BUT RUN-OF-THE-MILL. DAILY LIFE CONSISTS OF MANY EMOTIONAL PEAKS AND VALLEYS.

---

are active in sports and hobbies, for example, are stimulated to develop new levels of expertise and accomplishment, and their success then stimulates them to search for fresh challenges, slightly harder but still within reach. This growing involvement and ability, Csikszentmihalyi and Larson argue, is what takes adolescents beyond the impulsive, egocentric activities of childhood, into a more adult world of widely shared rules, symbols and communication.

But teenagers easily fall into less productive, less satisfying activities that take up large amounts of their time, like watching television and engaging in aimless conversation with friends. These are the paths of least resistance, often taken just to avoid boredom. Many teenagers have not learned to make the connection between a challenge and feelings of personal fulfillment, Csikszentmihalyi reports. Instead, they search for short-term pleasures, and they are disappointed in the results.

An even bigger problem is that teenagers' lives encompass many things that drag them down. Besides schoolwork or family chores, teens must contend with parents who "don't understand" them, breakups with girlfriends or boyfriends, invasions of privacy, worries about their personal appearance or popularity and a host of other "downers."

Coming to terms with the downs as well as the ups, Csikszentmihalyi says, is one of the greatest challenges that teenagers face. Rather than being thrown completely off stride by bad experiences, teenagers must find ways to maintain some balance in their lives. They need to find order amid the disorder—turn boredom into useful reflection, strive to see their parents' point of view after an argument instead of just brooding, and learn from mistakes.

This shift in perspective comes with age. The reality of bad or unpleasant experiences doesn't change; what changes is the way teenagers interpret them. When Csikszentmihalyi and Larson repeated the study two years later with some of the same students, now juniors and seniors, the students saw the world in a different way. "It used to get on my nerves, the time with my family," one girl recalled. "I still get mad but not that mad. A lot of my friends went off to school, so they're gone, but my family will be there; I think I accept them more."

Overall, older teenagers reported that their lives had become better in the two years. More than half reported they were happier with their families, while only 12 percent said they were less happy; the same pattern held true for teenagers' feelings about their friends and about time spent by themselves. "Instead of thinking I'm alone because others aren't with me, it is because I choose to be alone . . . ," one teenager said.

Not every teenager embraces this increasingly complex view of life. "Some kids simply split off work from play and enjoyment," Csikszentmihalyi says. "This splits their psychic life and is dangerous. They work, then they go off with friends to smoke pot." They rely on crutches, instead of developing skills that will serve them later on. "They stay adolescents forever."

Perhaps the best help adults can offer teenagers, Csikszentmihalyi and Larson conclude, is to provide models, "examples of how to choose among goals, how to persevere, how to have patience, how to recognize the challenges of life and enjoy meeting them . . . letting them see that achieving control over experience can bring serenity and enjoyment in its wake."

—Joshua Fischman

Three Days in the Life of Greg

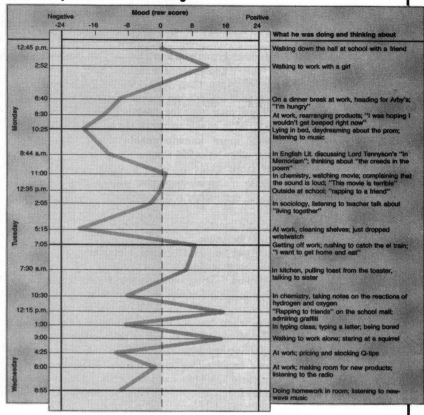

Three Days in the Life of Kathy

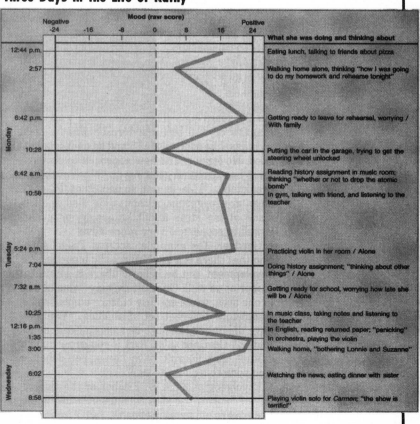

BY M. CSIKSZENTMIHALYI AND R. LARSON © 1984 BASIC BOOKS. REPRINTED BY PERMISSION OF THE PUBLISHER

ADAPTED FROM BEING ADOLESCENT

presses creative growth. Children get the message quickly, and spend a lot of energy learning how to do things "right"—the way adults do. "This is a time of rules, regulations, principles—a kind of narrow but dogged skill-building...," Gardner and his colleagues write.

The decline of creative ability with age is not, however, inevitable. Psychologist E. Paul Torrance, professor emeritus from the University of Georgia, has designed learning experiences that increase creativity in schoolchildren without leading to behavior problems. For instance, he asked teachers to be more tolerant of unusual questions (such as "Do rocks grow?"), to praise original ideas and to provide ungraded activities. It worked. "I gave a creativity test to one class of first-graders at the end of the school year," Torrance reports. The students who had received this training "were so well behaved, without any of the bickering, fighting and pushing that one so often sees in first-graders, that I thought, 'Surely, these kids must have had their creativity driven out of them.' But to my delight they produced far more original ideas than a group of children that had not had the training."

Gardner notes that creativity receives new stimulation as teenagers rebel against adult norms and once again seek to personalize the language, art, music and other cultural realms they had mastered earlier. Sometimes this rebellion leads to disruptive behavior, as many parents know, but it can also mark a return to the creative patterns of early childhood and yield a new burst of growth in creative ability. The teenager who rediscovers the joy of making clouds blue and grass red may go on to develop high-level artistic skills.

Social skills is another area that defies the neat stratification of stages. Psychologists are finding that becoming socially adept is partly a matter of learning that each kind of social interaction has its distinctive script. Psychologists Roger C. Schank and Robert P. Abelson of Yale University offer restaurant dining as a good example. Very young children have no clear idea of dining out as a distinct set of interactions, but this changes as they eat out more. Children come to recognize that they, their parents and others are performing in a kind of play, following a script that gives each person lines to say and acts to perform.

A child's first restaurant script is very crude, but as the child becomes more familiar with the roles and actions, the script becomes more sophisticated. For instance, when Schank asked his 3-year-old daughter, Hana, to tell him what happens in a restaurant, she replied, "You sit down and you, uh, eat food." On questioning, she was able to say that one got food from a waitress, but she was unclear what happens after a meal. Asked if one simply left the restaurant, she replied, "No, the waitress gives you

some money and you pass some money to her and she gives you some money back to you and then you leave." Ten months later, when Schank asked her what happens in a restaurant, Hana answered, "You come in and you sit down at a table. And then the waitress comes. And she gives you a menu. And then she takes it back and writes down your order. And then you eat what she gave you. And then, you get up from the table. And you pay the money and then you walk out...."

It isn't just by visiting restaurants that children learn such scripts. In early childhood, imaginative play provides opportunities to gain information from others as well as to practice what the child has observed firsthand. Through such social exchanges, children learn what happens and how they are to behave in schools, churches, at parties, in a dentist's office and so on. Schank notes that by age 4, Hana had a working knowledge of most scripts that are familiar to adults, though in less detail.

In middle childhood and adolescence, children learn to flesh out these scripts in much the same way, sharing what they have learned with friends. In adolescence, children learn some of the more subtle nuances, such as the different behavior that is expected in informal eating places and in "fancy" restaurants. Adolescence is also a time for experimenting in new situations—dating and working, for example—and developing a greater understanding of these scripts (see "The Ups and Downs of Teenage Life," this article). Developing social knowledge and skill, then, is a long, complex learning process.

Much of the improvement in these skills, as well as others such as abstract reasoning and language comprehension, depends upon the tremendous growth of knowledge. This is especially true in adolescence. Skills appear to have improved more than they really have, because increased knowledge makes many tasks easier.

Take memory, for instance. Since familiar material is easier to recall than unfamiliar material, more knowledgeable people have an edge when asked to remember something: They are more likely to have already encountered similar material. A person who speaks French should remember a list of French words better than someone who knows no French, even though neither has ever seen the particular words before.

"What the head knows," concludes Flavell, "has an enormous effect on what the head remembers." Young chess experts, for example, can remember chessboard positions much better than adults who don't play a lot of chess, even when seeing the positions for the first time. Michelene T.H. Chi, a psychologist at the University of Pittsburgh, compared six young chess experts, 10 years old on average, with graduate students and research assistants who

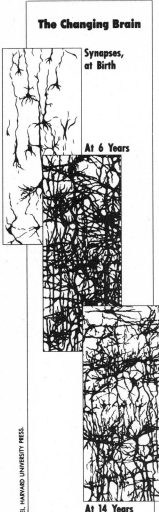

ADAPTED FROM *THE POSTNATAL DEVELOPMENT OF THE HUMAN CORTEX*, J.L. CONEL, HARVARD UNIVERSITY PRESS.

The Changing Brain

Synapses, at Birth

At 6 Years

At 14 Years

As a child grows into an adult, the maturing brain might lose 40 percent of its synapses, the connections between brain cells, in a particular region. This streamlining may follow childhood experience: The neural connections that are used most often are strengthened and the unused ones are stripped away.

knew much less about the game. When Chi showed both groups a series of eight midgame positions on a chessboard, the youngsters were markedly better at remembering these positions than the grown-ups were. The children didn't do as well as the adults, however, when both groups were asked to remember a random series of numbers. The children's expertise in chess gave them superior memories, but only in chess.

Growing up now looks quite different than it did 20 years ago. Then, one could gaze into Piaget's crystal ball and foresee what a child would be like at 2, at 4, at 8 and so on into adulthood. No longer. The new approach to development has revealed growing up to be much less predictable. This is not to say that Piaget's ideas are forgotten. Although he died in 1980, some of his ideas about stages, greatly narrowed in scope, are very much alive.

Gardner, for example, accepts Piaget's notion that there is a kind of genetic blueprint for development, but he believes there is a cultural blueprint as well. One culture may value ability in one area, while another culture emphasizes a completely different set of abilities. "You may have the innate talent to be a magnificent chess player," Gardner says, "but if you grow up in a culture that doesn't have chess, you'd better find something else to do."

Psychologist Kurt W. Fischer of the University of Denver has identified a series of Piaget-like stages of mental growth, but Fischer argues that a child's particular stage changes as the child's surroundings change. Fischer has been watching how children of various ages play with dolls, using this play to recreate different social situations. The striking thing, he says, is that a child's ability to do this depends not so much on age, but on receiving support and coaching. Without this help, 6-year-olds will act out scenes no more complex than scenes acted out by 4-year-olds or by even younger children. "The basic idea of Piagetian stage theory is that a child either has a competence or

*W*HILE THE POWER OF LEARNING AND EXPERIENCE TO SHAPE A CHILD IS GREAT, SO IS THE COMPLEXITY OF THE SHAPING PROCESS.

doesn't," Fischer says. "We're saying that competence depends on context. Change the context and you change the competence."

Parents and teachers have been quick to realize that if children's abilities depend largely on experience instead of age, child-rearing practices and education can have a tremendous impact. And as a result, childhood itself has changed for many kids: They are no longer locked into fixed stages of mental growth, but are pushed to develop advanced skills at ages that would have seemed ridiculously young a few years ago. Aggressive educational programs, even in preschool, are commonplace today.

Yet an enriched environment is no guarantee of a superkid, or of a superadult. In the words of Harvard psychologist Jerome Kagan, "Each person can be understood only as a coherence of many, many past events." While the power of learning and experience to shape a child is great, so is the complexity of the shaping process. Appreciating this complexity, and becoming aware of the many dimensions of a child's growth, may be the first step in understanding the magic of childhood.

Puberty and Parents

Understanding Your Early Adolescent

Dr. Bruce A. Baldwin

Dr. Baldwin is a practicing psychologist who heads Direction Dynamics, a consulting service specializing in promoting professional development and quality of life in achieving men and women. He responds to many requests each year for seminars on topics of interest to professional organizations and businesses.

For busy achievers and involved parents, Dr. Baldwin has authored a popular, positive parenting cassette series and a new book, It's All In Your Head: Lifestyle Management Strategies for Busy People! *Both are available in bookstores or from Direction Dynamics in Wilmington, N.C.*

In the large auditorium, concerned parents wait for the program to begin. The speaker appears to talk about the problems of parenting in the eighties. The program begins with a question to the audience: "How many of you would choose to live your adolescent years over if you had the chance?" Relatively few hands are raised and some of them waver indecisively. For just a few, the adolescent years are some of the best. The majority, however, are happy to have reached adulthood and put those tumultuous years behind them.

Then a second question: "How many of you would choose to live your adolescent years over if you had to do it *right now*?" This time, practically no hands are raised. The fact is that in any era, early adolescence is a most difficult time of life. On the other hand, there is ample evidence that this critical period of growth and change for young people is steadily becoming more difficult to negotiate emotionally. Caring parents seem to sense this and they are afraid for their children. Sadly,

their intuitive awareness is quite accurate: what they remember as the simpler world of their own youth has changed irrevocably.

Still, beyond the social environment and value system characteristic of this decade resides the basic adolescent. Understanding the changes that occur and the behaviors that are typical of a young man or woman growing up, regardless of time or place, provides parents with a backdrop of awareness that is most reassuring. It also provides the basis for the necessarily changed relationship with a child who is rapidly growing physically and emotionally. Armed with such understanding, parents can better cope with the many issues that are presented by the changes in their adolescent. At times, they can even manage a knowing smile at the many typical reactions they observe.

Parents who have survived the perils of puberty know, though, that dealing with one or more adolescents is not fun and games. Looking after the kids is relatively easy when they are small and dependent and the immediate neighborhood is their whole world. Three parental apprehensions, however, are forced into the forefront of consciousness by the onset of puberty and fueled daily by powerful adolescent strivings for independence.

Parental apprehension #1: "My adolescent will do the same things I did when I was young." With the wisdom of the years, parents look back at their adolescent antics with a bit of amusement tempered by a fair share of "only by the grace of God . . ." feelings. These parents, now mature individuals, simply don't want their children to take the same chances.

Parental apprehension #2: "The world my teen must live in is much more dangerous than it was years ago." This absolutely valid fear is constantly reinforced by public awareness of high suicide rates in adolescents, life-threatening sexually transmitted diseases and the easy availability of drugs. Mistakes and missteps can be much more serious than they were in the past.

Parental apprehension #3: "My child now has a private life that I can't directly control anymore." A reality is that teens force parents to trust them. Adolescence brings increased mobility and an expansion of time spent outside the sphere of direct family influence. Parents are forced to let go and hope that their teen will handle unknown and possibly dangerous situations well.

With impending puberty, the drama of early adolescence begins to unfold relentlessly. Responsible parents struggle to safeguard their teen's present and future. At the same time, their adolescent precociously lays claim to all adult prerogatives and privileges. In the background, a chosen peer group powerfully influences a child to do its immature bidding. Peers, parents and puberty all interact to produce the conflict-laden "adolescent triangle." It's normal but not easy.

The complex relationships of the adolescent triangle have been a perplexing part of the family life for centuries. It is incumbent upon parents to try to understand the developmental processes being experienced by their growing teen. Only then can they effectively modify their parenting relationship to their child-cum-adult in ways that will promote healthy growth toward maturity. And they must persevere without thanks in the face of active resistance by their teen. To set the stage for effective parental coping, here's an overview of the normal changes that occur during early adolescence.

THE STAGES OF ADOLESCENCE

If the typical individual is asked where adolescence begins and ends, the immediate response is "the teen years." Implicit in this response is the assumption that when the early twenties are reached, adolescence has ended and the individual has become an adult. Nothing could be further from the truth. True, in the past there has been an easy biological marker for the beginning of the adolescent years: puberty. And, in generations past, young men and women became financially and emotionally self-sufficient shortly after leaving home in their late teens.

However, the beginning and end of adolescence have become increasingly diffuse and difficult to define clear-ly. On one hand, we sometimes see a precocious beginning to adolescence that may predate overt signs of puberty. Children frequently begin to act like adolescents before physical changes begin. At the other end of this growth period, the difficulty is obvious. How do you define the moment when a child has become a true adult? Of course the best way is to use emotional maturity as a gauge rather than more obvious but often misleading criteria such as completing an education, earning a living, marrying or becoming a parent.

In short, adolescence in this society at present spans approximately 20 years. For almost two decades young people struggle to become emotionally mature adults. There are three basis stages of the adolescent experience as it exists today. However, for parents and children the most critical and dangerous is the first.

Stage I: Early adolescence (the rise of tribal loyalties). Age span: 10 or 11 through 17 years. In other words, this most tumultuous stage of growth begins in late fifth or sixth grade and typically ends at about the senior year of high school. During this time, your child joins a "tribe" of peers that is highly separate from the adult world. The peer group (tribe) clearly defines itself as a distinct subculture struggling for identity with its own dress codes, language codes, defined meeting places and powerfully enforced inclusion criteria.

During these most difficult years for both parent and child, the most pronounced changes of puberty occur. The core struggle of the child is to become independent—and that means emotionally separating from parents and forging a new adult identity. Initial attempts are awkward and emotionally naive. In three key areas, here's what the early adolescent is like.

A. Relationship to parents. Suspicious and distrustful, the adolescent begins actively to push parents away and resists their attempts to give advice. Life is conducted in a secretive world dominated by peers. Rebelling, pushing limits and constantly testing parental resolve are characteristic.

B. Relationship to peers. The youth experiences emotionally intense "puppy love" relationships with members of the opposite sex and "best friends" relationships with peers of the same sex. These relationships are often superficial, with undue emphasis placed on status considerations: participation in sports, attractiveness, belonging to an *in* group.

C. Relationship to career/future. Largely unrealistic in expectations of the adult world, the adolescent sees making a good living—and getting the training required—as easy and "no problem." Money made by working is often spent on status items such as cars or clothes or just on having a good time. The future is far away.

Stage II: Middle adolescence (testing adult realities). Age span: about 18 through 23 or 24. Beginning late in the high school years, a new awareness, with a subtle accompanying fear, begins to grow within the adolescent: "It's almost over. Soon I'll have to face the world on my own." A personal future and the hard realities it entails can no longer be completely denied. Shortly after high school, this young adult typically leaves home to attend college or technical school, join the service or enter the work force.

5. ADOLESCENCE AND EARLY ADULTHOOD

While on their own, but still basically protected by parents, middle adolescents are actively engaged in testing the self against the real world in ways not possible while living at home. More personal accountability is required and some hard lessons are learned. These sometimes painful experiences help the middle adolescent learn the ways of the world, but many signs of immaturity remain. In more specific terms, here is what's happening.

A. Relationship to parents. This dynamic is improved but still problematic at times. Middle adolescents still aren't really ready to be completely open with parents, but they are less defensive. During visits home, intense conflicts with parents will still erupt about lifestyle, career decisions and responsibility.

B. Relationships with peers. Frequent visits home may be made more with the intention to see the old gang from high school than to see parents. Good buddies remain at home, but new friends are being made in a work or school setting. A deeper capacity for caring is manifested in increasingly mature relationships with both sexes.

C. Relationship to career/future. The economics of self-support are steadily becoming more important. Sights may be lowered and changes in career direction are common. Meeting new challenges successfully brings a growing sense of confidence and self-sufficiency.

Stage III: Late adolescence (joining up). Age span: 23 or 24 to about 30 years. By the mid-twenties, early career experimentation has ended, as has protective parental involvement. The late adolescent is usually financially self-sufficient and remains quite social. Life is relatively simple because there is minimal community involvement, little property needing upkeep and usually an income at least adequate to meet basic needs. The late adolescent years tend to be remembered fondly as having been filled with hard work and good times.

At first glance, the late adolescent may appear to be fully adult, but this perception is deceptive. Significant adjustments to the adult world are still being made but are less obvious than they were. Many insecurities in relationships and at work continue to be faced and resolved. Spurred by a growing commitment to creating a personal niche in the adult world, the individual continues to change in the direction of true adult maturity. Here's how.

A. Relationship to parents. Over 20 years, the late adolescent has come full circle. Now that he or she is emotionally self-sufficient, a closer relationship with parents becomes possible. Mutual respect and acceptance grow. The late adolescent begins to understand parenting behaviors that were resisted earlier.

B. Relationship to peers. Most high school chums have been left behind and are seen only occasionally. A new group of work-related peers has been solidly established. Love relationships are more mature and show increased capacity for give and take. Commitment to a shared future and to a family grows.

C. Relationship to career/future. Active striving toward the good life and personal goals intensifies. At work, there is a continuing need to prove competency and get ahead. At home, a more settled lifestyle, one that is characteristic of the middle-class mainstream, slowly evolves. Limited community involvement is seen.

EARLY ADOLESCENT ATTITUDES

While adolescents struggle for nearly two decades to attain emotional maturity, the period of early adolescence is clearly the most striking. It is during this critical six or seven years that the growing young adult is most vulnerable to major mistakes. It is an emotional, intense, painful and confusing phase. It is also the time remembered by parents as most trying of their ability to cope.

Because the vulnerability of parents and their children is never so high as it is during early adolescence, it is well to define some of the characteristics of the normal teen during these years. Here are listed 15 of the most common adolescent attitudes that make life difficult for parents and children, but which are entirely normal for this age group. (NOTE: "Teen" in this discussion refers specifically to an early adolescent.)

Adolescent attitude #1: Conformity within nonconformity. The early adolescent attempts to separate from parents by rejecting their standards. At the same time there is an absolute need to conform to peer group standards. It is very important to be like peers and unlike parents.

Adolescent attitude #2: Open communication with adults diminishes. The early adolescent doesn't like to be questioned by parents and reveals little about what is really going on. Key items may be conveniently forgotten as a personal life outside the family is protected.

Adolescent attitude #3: Withdrawal from family altogether. With the advent of puberty, there is increasing resistance to the family. The teen would much rather stay home merely in order to be available or spend time doing nothing with friends than participate in anything with the family.

Adolescent attitude #4: Acceptability is linked to externals. Personal acceptability is excessively linked to having the right clothes, friends and fad items in teen culture. Parents are badgered constantly to finance status needs deemed necessary for acceptance by a chosen peer group.

Adolescent attitude #5: Spending more time alone. Ironically, although early adolescents are quite social most of the time, they also like to spend time by themselves. Often, teens will retreat for hours to a bedroom and tell parents in no uncertain terms to respect their privacy and let them alone.

Adolescent attitude #6: A know-it-all pseudo-sophistication. Attempts by parents to give helpful advice are usually met with a weary "I already know that!" More often than not, however, a teen's information about topics important to health and well-being is incomplete, full of distortions or patently false.

Adolescent attitude #7: Rapid emotional changes. One of the most difficult aspects of early adolescence for parents to cope with is rapid mood changes. A teen is on top of the world one minute and sullen or depressed the next. The emotional triggers for such changes are frequent but unclear and unpredictable.

Adolescent attitude #8: Instability in peer relationships. Early adolescent relationships are marked by intensity and change. Overnight, a best friend may become a mortal enemy because of a real or imagined betrayal. Changing loyalties are often triggered by the incessant gossiping characteristic of teen culture.

Adolescent attitude #9: Somatic sensitivity. In other words, a teen's rapidly changing body is cause for great concern. Frequently, an early adolescent will become obsessed with and distraught over a perceived major physical deformity (an asymmetrical nose, not-quite-right ears, two pimples).

Adolescent attitude #10: Personal grooming takes a spectacular upturn. To parents' astonishment, a lackadaisical preadolescent turns practically overnight into a prima donna who spends hours grooming and checking the mirror to make sure that every feature of personal appearance is letter-perfect.

Adolescent attitude #11: Emotional cruelty to one another. Early teens can be incredibly insensitive to one another. Malicious gossip, hurtful teasing, and descriptive nicknames, outright rejection by the peer group—all are reasons why early adolescence is a time of great pain for so many.

Adolescent attitude #12: A highly present-oriented existence. Parents often learn the hard way that seriously discussing the future with an early adolescent is an exercise in futility. Conflict results when an unconcerned teen insists on continuing a day-to-day, pleasure-oriented way of life.

Adolescent attitude #13: A rich fantasy life develops. The adolescent's world is filled with hopes and dreams: knights in shining armor, great achievements, plenty of money, a life of freedom and fun—all without much personal effort. Such fantasies often help deny true realities.

Adolescent attitude #14: There is a strong need for independence. Translation: "I can make my own decisions by myself." Teens take it as an insult to their maturity to have to ask permission for anything. This leads to circumventing established rules or making decisions without parental knowledge.

Adolescent attitude #15: A proclivity for experimentation. With a new body and new feelings, the early adolescent develops an unwarranted sense of personal maturity. This leads to covert experimentation wth adult behaviors (smoking, drug use, sexual activity) aimed at the achievement of status and the satisfaction of curiosity. The knowledge that this is taking place leads to another legitimate parental fear.

THE EMOTIONAL AROUSAL OF PARENTS

It is a given that early adolescence is difficult for parents. To a degree this is attributable to the erratic and challenging behavior of their teens. It is also true that as long as the child is clearly a child, the parents remain weak or lie dormant. However, once puberty begins, a myriad of powerful feelings wells up in the parents.

In many respects, a child's puberty forces parents to deal actively with emotional issues that promote *their* growth and development if handled well. It is as important for parents to understand their suddenly aroused feelings as it is for them to understand what is happening emotionally within their teen.

Aroused emotion #1: Unadulterated fear. I would not be going too far to say that the parents of teens live with fear and constant worry. "What's happened now?" "What am I going to find out about next?" A child's world is quite small. At puberty, it suddenly expands and the teen is gone much of the time. This occurs at about the same time that a teen becomes evasive about what is going on in his or her world. Fears grow.

Aroused emotion #2: A deep sense of helplessness. Parents grow very uncomfortable as they watch their teen experience all the pain and turmoil that early adolescence usually brings. Because adolescents perceive adults as unable *really* to understand anything of importance to themselves, parents may be pushed away when problems occur. The kids don't realize how helpless parents feel when they see their child suffering emotionally but are relegated to the sidelines.

Aroused emotion #3: High levels of frustration. It is a given that many of the behaviors of an early adolescent trigger parental anger. One of a teen's strongest emotional needs is to be separate emotionally. This need is expressed by constantly confronting parents verbally, violating rules and pushing limits right to the brink. This entirely normal adolescent response pattern takes its toll on parents who become highly stressed, frustrated and tired.

Aroused emotion #4: A growing awareness of loss. With the onset of puberty, parents are forced to recognize that in just a few years, their teen will be going into the world and lost forever to the nuclear family. The undeniable fact that "our little girl/boy is growing up" triggers this deepening sense of loss on the part of parents: the sadness is compounded by the withdrawal of the teen from family life. Often this particular feeling is overwhelmed by fleeting wishes that the child would hurry and grow up so parents can have some peace of mind.

Aroused emotion #5: Personal hurt. Parents of teens struggle to do their very best to guide and protect their children. However, no thanks are forthcoming. In fact, parents' efforts are often resented and they are labeled as old-fashioned or Victorian or old fogies who are obviously completely out of touch with reality. Angry confrontations are the norm. Sullen withdrawal is an everyday occurrence. Continued rejection and hurt feelings make it difficult for parents to continue giving their personal best to an unappreciative teen.

IN THE EYE OF THE HURRICANE

At the center of every hurricane is the eye. That's where there is calm despite the intensity of the storm that swirls around it. This is an excellent way to conceptualize the relationship of effective parents to their children during the tumultuous early adolescent years. At puberty, a teen becomes inexorably swept up in the swift winds of change. To help themselves and their child, parents must remain calm and aware in the eye of the hurricane.

In recent years, much has been written about the changing nature of growing up in America. Some authorities emphasize the group's premature sophistication consequent on the fact that teens these days are ex-

5. ADOLESCENCE AND EARLY ADULTHOOD

posed to much more at an earlier age than their parents were. Others who study this special group find that beyond the surface precocity of teens, attaining emotional maturity is steadily becoming a more prolonged and difficult process than ever before. The reality that parents must understand is that these seemingly divergent points of view are both absolutely valid and in no way contradict one another.

To be effective, parents must not be fooled by the misleading sophistication of teens and instead respond to the more complex developmental problems that lie beneath this surface veneer. To be of maximum aid in promoting healthy growth toward maturity, parents must make sure that their responses reflect three important teenage needs.

Teen need #1: "Depth perception" by parents. Basically, parents must be able to see accurately beyond the often erratic surface behaviors of an adolescent to the real issues that simply can't be articulated by a teen. Then parents must respond in caring ways to those emotional needs despite protests, confrontations and denials.

Teen need #2: Consistency of parental responses. Teens are notorious for their inconsistency. One of their deepest needs during these years of turmoil is to have parents who are steady and consistent. Such parents become a stabilizing influence and a center of strength—this helps a teen cope effectively with rapid change in every part of life.

Teen need #3: Strength of parental conviction. At no other time during the entire child-rearing process must parents be surer of their values. Teens focus tremendous pressure on parents to convince them they are wrong or that their values are irrelevant. Far too often the kids succeed in compromising solid parental values, to the detriment of the family and themselves.

One of the most emotionally rigorous tasks that parents face during the adolescent years is to keep doing what is right with very little encouragement and without becoming too insecure. And, after all is said and done and those difficult years are over, most teens do mature to join the ranks of respectable adults. Didn't you? And if you parented well, you will be rewarded eventually when your adult son or daughter thanks you directly for all the sacrifices you made in the face of all the obstacles.

But the progress toward the goal is a nightmare. One frustrated parent put up a sign in the kitchen: "NOTICE TO ALL TEENS! If you are tired of being hassled by unreasonable parents, NOW IS THE TIME FOR ACTION. Leave home and pay your own way WHILE YOU STILL KNOW EVERYTHING!"

These days puberty has perils for parents and for the children in grown-up bodies who are in their charge. And adolescence is no time to cut corners and take the easy road. Perhaps it was a wise parent who remarked that "a shortcut is often the quickest way to get somewhere you weren't going." With adolescents, the best road is always difficult but eventually rewarding. Shortcuts too often lead to dead ends. Or dangerous precipices. Sometimes to places you never expected to visit.

JEALOUSY & ENVY

The Demons Within Us

JON QUEIJO

Jon Queijo *is a free-lance writer who resides in West Roxbury, Massachusetts.*

Rick and Liz seemed to have a wonderful marriage; they did everything together. This changed suddenly, however, when Liz's ex-boyfriend began working at her law firm. Besieged by insecurity, Rick began calling Liz's office at odd hours and at night questioned her suspiciously. In a coup de grace, *he burst in on her during a business luncheon and falsely accused her of having an affair.*

Ann worked extremely hard to achieve success as a real estate agent. Her satisfaction turned sour, however, when a new agent was hired who managed to work less, yet made more sales. Ann hid her dislike of the new agent by offering to take her phone messages while she was out. When Ann began making more sales than her rival, no one made the connection between this turn of events and Ann's tendency to "accidentally" forget to deliver certain phone messages.

The jealous rage of a lover. The shameful actions of envy. Despite our better intentions, most of us feel these emotions dozens of times in our lifetimes. Pulling us apart from lovers, friends, family members, co-workers and even perfect strangers, jealousy and envy can devastate our lives and cause effects ranging from sadness, anger and depression, to estrangement, abuse and even violence.

Beyond our own lives, the power of these emotions has spawned countless works of poetry and prose and triggered numerous historical events. Perhaps for this reason society proclaimed judgment on jealousy and envy thousands of years ago, with the verdict coming down harder on envy. For example, while the pain of jealousy has been forever immortalized in poetry and song, the shame of envy emerges as early as the Ten Commandments: "Thou shalt not covet thy neighbor's house, field, wife or anything that is thy neighbor's." In fact, envy is despicable enough to be considered one of the "Seven Deadly Sins," taking its place alongside pride, gluttony, lust, sloth, anger and greed.

Although we see jealousy and envy arise in numerous situations, their basic definitions are fairly simple: jealousy is "the fear of losing a relationship" (romantic, parental, sibling, friendship); and envy is "the longing for something someone else has" (wealth, possessions, beauty, talent, position).

Despite these definitions and the numerous philosophers, poets and scientists who have pondered these emotions, some remarkably fundamental questions remain: What causes jealousy and envy? Are the emotions actually different? What do they feel like? What are the best ways to cope with these feelings? Why do we often use the terms interchangeably? And what are their implications for society?

Researchers have taken various approaches to answer these questions. The biological view, for example, says that jealousy and envy serve a basic purpose — the emotions lead to biochemical changes that spur the individual to take action and improve the situation. The evolutionary view holds that jealousy may enhance survival by keeping parents together, thus increasing protection of the offspring.

From *Bostonia*, May/June 1988, pp. 31-36. Reprinted with permission of *Bostonia Magazine*.

"People who are dissatisfied with themselves are primed for having other people's talents impinge on them. If, on the other hand, you're satisfied with yourself, then what other people have or do won't unduly raise your expectations, and you should be less likely to feel envy."

RICHARD SMITH

Other explanations range from the reasonable — envy stems from parental attitudes that make a child feel inferior; to the bizarre — the emotions begin in infants when the mother withholds breast-feeding.

Probably the most practical understanding of jealousy and envy, however, emerges from the work of social psychologists — researchers who look at the way people react to each other and society. To them, jealousy and envy arise when the right mix of internal *and* external ingredients are present in society.

"I tend to look at jealousy and envy in terms of motivation and self-esteem," explains Peter Salovey, a social psychologist at Yale University. "It's the interaction between what's important to you and what's happening in the environment. The common denominator is this threat to something that's very important to the person — something that defines self-worth."

Richard Smith, a social psychologist at Boston University who has conducted several studies on jealousy and envy, emphasizes external factors, such as how society affects our view of ourselves. "My perspective is from social comparisons," he explains, "which says we have no objective opinion for evaluating our abilities, so we look at others."

Smith, like Salovey, also stresses internal factors — the role of self-esteem, for example — in determining whether we will feel jealous or envious in any given situation. "People who are dissatisfied with themselves are primed for having other people's talents and possessions impinge on them," Smith points out. "If, on the other hand, you're satisfied with yourself, then what other

people have or do won't unduly raise your expectations for yourself, and you should be less likely to feel envy."

Embarrassed by his display of jealousy, Rick apologizes to Liz and they discuss the problem. Soon they realize that while Rick loves Liz and fears losing her, something else is at work here. Because Liz's ex-boyfriend is a lawyer, he possesses skills Rick does not. While Rick is proud of his ability as a store manager, he fears Liz's ex-boyfriend could lure her away with other skills.

Feeling guilty about her actions, Ann calls a friend for support. Ann knows she feels inadequate because the new agent is succeeding in a career that is very important to her, but that doesn't explain everything; others have done better whom Ann has not envied. Then it occurs to her: What bothers her is the way the woman was bettering Ann. She was more outgoing and self-confident — two skills about which Ann has always felt insecure.

As Rick and Ann's situations illustrate, if someone is unsure about an ability — such as Rick and his law knowledge or Ann and her communication skills — then a social situation can bring out that insecurity. "In envy," notes Salovey, "the threat may come from someone else's possessions or attributes. In jealousy the situation is the same, except that the other person's possessions or attributes cause you to fear losing the relationship. Either way, somebody else threatens your self-esteem."

Yet Salovey emphasizes that it is not as simple as saying someone is at risk for these emotions if they have a low opinion of themselves. "It's low self

esteem in a specific *area*," he explains. "If you have a low opinion about your physical looks or occupation, then that's the area in which you're more likely to be vulnerable. You feel it when you confront somebody else who is superior to you in that respect."

From Smith's point of view, the key is how that person compares him or herself to others. In one study, for example, he found that envy was strongest among people who performed below their expectation in an area that was important to them and then confronted someone who functioned better. In a related study, Smith also found that a person's "risk" for feeling jealous or envious increases with the increased importance they put on the quality.

While much of this may sound like common sense, in fact little research has been done to establish even the most basic ground rules of jealousy and envy. For example, are the two emotions actually different? What do people feel when they are jealous or envious? Despite centuries of long-held assumptions, only recently have researchers begun to answer these questions scientifically.

Smith and his colleagues, for example, recently conducted a study to see if the classic distinctions between the two emotions are actually true. Their findings — presented last August at the annual convention of the American Psychological Association — validated what we have always suspected. Jealous people tend to feel a fear of loss, betrayal, loneliness, suspicion and uncertainty. Envious people, on the other hand, tend to feel inferior, longing for what another has, guilt over feeling ill will towards someone, shame and a tendency to deny the emotion.

The study was not an idle exercise in stating the obvious. It was designed to help clear an ongoing debate about jealousy and envy and our curious tendency not only to mix up the terms, but to experience an overlap of both emotions.

Consider, for example, the following uses of the word jealousy: Bruce was jealous when his girlfriend began talking to another man at the party; the boy cried in a fit of jealousy when his parents paid attention to the new baby; Ellen became jealous when her friend began spending more time at the health club. Nothing unusual with any of these uses of jealousy—they all refer to someone's "fear of losing a relationship."

Now, however, consider these uses: The professor, jealous of his colleague's success, broke into his lab and ruined his experiment; Mary is always complaining of being jealous of her sister's beautiful blonde hair; Mark admits that he is jealous of John's athletic ability. All of these situations actually refer to envy, "the longing for something someone else has." Researchers have noticed the mix-up and it has led them to question how different the feelings really are.

"In everyday language it's clear that people use the terms interchangeably," notes Smith, adding, "For that reason, there's naturally some confusion about whether they're different." In a recent study, however, Smith and his colleagues found the mix-up only works one way, with jealousy being the broader term. That is, jealousy is used sometimes in place of envy, but envy is rarely used when referring to jealousy. So while you might say, "I'm jealous of Paul's new Mercedes," you would never say, "When she left her husband, he flew into an *envious* rage."

Is there an underlying reason for why the terms are used interchangeably? One reason people may use jealousy in place of envy, and not vice versa, is because of the social stigma attached to envy. But Salovey and Smith both point to another reason for the overlap.

Dave and Marcia had been dating for a year when Marcia decided she wanted to see other men. Dave was devastated—not only because he cared for her, but because he was older and feared his age was working against him. One evening he bumped into Marcia, arm-in-arm with another man. As Dave talked to the couple, sarcasm led to verbal abuse, until finally Dave took a swing at Marcia's date—a man at least 10 years his junior.

In case you haven't guessed, Dave wanted back more than his relationship with Marcia; he also wanted the return of his youth. He was feeling a painful mixture of jealousy *and* envy. Explains Salovey, "The same feelings emerge when your relationship is threatened by someone else as when you'd like something that person has. I think one reason is that in most romantic situations, envy plays a role. You're jealous

because you're going to lose the relationship, but you're also envious because there is something the other person has that allows him to be attractive to the person you care about."

Because this overlap occurs so frequently, Salovey has found the best way to understand jealousy and envy is to examine each *situation*. In addition, because jealousy is the more encompassing term, he views envy as a form of jealousy and distinguishes the two by the terms "romantic jealousy," the fear of losing a relationship; and "social-comparison jealousy," the envy that arises when people compare traits like age, intelligence, possession and talent.

Smith agrees with Salovey that one reason people confuse the terms may be that envy is present in most cases of jealousy. Nevertheless, he takes issue with Salovey's use of the term "social-comparison jealousy." "It may be true that there's almost invariably envy in every case of jealousy," he notes, "but it doesn't mean there's no value in distinguishing the two feelings. The overlap in usage only goes one way, so there's no reason to throw out the term 'envy.'"

Salovey counters, "I'm not saying we need to stop using 'envy.' The reason we use 'social-comparison jealousy' is to emphasize that the situation that creates the feeling is important." And one reason

---

COMPARING OURSELVES TO OTHERS — In Sickness and In Health

Envy, according to Richard Smith, arises when we compare ourselves to others and can't cope with what we see. Indeed, he believes that the way in which we cope with "social comparisons" plays an important role in our physical as well as mental health.

Smith theorizes that we use one of four "comparison styles" to cope with social differences. Two of these styles are "constructive" to well-being, while two are "destructive." "It's a difficult problem to tackle," says Smith, "but we're trying to measure these styles and see if they predict a person's general satisfaction with life or ability to cope with illness." For example, he notes, "There's considerable evidence" that one way people cope with serious illness is by focusing on others who are not doing as well.

Smith has arranged the four comparison styles in a matrix, with the descriptions in each box referring to the characteristics of that style. "*Upward*-Constructive," for example, represents those who compare themselves to others who are better off, and use it as a healthy stimulus. In this category, "You don't feel hostile to others who are better," says Smith, "because you hope to be like them. It suggests that upward comparisons are not necessarily bad."

"*Upward*-Destructive," however, shows how comparing yourself to those doing better can be unhealthy.

Envy, resentment, Type A behavior and poor health all fit into this category. Smith points to a study that looked at personalities of people who had heart attacks, "and the only dimension that predicted heart disease was this jealousy-suspicion trait."

What Smith calls "*Downward*-Constructive," refers to people who compare themselves to those worse off, and use it to feel better about themselves. Such people, says Smith, "realize that others aren't doing so well and how lucky they are. There's some solid evidence in the health literature showing the value of that kind of comparison."

Finally, "*Downward*-Destructive," describes those who get pleasure out of comparing themselves to others who are worse off. "I'd call the effect 'schadenfreude,' or joy at the suffering of others," says Smith. "It's akin to sadism and it's probably not conducive to health."

"What's interesting about all four comparison styles," notes Smith, "is that they don't necessarily have any relation to reality. They reflect what people construe and focus on." While Smith stresses, "This is all speculative," he adds that "my feeling is that 'Upward-Destructive' explains people's hostility in terms of their social comparison context. It shows why their relation to people doing better leads to envy and why they'd feel hostile to begin with." J.Q.

Salovey stresses the situation—rather than other mood differences—is that romantic jealousy, since it includes envy, is usually very intense, making it difficult to separate distinct feelings.

Nevertheless, Smith believes the distinction should be made, especially because "In its traditional definition, envy has a hostile component to it." Not everyone would agree with that. After all, in envying others, we can also admire *them* and even use *them* as role models to spur ourselves to greater abilities. There is no hostility in that, yet these cases, Smith contends, are not precisely envy. Indeed, Smith believes envy differs from jealousy not only because of its hostility, but because of another distinct ingredient: privacy.

By the time Bill was 40, he was vice president at his firm and owned a luxurious house in an affluent neighborhood. Nevertheless, Bill had never married and was lonely. He envied his brother Jim, who lived a modest but happy life with his wife and children. One day Jim asked Bill to write a reference letter to help him get a bank loan for a new home. Bill said he'd be delighted, but soon realized he could send the letter to the bank without Jim ever seeing it. Bill wrote the letter and his brother, never knowing why, was refused the loan.

Although this anecdote is fictional, Smith has shown in his research that the principles illustrated are probably true. In a recent study, Smith had subjects identify with an envious person. He then gave them the option of dividing a "resource" between themselves and the envied person. Among the many options were: dividing the resource equally; dividing it so the subject kept the most and gave the least to the envied person; and dividing it so they sacrificed the amount they could otherwise keep for themselves if it meant giving the least to the envied person. Most subjects chose the last option, but only when they could do so in private, rather than public, circumstances.

Although Smith admits the findings need to be verified, he believes the results are strongly "suggestive." The envious person's choice, he says, "was unambiguously hostile. Not selfish, but hostile. And the findings verified the conventional wisdom that envy has a secretive quality about it that you wouldn't admit to the person you envy. And under the right circumstances it will lead to actual hostile behavior."

The reason people would be hostile in private seems obvious, given that envy is socially unacceptable. But why the hostility in the first place? Smith theorizes that the envied person's "superiority" emphasizes the envious person's low self-esteem in a specific area — the way the new real estate agent's communication skills affected Ann; Marcia's young date affected Dave; and Jim's happy family life affected Bill. Hostility is a way of putting down the envied person and devaluing his or her "superiority." Whether it takes the form of thought, word or deed, hostility pushes the envied person away, thereby allowing the envious person to restore his self-esteem.

Smith is looking at other implications of envy and hostility in society. For example, "We don't know much about why people are hostile to begin with, but since envy is related to hostility, maybe the way people respond to the way they compare themselves to others is at the root of hostility." And there are other subtle—and even more frightening—implications. For one thing, Smith notes, because envy is socially unacceptable, "It's often not conscious, and as a result people will arrange the details of their situation and their perception of the other person to make envy something they can label as 'resentment' — resentment in the sense of righteous indignation."

Smith goes as far as to propose that many intergroup conflicts in the world —between countries, races and religious groups, for example—may begin with envy. One group is better off economi-

---

RANKING JEALOUSY & ENVY

What situations are most likely to evoke feelings of jealousy and envy? In a study published in the *Journal of Personality and Social Psychology*, Peter Salovey and Judith Rodin asked subjects to rank 53 situations according to the degree of emotion each would evoke. Below are 25 of those situations, listed in decreasing order, that received the highest "jealousy/envy" ratings:

1 You find out your lover is having an affair.

2 Someone goes out with a person you like.

3 Someone gets a job that you want.

4 Someone seems to be getting closer to a person to whom you are attracted.

5 Your lover tells you how sexy his/her old girl/boyfriend was.

6 Your boyfriend or girlfriend visits the person he or she used to date.

7 You do the same work as someone else and get paid less than he or she.

8 Someone is more talented than you.

9 Your boyfriend or girlfriend would rather be with his/her friends than with you.

10 You are alone while others are having fun.

11 Your boyfriend or girlfriend wants to date other people.

12 Someone is able to express himself or herself better than you.

13 Someone else has something you wanted and could have had but don't.

14 Someone else gets credit for what you've done.

15 Someone is more intelligent than you.

16 Someone appears to have everything.

17 Your steady date has lunch with an attractive person of the opposite sex.

18 Someone is more outgoing and self-confident than you.

19 Someone buys something you wanted but couldn't afford.

20 You have to work while your roommate is out partying.

21 An opposite-sex friend gives another friend a compliment, but not you.

22 Someone has more free time than you.

23 You hear that an old lover of yours has found a new lover.

24 Someone seems more self-fulfilled than you.

25 You listen to someone tell a story about things he did without you. J.Q.

cally, for example, than another, and the "inferior" group feels envy as a result. "But if they're just envious," he explains, "no one is going to give them any sympathy. So they tend to see their situation as unfair and unjust. In this way, envy becomes righteous resentment, which in turn gives them the 'right' to protest and conduct hostile — or even terrorist —activity."

This topic raises questions about what is fair or unfair in our society, how people cope with differences and whether, as a result, they feel envy, resentment or acceptance. Smith points out that coping with envy may depend on how well we learn to accept our inequalities. "I think as people mature, they learn to cope with differences by coming to terms with the fact that life *isn't* fair and that it's counterproductive to dwell on things you can't do anything about."

While some of Smith's ideas on coping with envy are speculative, Salovey has found that there are specific strategies —illustrated in the following anecdotes — that work in preventing and easing jealousy and envy. . . .

Rick and Liz are getting along much better these days even though Liz still works with her ex-boyfriend at the law firm. Rick isn't thrilled by this, but he overcame his jealousy by focusing instead on his relationship with Liz: spending time with her, planning vacations, discussing their future.

Ann no longer feels envy or hostility towards her co-worker at the real estate agency. She put an end to those negative feelings by simply ignoring her rival's superior communication skills and concentrating instead on her own achievements.

As illustrated here and isolated in a survey Salovey and Yale associate Judith Rodin conducted with *Psychology Today* readers, there are three major coping strategies: *Self-reliance*, in which a person does not give into the emotion, but continues to pursue the goals in the relationship; *Selective ignoring*, or simply ignoring the things that cause the jealousy or envy; and *Self-bolstering*, or concentrating on positive traits about yourself.

Surprisingly, "We found that the first two coping strategies are very effective in helping a person not feel jealousy," reports Salovey. "We thought that self-bolstering would also be good, since if something that's important to you is threatened, then maybe you should think of things in which you do well."

Although self-bolstering was not helpful in preventing jealousy. "Once you were *already* jealous, it was the only thing that kept you from becoming depressed and angry," notes Salovey. "So the first two keep jealousy in check, but if jealousy does emerge, self-bolstering keeps jealousy from its worst effects."

In the same study, Salovey and Rodin also uncovered some interesting data about how men and women experience jealousy and envy. "Men tend to be more envious in situations involving wealth and fame, and women more so in beauty and friendship," reports Salovey, but he emphasizes, "I should put that finding in context. We looked at a lot of variables and very rarely found differences. Men and women were very similar on nearly everything you could measure except that one difference."

Smith and Salovey do agree that while jealousy and envy can be devastating to those experiencing them, in milder forms they can actually be helpful. "I tend to think of jealousy and envy as normal," says Salovey. "In any relationship where you really care about the other person, when your relationship is threatened by someone, you're going to feel negative emotions. If you don't, maybe you don't care that much."

As for envy, Smith points out that, "I don't think it's a bad thing, necessarily. It's a motivator when it's in the form of admiration and hero worship." He does add, however, that at those levels the emotion may not be envy since envy, by definition is hostile. "It's hard to know where one stops and the other begins," he notes.

Nevertheless, Smith stresses that "Coping with differences is something we all do. Some of us do it in constructive ways and others in destructive, and that has implications for who is going to be happy or unhappy. Envy is a sign of not coping well—maybe." He adds that while "some people have a right to recognize that a situation is unfair, the next question is, what do you do? It may be best to recognize the unfairness and cope with it before it leads to more painful feelings."

Jealousy and envy can have an unpleasant knack of cropping up between the people who care most about each other. Our first reaction is often to blame the other person — my *wife* is the one who is lunching with her ex-boyfriend; my *co-worker* is undeservedly making more sales; why should my *brother* have a happy family *and* a big house? Part of our blame is understandable: life *is* unfair; society and circumstance *do* create differences between us beyond our control.

Nevertheless, the bottom line is not how we view other people, but how we view ourselves. When jealousy or envy become overwhelming, it is as much from passing judgment on ourselves as on others. That's when we owe it to everyone to talk it out, change what needs to be changed and—perhaps most importantly—accept ourselves for what we are.

The Prime of Our Lives

WHAT SEEMS TO MARK OUR ADULT YEARS MOST IS OUR SHIFTING PERSPECTIVE ON OURSELVES AND OUR WORLD. IS THERE A COMMON PATTERN TO OUR LIVES?

Anne Rosenfeld and Elizabeth Stark

Anne Rosenfeld and Elizabeth Stark, both members of Psychology Today's *editorial staff, collaborated across cohorts to write this article.*

My parents had given me everything they could possibly owe a child and more. Now it was my turn to decide and nobody ... could help me very far...." That's how Graham Greene described his feelings upon graduation from Oxford. And he was right. Starting on your own down the long road of adulthood can be scary.

But the journey can also be exciting, with dreams and hopes to guide us. Maybe they're conventional dreams: getting a decent job, settling down and starting to raise a family before we've left our 20s. Or maybe they're more grandiose: making a million dollars by age 30, becoming a movie star, discovering a cure for cancer, becoming President, starting a social revolution.

Our youthful dreams reflect our unique personalities, but are shaped by the values and expectations of those around us—and they shift as we and our times change. Twenty years ago, college graduates entered adulthood with expectations that in many cases had been radically altered by the major upheavals transforming American society. The times were "a-changin'," and almost no one was untouched. Within a few years many of the scrubbed, obedient, wholesome teenagers of the early '60s had turned into scruffy, alienated campus rebels, experimenting with drugs and sex and deeply dissatisfied with their materialistic middle-class heritage.

Instead of moving right on to the career track, marrying and beginning families, as their fathers had done, many men dropped out, postponing the obligations of adult life. Others traveled a middle road, combining "straight" jobs with public service rather than pursuing conventional careers. And for the first time in recent memory, large numbers of young men refused to serve their country in the military. In the early 1940s, entire fraternities went together to enlist in World War II. In the Age of Aquarius, many college men sought refuge from war in Canada, graduate school, premature marriages or newly discovered medical ailments.

Women were even more dramatically affected by the social changes of the 1960s. Many left college in 1967 with a traditional agenda—work for a few years, then get married and settle down to the real business of raising a family and being a good wife—but ended up following a different and largely unexpected path. The women's movement and changing economics created a whole new set of opportunities. For example, between 1967 and 1980, women's share of medical degrees in the United States rocketed from 5 percent to 26 percent, and their share of law degrees leaped from 4 percent to 22 percent.

From *Psychology Today,* May 1987, pp. 62-64, 66, 68-72. Copyright © 1987 by PT Partners L. P. Reprinted by permission.

A group of women from the University of Michigan class of 1967 who were interviewed before graduation and again in 1981 described lives very different from their original plans. Psychologists Sandra Tangri of Howard University and Sharon Jenkins of the University of California found that far more of these women were working in 1981 than had expected to, and far more had gotten advanced degrees and were in "male" professions. Their home lives, too, were different from their collegiate fantasies: Fewer married, and those who did had much smaller families.

Liberation brought problems as well as opportunities. By 1981, about 15 percent of the women were divorced (although some had remarried), and many of the women who "had it all" told Tangri and Jenkins that they felt torn between their careers and their families.

Living out our dreams in a rapidly changing society demands extreme flexibility in adjusting to shifting social realities. Our hopes and plans, combined with the traditional rhythms of the life course, give some structure, impetus and predictability to our lives. But each of us must also cope repeatedly with the unplanned and unexpected. And in the process, we are gradually transformed.

For centuries, philosophers have been trying to capture the essence of how people change over the life course by focusing on universally experienced stages of life, often linked to specific ages. Research on child development, begun earlier in this century, had shown that children generally pass through an orderly succession of stages that correspond to fairly specific ages. But recent studies have challenged some of the apparent orderliness of child development, and the pattern of development among adults seems to be even less clear-cut.

When we think about what happens as we grow older, physical changes leap to mind—the lessening of physical prowess, the arrival of sags, spreads and lines. But these take a back seat to psychological changes, according to psychologist Bernice Neugarten of Northwestern University, a pioneer in the field of human development. She points out that although biological maturation heavily influences childhood development, people in young and middle adulthood are most affected by their own experiences and the timing of those experiences, not by biological factors. Even menopause, that quintessentially biological event, she says, is of relatively little psychological importance in the lives of most adult women.

In other words, chronological age is an increasingly unreliable indicator of what people will be like at various points. A group of newborns, or even 5-year-olds, shows less variation than a group of 35-year-olds, or 50-year-olds.

What seems to mark our adult years most is our shifting perspective on ourselves and our

STAGE THEORIES ARE A LITTLE LIKE HOROSCOPES— VAGUE ENOUGH TO LET EVERYONE SEE SOMETHING OF THEMSELVES IN THEM. THAT'S WHY THEY'RE SO POPULAR.

world—who we think we are, what we expect to get done, our timetable for doing it and our satisfactions with what we have accomplished. The scenarios and schedules of our lives are so varied that some researchers believe it is virtually impossible to talk about a single timetable for adult development. However, many people probably believe there is one, and are likely to cite Gail Sheehy's 1976 best-seller *Passages* to back them up.

Sheehy's book, which helped make "midlife crisis" a household word, was based on a body of research suggesting that adults go through progressive, predictable, age-linked stages, each offering challenges that must be met before moving on to the next stage. The most traumatic of these transitions, Sheehy claimed, is the one between young and middle adulthood—the midlife crisis.

Sheehy's ideas were based, in part, on the work of researchers Daniel Levinson, George Vaillant and Roger Gould, whose separate studies supported the stages of adult development Erik Erikson had earlier proposed in his highly influential model (see "Erikson's Eight Stages," next page).

Levinson, a psychologist, had started his study in 1969, when he was 49 and intrigued with his own recent midlife strains. He and his Yale colleagues intensively interviewed 40 men between the ages of 35 and 45 from four occupational groups. Using these interviews, bolstered by the biographies of great men and the development of memorable characters in literature, they described how men develop from 17 to 65 years of age (see "Levinson's Ladder," this article).

At the threshold of each major period of adulthood, they found, men pass through predictably unstable transitional periods, including a particularly wrenching time very close to age 40. At each transition a man must confront issues that may involve his career, his marriage, his family and the realization of his dreams if he is to progress successfully to the next period. Seventy percent to 80 percent of the men Levinson interviewed found the midlife transition (ages 40 to 45) tumultuous and psychologically painful, as most aspects of their lives came into question. The presumably universal timetable Levinson offered was very rigid, allowing no more than four years' leeway for each transition.

Vaillant's study, although less age-bound than Levinson's, also revealed that at midlife men go through a period of pain and preparation—"a time for reassessing and reordering the truth about adolescence and young adulthood." Vaillant, a psychiatrist, when he conducted his study at Harvard interviewed a group of men who were part of the Grant Study of Adult Development. The study had tracked almost 270 unusually accomplished, self-reliant and healthy Harvard freshmen

sage with its emphasis on orderly and clearly defined transitions. According to Cornell historian Michael Kammen, "We want predictability, and we desperately want definitions of 'normality.'" And (drawn mostly from the classes of 1942 to 1944) from their college days until their late 40s. In 1967 and 1977 Vaillant and his team interviewed and evaluated 94 members of this select group.

Erikson's Eight Stages

According to Erik Erikson, people must grapple with the conflicts of one stage before they can move on to a higher one.

BONNIE SCHIFFMAN

	1	2	3	4	5	6	7	8
Old Age								Integrity vs. Despair, Disgust
Maturity							Generativity vs. Self-absorption	
Young Adulthood						Intimacy vs. Isolation		
Adolescence					Identity vs. Identity Confusion			
School Age				Industry vs. Inferiority				
Play Age			Initiative vs. Guilt					
Early Childhood		Autonomy vs. Shame, Doubt						
Infancy	Trust vs. Mistrust							

SOURCE: ADAPTED FROM "REFLECTION ON DR. BORG'S LIFE CYCLE": ERIK H. ERIKSON, DAEDALUS, SPRING 1976.

They found that, despite inner turmoil, the men judged to have the best outcomes in their late 40s "regarded the period from 35 to 49 as the happiest in their lives, and the seemingly calmer period from 21 to 35 as the unhappiest." But the men least well adapted at midlife "longed for the relative calm of their young adulthood and regarded the storms of later life as too painful."

While Levinson and Vaillant were completing their studies, psychiatrist Roger Gould and his colleagues at the University of California, Los Angeles, were looking at how the lives of both men and women change during young and middle adulthood. Unlike the Yale and Harvard studies, Gould's was a one-time examination of more than 500 white, middle-class people from ages 16 to 60. Gould's study, like those of Levinson and Vaillant, found that the time around age 40 was a tough one for many people, both personally and maritally. He stressed that people need to change their early expectations as they develop. "Childhood delivers most people into adulthood with a view of adults that few could ever live up to," he wrote. Adults must confront this impossible image, he said, or be frustrated and dissatisfied.

The runaway success of *Passages* indicated the broad appeal of the stage theorists' mes-

almost everyone could find some relationship to their own lives in the stages Sheehy described. Stage theories, explains sociologist Orville Brim Jr., former president of the Russell Sage Foundation, are "a little like horoscopes. They are vague enough so that everyone can see something of themselves in them. That's why they're so popular."

But popularity does not always mean validity. Even at the time there were studies contradicting the stage theorists' findings. When sociologist Michael Farrell of the State University of New York at Buffalo and social psychologist Stanley Rosenberg of Dartmouth Medical School looked for a crisis among middle-aged men in 1971 it proved elusive. Instead of finding a "universal midlife crisis," they discovered several different developmental paths. "Some men do appear to reach a state of crisis," they found, "but others seem to thrive. More typical than either of these responses is the tendency for men to bury their heads and deny and avoid all the pressures closing in on them."

Another decade of research has made the picture of adult development even more complex. Many observations and theories accepted earlier as fact, especially by the general public, are now being debated. Researchers have espe-

Oh, God, I'm only twenty and I'll have to go on living and living and living.
—Jean Rhys, *Diary*

At thirty a man should know himself like the palm of his hand, know the exact number of his defects and qualities, know how far he can go, foretell his failures—be what he is. And above all accept these things.
—Albert Camus *Carnets.*

cially challenged Levinson's assertion that stages are predictable, tightly linked to specific ages and built upon one another.

In fact, Gould, described as a stage theorist in most textbooks, has since changed his tune, based upon his clinical observations. He now disagrees that people go through "formal" developmental stages in adulthood, although he says that people "do change their ways of looking at and experiencing the world over time." But the idea that one must resolve one stage before going on to the next, he says, is "hogwash."

Levinson, however, has stuck by his conceptual guns over the years, claiming that no one has evidence to refute his results. "The only way for my theory to be tested is to study life structure as it develops over adulthood," he says. "And by and large psychologists and sociologists don't study lives, they study variables."

Many researchers have found that changing times and different social expectations affect how various "cohorts"—groups of people born in the same year or time period—move through the life course. Neugarten has been emphasizing the importance of this age-group, or cohort, effect since the early 1960s. Our values and expectations are shaped by the period in which we live. People born during the trying times of the Depression have a different outlook on life from those born during the optimistic 1950s, according to Neugarten.

The social environment of a particular age group, Neugarten argues, can influence its so-

WHAT WAS TRUE FOR PEOPLE BORN IN THE DEPRESSION ERA MAY NOT HOLD FOR TODAY'S 40-YEAR-OLDS, BORN IN THE UPBEAT POSTWAR YEARS.

cial clock—the timetable for when people expect and are expected to accomplish some of the major tasks of adult life, such as getting married, having children or establishing themselves in a work role. Social clocks guide our lives, and people who are "out of sync" with them are likely to find life more stressful than those who are on schedule, she says.

Since the 1960s, when Neugarten first measured what people consider to be the "right" time for major life events, social clocks have changed (see "What's the Right Time?" this article), further altering the lives of those now approaching middle age, and possibly upsetting the timetable Levinson found in an earlier generation.

As sociologist Alice Rossi of the University of Massachusetts observes, researchers trying to tease out universal truths and patterns from

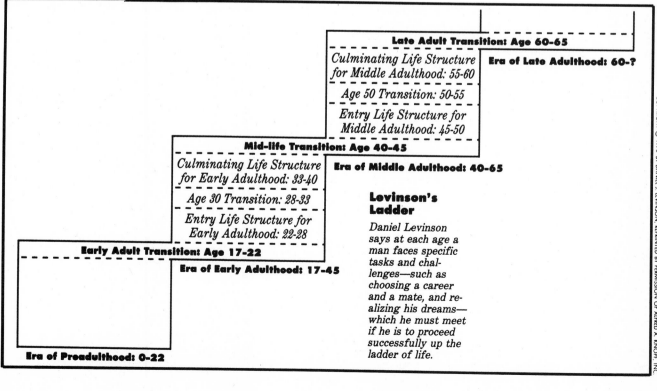

Late Adult Transition: Age 60-65

Culminating Life Structure for Middle Adulthood: 55-60

Era of Late Adulthood: 60-?

Age 50 Transition: 50-55

Entry Life Structure for Middle Adulthood: 45-50

Mid-life Transition: Age 40-45

Culminating Life Structure for Early Adulthood: 33-40

Era of Middle Adulthood: 40-65

Age 30 Transition: 28-33

Entry Life Structure for Early Adulthood: 22-28

Levinson's Ladder

Daniel Levinson says at each age a man faces specific tasks and challenges—such as choosing a career and a mate, and realizing his dreams—which he must meet if he is to proceed successfully up the ladder of life.

Early Adult Transition: Age 17-22

Era of Early Adulthood: 17-45

Era of Preadulthood: 0-22

the lives of one birth cohort must consider the vexing possibility that their findings may not apply to any other group. Most of the people studied by Levinson, Vaillant and Gould were born before and during the Depression (and were predominantly male, white and upper middle class). What was true for these people may not hold for today's 40-year-olds, born in the optimistic aftermath of World War II, or the post baby-boom generation just approaching adulthood. In Rossi's view, "The profile of the midlife men in Levinson's and Vaillant's studies may strike a future developmental researcher as burned out at a premature age, rather than reflecting a normal developmental process all men go through so early in life."

Based on her studies of women at midlife, Nancy Schlossberg, a counselor educator at the University of Maryland, also disagrees that there is a single, universal timetable for adult development—or that one can predict the crises in people's lives by knowing their age. "Give me a roomful of 40-year-old women and you have told me nothing. Give me a case story about what each has experienced and then I can tell if one is going to have a crisis and another a tranquil period." Says Schlossberg: "What matters is what transitions she has experienced. Has she been 'dumped' by a husband, fired from her job, had a breast removed, gone back to school, remarried, had her first book published. It is what has happened or not happened to her, not how old she is, that counts. . . . There are as many patterns as people."

Psychologist Albert Bandura of Stanford University adds more fuel to the anti-stage fire by pointing out that chance events play a big role in shaping our adult lives. Careers and marriages are often made from the happenstance of meeting the right—or wrong—person at the right—or wrong—time. But, says Bandura, while the events may be random, their effects are not. They depend on what people do with the chance opportunities fate deals them.

The ages-and-stages approach to adult development has been further criticized because it does not appear to apply to women. Levinson claims to have confirmed that women do follow the same age-transition timetable that men do. But his recent study of women has yet to be published, and there is little other evidence that might settle the case one way or the other.

Psychologists Rosalind Barnett and Grace Baruch of the Wellesley Center for Research on Women say, "It is hard to know how to think of women within this [stage] theory—a woman may not enter the world of work until her late 30s, she seldom has a mentor, and even women with lifelong career commitments rarely are in a position to reassess their commitment pattern by age 40."

But University of Wisconsin-Madison psychologist Carol Ryff, who has directly compared the views of men and women from different age groups, has found that the big psychological issues of adulthood follow a similar developmental pattern for both sexes.

Recently she studied two characteristics highlighted as hallmarks of middle age: Erikson's "generativity" and Neugarten's "complexity." Those who have achieved generativity, according to Ryff, see themselves as leaders and decision makers and are interested in helping and guiding younger people. The men and women Ryff studied agreed that generativity is at its peak in middle age.

Complexity, which describes people's feeling that they are in control of their lives and are actively involved in the world, followed a somewhat different pattern. It was high in young adulthood and stayed prominent as people matured. But it was most obvious in those who are now middle-aged—the first generation of middle-class people to combine family and work in dual-career families. This juggling of roles, although stressful, may make some men and women feel actively involved in life.

Psychologist Ravenna Helson and her colleagues Valory Mitchell and Geraldine Moane at the University of California, Berkeley, have recently completed a long-term study of the lives of 132 women that hints at some of the forces propelling people to change psychologically during adulthood. The women were studied as seniors at Mills College in California in the late 1950s, five years later and again in 1981, when they were between the ages of 42 and 45.

Helson and her colleagues distinguished three main groups among the Mills women: family-oriented, career-oriented (whether or not they also wanted families) and those who followed neither path (women with no children who pursued only low-level work). Despite their different profiles in college, and their diverging life paths, the women in all three groups underwent similar broad psychological changes over time, although those in the third group changed less than those committed to career or family.

Personality tests given through the years revealed that from age 21 to their mid-40s, the Mills women became more self-disciplined and committed to duties, as well as more independent and confident. And between age 27 and the early 40s, there was a shift toward less traditionally "feminine" attitudes, including greater dominance, higher achievement motivation, greater interest in events outside the family and more emotional stability.

To the Berkeley researchers, familiar with the work of psychologist David Gutmann of Northwestern University, these changes were not surprising in women whose children were mostly grown. Gutmann, after working with Neugarten and conducting his own research, had theorized that women and men, largely locked into traditional sex roles by parenthood, become less rigidly bound by these roles once the major duties of parenting decline; both are then freer to become more like the opposite sex—and do. Men, for example, often become more willing to share their feelings. These changes in both men and women can help older couples communicate and get along better.

During their early 40s, many of the women Helson and Moane studied shared the same midlife concerns the stage theorists had found in men: "concern for young and old, introspectiveness, interest in roots and awareness of limitation and death." But the Berkeley team described the period as one of midlife "consciousness," not "crisis."

In summing up their findings, Helson and Moane stress that commitment to the tasks of young adulthood—whether to a career or family (or both)—helped women learn to control impulses, develop skills with people, become independent and work hard to achieve goals. According to Helson and Moane, those women who did not commit themselves to one of the main life-style patterns faced fewer challenges and therefore did not develop as fully as the other women did.

The dizzying tug and pull of data and theories about how adults change over time may frustrate people looking for universal principles or certainty in their lives. But it leaves room for many scenarios for people now in young and middle adulthood and those to come.

People now between 20 and 60 are the best-educated and among the healthiest and most fit of all who have passed through the adult years. No one knows for sure what their lives will be like in the years to come, but the experts have some fascinating speculations.

For example, Rossi suspects that the quality of midlife for baby boomers will contrast sharply with that of the Depression-born generation the stage theorists studied. Baby boomers, she notes, have different dreams, values and opportunities than the preceding generation. And they are much more numerous.

Many crucial aspects of their past and future lives may best be seen in an economic rather than a strictly psychological light, Rossi says. From their days in overcrowded grade schools, through their struggles to gain entry into college, to their fight for the most desirable jobs, the baby boomers have had to compete with one another. And, she predicts, their competitive struggles are far from over. She foresees that many may find themselves squeezed out of the workplace as they enter their 50s—experiencing a crisis at a time when it will be difficult to redirect their careers.

But other factors may help to make life easier for those now approaching midlife. People are on a looser, less compressed timetable, and no longer feel obliged to marry, establish their careers and start their families almost simulta-

The first forty years of life furnish the text, while the remaining thirty supply the commentary.
—Schopenhauer, *Parerga and Paralipomena.*

neously. Thus, major life events may not pile up in quite the same way they did for the older generation.

Today's 20-year-olds—the first wave of what some have labeled "the baby busters"—have a more optimistic future than the baby boomers who preceded them, according to economist Richard Easterlin of the University of Southern California. Easterlin has been studying the life patterns of various cohorts, beginning with the low-birthrate group born in the 1930s—roughly a decade before the birthrate exploded.

WHAT'S THE RIGHT TIME?

Two surveys asking the same questions 20 years apart (late 1950s and late 1970s) have shown a dramatic decline in the consensus among middle-class, middle-aged people about what's the right age for various major events and achievements of adult life.

Activity/Event	Appropriate Age Range	Late '50s Study % Who Agree Men	Women	Late '70s Study % Who Agree Men	Women
Best age for a man to marry	20-25	80%	90%	42%	42%
Best age for a woman to marry	19-24	85	90	44	36
When most people should become grandparents	45-50	84	79	64	57
Best age for most people to finish school and go to work	20-22	86	82	36	38
When most men should be settled on a career	24-26	74	64	24	26
When most men hold their top jobs	45-50	71	58	38	31
When most people should be ready to retire	60-65	83	86	66	41
When a man has the most responsibilities	35-50	79	75	49	50
When a man accomplishes most	40-50	82	71	46	41
The prime of life for a man	35-50	86	80	59	66
When a woman has the most responsibilities	25-40	93	91	59	53
When a woman accomplishes most	30-45	94	92	57	48

SOURCE: ADAPTED FROM "AGE NORMS AND AGE CONSTRAINTS TWENTY YEARS LATER," P. PASSUTH, D. MAINES AND B.L. NEUGARTEN. PAPER PRESENTED AT THE MIDWEST SOCIOLOGICAL SOCIETY MEETING, CHICAGO, APRIL 1984.

SUDDENLY I'M THE ADULT?

BY RICHARD COHEN

Several years ago, my family gathered on Cape Cod for a weekend. My parents were there, my sister and her daughter, too, two cousins and, of course, my wife, my son and me. We ate at one of those restaurants where the menu is scrawled on a blackboard held by a chummy waiter and had a wonderful time. With dinner concluded, the waiter set the check down in the middle of the table. That's when it happened. My father did not reach for the check.

In fact, my father did nothing. Conversation continued. Finally, it dawned on me. Me! I was supposed to pick up the check. After all these years, after hundreds of restaurant meals with my parents, after a lifetime of thinking of my father as the one with the bucks, it had all changed. I reached for the check and whipped out my American Express card. My view of myself was suddenly altered. With a stroke of the pen, I was suddenly an adult.

Some people mark off their life in years, others in events. I am one of the latter, and I think of some events as rites of passage. I did not become a young man at a particular year, like 13, but when a kid strolled into the store where I worked and called me "mister." I turned around to see whom he was calling. He repeated it several times—"Mister, mister"—looking straight at me. The realization hit like a punch: Me! He was talking to me. I was suddenly a mister.

There have been other milestones. The cops of my youth always seemed to be big, even huge, and of course they were older than I was. Then one day they were neither. In fact, some of them were kids—short kids at that. Another milestone.

The day comes when you suddenly realize that all the football players in the game you're watching are younger than you. In-

Richard Cohen is a syndicated columnist for The Washington Post.

stead of being big men, they are merely big kids. With that milestone goes the fantasy that someday, maybe, you too could be a player—maybe not a football player but certainly a baseball player. I had a good eye as a kid—not much power, but a keen eye—and I always thought I could play the game. One day I realized that I couldn't. Without having ever reached the hill, I was over it.

For some people, the most momentous milestone is the death of a parent. This happened recently to a friend of mine. With the burial of his father came the realization that he had moved up a notch. Of course, he had known all along that this would happen, but until the funeral, the knowledge seemed theoretical at best. As long as one of your parents is alive, you stay in some way a kid. At the very least, there remains at least one person whose love is unconditional.

For women, a milestone is reached when they can no longer have children. The loss of a life, the inability to create one—they are variations on the same theme. For a childless woman who could control everything in life but the clock, this milestone is a cruel one indeed.

I count other, less serious milestones—like being audited by the Internal Revenue Service. As the auditor caught mistake after mistake, I sat there pretending that really knowing about taxes was for adults. I, of course, was still a kid. The auditor was buying none of it. I was a taxpayer, an adult. She all but said, Go to jail.

There have been others. I remember the day when I had a ferocious argument with my son and realized that I could no longer bully him. He was too big and the days when I could just pick him up and take him to his room/isolation cell were over. I needed to persuade, reason. He was suddenly, rapidly,

---

The size of a birth cohort, Easterlin argues, affects that group's quality of life. In its simplest terms, his theory says that the smaller the cohort the less competition among its members and the more fortunate they are; the larger the cohort the more competition and the less fortunate.

Compared with the baby boomers, the smaller cohort just approaching adulthood "will have much more favorable experiences as they grow up—in their families, in school and finally in the

labor market," he says. As a result, they will "develop a more positive psychological outlook."

The baby busters' optimism will encourage them to marry young and have large families—producing another baby boom. During this period there will be less stress in the family and therefore, Easterlin predicts, divorce and suicide rates will stabilize.

Psychologist Elizabeth Douvan of the University of Michigan's Institute for Social Re-older. The conclusion was inescapable: So was I.

One day you go to your friends' weddings. One day you celebrate the birth of their kids. One day you see one of their kids driving, and one day those kids have kids of their own. One day you meet at parties and then at weddings and then at funerals. It all happens in one day. Take my word for it.

I never thought I would fall asleep in front of the television set as my father did, and as my friends' fathers did, too. I remember my parents and their friends talking about insomnia and they sounded like members of a different species. Not able to sleep? How ridiculous. Once it was all I did. Once it was what I did best.

I never thought that I would eat a food that did not agree with me. Now I meet them all the time. I thought I would never go to the beach and not swim. I spent all of August at the beach and never once went into the ocean. I never thought I would appreciate opera, but now the pathos, the schmaltz and, especially, the combination of voice and music appeal to me. The deaths of Mimi and Tosca move me, and they die in my home as often as I can manage it.

I never thought I would prefer to stay home instead of going to a party, but now I find myself passing parties up. I used to think that people who watched birds were weird, but this summer I found myself watching them, and maybe I'll get a book on the subject. I yearn for a religious conviction I never thought I'd want, exult in my heritage anyway, feel close to ancestors long gone and echo my father in arguments with my son. I still lose.

One day I made a good toast. One day I handled a headwaiter. One day I bought a house. One day—what a day!—I became a father, and not too long after that I picked up the check for my own. I thought then and there it was a rite of passage for me. Not until I got older did I realize that it was one for him, too. Another milestone.

search shares Easterlin's optimistic view about the future of these young adults. Surprisingly, she sees as one of their strengths the fact that, due to divorce and remarriage, many grew up in reconstituted families. Douvan believes that the experience of growing up close to people who are not blood relatives can help to blur the distinction between kinship and friendship, making people more open in their relationships with others.

Like many groups before them, they are likely to yearn for a sense of community and ritual, which they will strive to fulfill in many ways, Douvan says. For some this may mean a turn toward involvement in politics, neighborhood or religion, although not necessarily the religion of their parents.

In summing up the future quality of life for today's young adults and those following them, Douvan says: "Life is more open for people now. They are judging things internally and therefore are more willing to make changes in the external aspects. That's pretty exciting. It opens up a tremendous number of possibilities for people who can look at life as an adventure."

For thousands of mental-health researchers and clinicians, "suicidology" is now a full-time specialty. ✓ *At some three hundred prevention centers, volunteers man hotlines.* ✓ *Yet experts still know too little about suicide, and its incidence in the U.S. increases each year.*

George Howe Colt

George Howe Colt is a contributing editor of Harvard Magazine. *He graduated from Harvard College in 1976.*

The rate remains highest among older people, but adolescent suicide has trebled over the past 25 years—a period that has also seen expanding prevention efforts, an outpouring of books and papers, and the creation of "suicidology" as a field of specialization. These developments are forcing a new awareness of the problem of self-destruction in American life.

At a recent high-school graduation in Weymouth, Massachusetts, a seventeen-year-old senior stepped up to the podium, said "This is the American way," and shot himself, though not fatally. His words provided a haunting corollary to Durkheim's theory that suicides tend to reflect not only personal failures but failures of society.

This year more than 30,000 Americans, an average of eighty per day, will take their own lives. The murder rate in this country is high, yet more people kill themselves than kill others. Suicide ranks ninth among major causes of death, and might rank as high as fifth if reporting procedures were stricter. Driven up by the alarming increase among adolescents, the U.S. suicide rate rose from 10.2 per hundred thousand in 1955 to 12.5 in 1980. Suicide now accounts for at least one percent of deaths in this country.

What are we doing about it? As a nation, almost nothing. Apparently satisfied with our rank near the middle of world statistics, the government no longer funds research on suicide. Funding for research on depression, such as it is, is biomedical, not socioeconomic.

As individuals and institutions, we are trying—but not concertedly. For the state of the art of suicide prevention is fragmented, controversial, and (worst of all) largely ignored. For more than two years I have been talking with men and women who have attempted suicide, with families of suicides, and with psychiatrists, social workers, sociologists, psychologists, and clergy. In these conversations I frequently felt we were talking about a dozen subjects at once. But often the talk was revealing, opening new perspectives on life as well as death.

EMMIE

Emmie* stares straight ahead as if at a drive-in movie. The windshield frames the kind of property advertised in the back pages of the New York Times Sunday Magazine. The six-bedroom house has enough angles and ivy to give it character. The yard grows into a meadow along the edge of a lake. Two blonde children in down vests chase each other in and out of view. The car windows are rolled up against the January chill and the ears of her children. I'm afraid to ask whether this car, the car she'll

*Asterisked names in this article are pseudonyms.

soon use to pick up a third blonde daughter from a piano lesson, is the car that two years ago, at age 47, she drove into a field at four a.m.; hooking up a vacuum-cleaner hose to the exhaust pipe, she zipped herself into a sleeping bag, stuck the hose in her mouth, and turned the key in the ignition.

When Emmie talks about it now—and she rarely does—she wonders how she could have reached that point, but the sadness in her voice suggests the forces aren't yet foreign to her. "It's easy be glib about it now," she interrupts herself, "but it wasn't that way at the time. I didn't realize what was happening to me. I was like a frog that can heat up so slowly, just adapting, adapting, and adapting. About the time he boils to death, he begins to wake up." She closes her eyes. "That's what happened to me, except unconsciously, I knew it."

Emmie had one of those "everything to live for" lives: good schools, an elite women's college, and a publishing job that sent her around the world from her Manhattan apartment. "But I had a few problems. I kept trying more things—flying an airplane, scuba diving, ballooning—to prove to myself I *was* somebody. Deep down, I didn't think I was very worthwhile."

She didn't marry till she was 36. Stan* was a paper manufacturer, divorced with two kids, a nice house in the suburbs, and a prefabricated life for her. "The whole situation changed," Emmie says. "I didn't have my job anymore, I didn't have my friends, I didn't have my name." She and Stan had three children, but as years passed, the marriage faltered. "I had trouble with Stan. He was not always taking my feelings into account." Emmie's voice rises in pitch but not volume. "I wasn't very assertive. I just let everything get eroded away, including my self-esteem."

When Emmie's father died, she was too busy with details of the funeral to confront his death. "I started to have what I now know to be anxiety attacks," she says. "It's as if you walked into the kitchen and your mother was on the floor in a pool of blood, you'd panic. I felt like that when the phone rang or I heard a car in the driveway." She saw several doctors, hoping for a diagnosis of physical illness. A psychiatrist told her she was having a normal reaction to her father's death—but that didn't stop the feelings. "I went through all the classic symptoms of depression," she says. "I couldn't sleep and would slip outside for walks at three in the morning. In the day I was doing the housework, but I wasn't able to concentrate. Our marriage had a lot to do with it—in many ways I was simply being ignored. I'd ask a question and the answers weren't forthcoming and it was as if I didn't exist. So I began to believe I didn't exist."

Emmie convinced Stan they should see a marriage counselor, but work kept him busy and appointments were irregular. "I had these wild dreams—'I'll get an M.B.A., then I'll be able to talk to him about business and he'll respect me.'" She took courses in the human-potential movement. "I kept thinking if I could just change, things would be better."

Things got worse. "We were on vacation when I first had a feeling I was going to kill myself. I was very vocal about it. I told Stan and I told my mother. They said, 'How terrible.' I saw a psychiatrist and I remember say-

ing, 'I understand people who talk about suicide never do it.' That was my great out. I must be okay because I was talking about it. But he said, 'No, people who talk about it usually do it.' That sort of cranked something into my subconscious." Her doctor suggested the hospital. "I had wanted to go six months before," says Emmie, "but the marriage counselor pooh-poohed it. She thought I was too intelligent." The doctor also sug-

> **E**MMIE: "I wanted something that was final but wouldn't be messy. I didn't want to blow my head off . . . I kept thinking about what would be easiest for everybody else."

gested a meeting with her husband and the counselor. "The more I thought about it, the more I thought, 'They're going to gang up on me and I don't want any part of that.' That's when I really decided I was going to have the last word.

"You have mixed feelings when you come to that point because you're living half in this world and half in the next. I lived for a month in that double state, knowing I was going to do it but wondering when, and unable to organize my life because I didn't know whether I'd be dead or alive. It's curious—if someone calls you up for lunch, you think, 'No, I can't make it Wednesday, I'm going to be dead.'" She purses her lips. "I didn't have any control over it—that's the interesting part. It all seemed inevitable, as if I had no choice but to carry out the impulse officially.

"The difference between having thoughts of doing it and actually doing it is tremendous. It takes an amazing amount of energy to figure out how to end your life, if you have certain criteria. I wanted something that was final but wouldn't be messy. I didn't want to blow my head off . . . I kept thinking about what would be easiest for everybody else." She smiles. "Of course, the easiest thing for everybody else would be if I lived. But that didn't occur to me.

"I was very matter-of-fact about it. I tried it one night in the garage but chickened out. I got the whole deal set up and then said, 'Well, this is just a dry run. Now I know how to do it.'" Two nights later, her husband woke at four to find Emmie gone. The police discovered her at eight the next morning, curled in her sleeping bag, unconscious. She revived at the hospital. No one could figure out how she survived. Three weeks later, a friend realized the car had a catalytic converter, which screened out enough carbon monoxide to keep Emmie alive. She'd taken the wrong car.

THE X-FACTOR

In the last 25 years, more time, energy, and money have been devoted to the study of suicide than in the previous two centuries. A computer check on the literature yields a five pound, seven ounce pile of printout. "Every psy-

5. ADOLESCENCE AND EARLY ADULTHOOD

chiatrist has to write a paper on suicide,'' jokes a psychiatrist who's written fourteen. The Encyclopedia Britannica gives the subject 25 times more space than it did 25 years ago. In the interim, an entire professional movement has sprung up: thousands of ''suicidologists'' now devote themselves to the study of suicide and its prevention, while thousands of trained volunteers tend hotlines at three hundred prevention centers across the country. Why, then, has the suicide rate continued to rise?

No one has a convincing answer. No one even has convincing statistics for the actual number of suicides in this country, because reporting procedures aren't uniform.* The accepted estimate for 1982 is 30,000, but some think 100,000 a year is more probable. Even the government-certified figures are only the tip of the ''suicide iceberg'': for every 30,000 suicides, runs the algorithm, there are 300,000 attempts, 3 million ''ideators'' (who threaten suicide), 30 million engaged in self-destructive behavior (from alcoholism, anorexia, and smoking to skydiving and compulsive gambling), and 120 million family members or ''survivors.''

Beyond the confetti of numbers, we know little about suicide, and less about preventing it. If we fail to understand why 30,000 people a year take their lives, it may be because we have such a limited understanding about one person's decision to do so. ''The problem of suicide cuts across all diagnoses,'' says John Mack, a psychiatrist at Cambridge Hospital. ''Some are mentally ill, most are not. Some are psychotic, most are not. Some are impulsive, most are not.''

Clinicians once tended to link suicide almost exclusively to depression, as if there were a threshold one could not sink below. This did not explain why many people who aren't depressed kill themselves, nor why the majority of depressed people do not take their own lives. For Aaron Beck, father of cognitive therapy, ''hopelessness'' is the key. Calvin Frederick, former chief of NIMH's Mental Health Disaster Assistance Program, goes two h's further: ''helplessness, hopelessness, haplessness.'' Psychiatrist Joseph Richman, of the Albert Einstein College of Medicine, declares that ''suicide is almost always the expression of some malfunction in the family system.'' Says Harvard Medical School psychiatrist Leon Eisenberg, ''A deficiency of social connections.'' Biologically oriented clinicians talk of chemical imbalance: recent research has linked suicide to low serotonin levels in the brain.

Nonclinicians are also divided. Some in the clergy cite erosion of faith, while philosophers speak of alienation and existential angst. To sociologists it is ''lack of connection,'' the breakdown of the family, changing sex roles, unemployment. Yet ''if life, or life in our culture provides the stress, the vulnerability to a response by suicide usually has a lifelong history,'' writes psychiatrist Herbert Hendin. Some clinicians see a predisposition to suicide that may have its roots in a childhood loss, past

* Some coroners classify a death as suicide only when a note is found (about fifteen percent of all suicides); others report only hangings and overdoses, overlooking shooting, jumping, drowning, and a dozen other forms of exit that can be viewed as accidents. Single-car crashes are rarely included, though some claim that as many as five percent are deliberate.

suicide in the family, unexpected separations from parents, or even the quality of mothering during the first three months of life.

While it is often said that suicide may be committed by twelve different people for twelve different reasons, it may be just as true that suicide may be committed by one individual for twelve different reasons. No single theory can account for all forms and motivations of suicidal behavior, and there can be no single preventative. ''There is no pill for suicide,'' says psychologist Edwin Shneidman, whose definition of suicide may be the best we've got: ''Suicide is a biological, sociocultural, interpersonal, dyadic existential malaise.''

How do we prevent *that*?

THE PREVENTION MOVEMENT

When Emile Durkheim published *Suicide* in 1897, demonstrating that suicide rates varied with social structure and with social change, he did not hope to save suicidal people; suicide was merely a mannikin for sociological theory. When Freud suggested that suicide is anger toward a lost love object turned back on the self, his interest was largely abstract. Suicide was not seen as something that could or should be prevented until Baptist minister Harry Warren formed the National Save-a-Life League in New York City in 1906. About the same time, the Salvation Army founded the London Antisuicide Bureau. Both used lay volunteers as counselors. Physicians refused to treat suicidal people, who were believed to be insane, or doomed by heredity; suicide was a crime and a sin, and the medical profession did not wish to contaminate itself with such cases. Today, almost all medical authorities regard suicide as a medical problem.

''Call me Ishmael,'' says Ed Shneidman, explaining the nest of neckties, each sporting thirteen spouting whales, on the desk of his Manhattan hotel room. Coat and tie slung over his chair, shirt unbuttoned, Shneidman looks out at the Staten Island ferry as he waits for a call from his wife in Los Angeles. He's in town for the annual meeting of the American Association of Suicidology. On the bureau, a pair of dark glasses, a copy of the New York Times, and a box of Vanilla Wafers complete what could be any traveling salesman's still life, only Ed Shneidman is never still. He's antic as a roly-poly third base coach giving all his signs at once. ''Okay.'' He claps his hands. ''Now.'' Another clap. ''The *first* paragraph is entirely about suicide. It says 'Some time ago, never mind how long ago exactly,' and blah-blah-blah, 'when I feel as though I'm going to commit homicide—knock somebody's hat off—I consider it high time to take to the ship.' '' Shneidman squirms to the edge of his seat and back. ''This is his substitute for pistol and ball, he says, for putting a gun to his head. 'Cato, with a philosophic flourish, throws himself on his sword.' *He* kills himself, he says, *I* take to the ship.'' He furls and unfurls his shirtsleeves. ''The whole book is about suicide, direct and indirect. Ahab *makes* the whale kill him. Ahab *nudgs* the whale: the whale wants to leave Ahab alone. Starbuck sees that, he says 'Look, fellow, you're making him kill you.' But Ahab *has* to pursue him. On the other hand, Ahab could not come back from that

192

voyage. He had to be killed. A substitute for suicide—a subintentioned death.''

"Subintentioned death" is one of dozens of words invented by Shneidman for the study of suicide. "Suicidology" is another. The growth of the field he named is difficult to separate from Shneidman's own career as a suicidologist—which began in 1949, when Shneidman, a psychologist working at the Los Angeles Veterans' Hospital, was asked to draft letters of condolence to the widows of two suicides. As preparation he visited the county coroner's office, where he found himself alone in an underground vault with two thousand suicide notes. He borrowed 721 of them and with psychologist Norman Farberow set out to see what last words might tell about people who kill themselves. Published articles led to grants, then to books, and then to the Los Angeles Suicide Prevention Center, a consortium of psychologists, social workers, psychiatrists, and trained volunteers, combining research and treatment. Opened in 1958, its guiding precept was that a suicidal person has a side that wants to live as well as a side that wants to die, and that a suicide attempt is a cry for help expressing that ambivalence.

The prevention movement gained federal sanction in 1966, when Shneidman was named first director of the Center for Studies of Suicide Prevention, a branch of NIMH: its goal was "to effect a reduction of the suicide rate in this country." CSSP spurred the growth of prevention centers. There were three when Shneidman arrived in Washington; three years later there were 200 and counting. The centers offered caring voices and ears (by telephone or in person) and perhaps "crisis intervention"—reality-oriented "short cut" therapy delivered by trained paraprofessionals.

Suicidology had been born. In 1967, CSSP sponsored a three-year postgraduate fellowship program in suicidology at Johns Hopkins. Social workers, sociologists, and psychologists took courses in crisis intervention, psychology of suicide, law and suicide, and did field work in prevention centers. In 1968, Shneidman promoted a "supergalactic symposium on suicide" in Chicago, with psychiatrists Karl Menninger and Erwin Stengel, and philosopher Jacques Choron ("the grand old men of suicidology"). The conference evolved into the American Association of Suicidology, an alliance of mental-health professionals, sociologists, clergy, and prevention-center volunteers. Shneidman was founding president.

Those were halcyon days. Regional centers were organized, symposia held, and ambitious ten-year plans outlined. At a 1970 CSSP summit conference in Phoenix, a task force proposed that government funding be apportioned among three groups: "Discoverers" (researchers), "disseminators" (playwrights, novelists, filmmakers, and journalists), and "gatekeepers" (clergy, police, bartenders, teachers, skid-row hotel managers, and anyone else who might usher a suicidal individual through a "gate" to prevention). The message was clear: suicide prevention is everybody's business.

But the growth of knowledge about suicide wasn't keeping pace with the zeal to prevent it. True, it was often pointed out that the rate of publications had more than tripled over ten years; in 1970, close to 500 articles

on suicide appeared, prompting that same CSSP task force to declare that "we are far from a state of ignorance in suicidology." But a review of these publications reveals little of practical use. AAS journal articles included "Lunar Associations with Suicide" (reviewing four years of Cuyahoga County suicides, the authors found a slight increase during the full moon, which they could not explain) and "Suicide on the Subway," which found "important differences" between those who lay in the train's path ("traumatic death") and those who touched the third rail ("nontraumatic death"). "One of the problems in the study of those who kill themselves is that the object of the study is deceased and hence not available for study," began an article on "The Role of Spiritualism in Suicide."

Research *is* limited by the fact that studies are post facto. "There is very little writing about completed suicides," says John Mack, whose *Vivienne* was a recent exception. "Clinicians are reluctant to write about it because it's so disturbing." Suicide is a statistically rare event, and it can take years to accumulate a statistically significant sample. Much research involves as few as six subjects, and the same case studies reappear like prize pupils, often to illustrate different points.

The movement's initial optimism underscored its apparent failure. In Los Angeles, site of the flagship prevention center, the suicide rate jumped from 14.8 in 1961 to 17.6 in 1965; in California, the rate rose from 15.9 to 18.8 during the Sixties. In the nation, the rate rose from 10.6 to 11.6. "The whole idea was sold as though it was going to change the suicide rate," says psychiatrist Herbert Hendin. "The trouble with the hypersell was that there was bound to be a reaction." The government, which in ten years had spent more than $10 million on suicide research, decided it was a bad investment. In 1975 the CSSP was dissolved.

Since then, "the commitment of the federal government to the cause of suicide prevention has been minimal," said psychologist James Selkin, former president of the American Association of Suicidology, in 1982. "A loss that approaches 30,000 lives per year is practically unrecognized." There is no federally funded research on suicide; organized efforts in prevention rest largely with AAS. But as an interdisciplinary organization, AAS reflects continuing conflicts in the field. "Suicidologists" are psychologists, social workers, psychiatrists, and others whose perspectives are often at odds. At one point, mental-health professionals and prevention-center volunteers threatened to split the association in two. But the most serious rift may be between members of the AAS and mental-health professionals who remain outside the movement. There is an unspoken feeling that the movement is too lay; it doesn't help that the AAS journal is scorned for its lack of academic rigor. "I don't want to be described as someone who's *only* interested in suicide," says one psychologist. Some express embarrassment over the word *suicidology* and accuse the leadership of self-interest. "The suicide industry is Shneidman's baby—he popularized it," says one well-known psychiatrist. "Shneidman's empire," scoffs another, who has his own theories about Shneidman's devotion to the subject. "People who think they know

*T*HE MOVEMENT'S initial optimism underscored its apparent failure. The government, which had spent more than $10 million on suicide research, decided it was a bad investment.

me say I'm running from death," says Shneidman. "I don't think so. . . . If you'd let me discover a roomful of personal documents of alcoholics or delinquents, I would have spent my life studying alcoholics or delinquents." He looks tired; the next three days he'll be pushing suicide prevention in three states. "It was a virgin field. My brightness was in recognizing what powerful possibilities lay there."

THE NUMBERS GAME

The main approach to suicide prevention has been demographic: predicting who's going to do it by analyzing who has done it. And who does it? "Any of us, if hit hard enough," says Douglas Jacobs, former director of Emergency Room Services at the Cambridge Hospital. Some are more likely than others. Three times more men than women kill themselves, though women make three times as many attempts. Whites commit suicide more than nonwhites. Suicide rates increase rapidly in the teen years, peak in the twenties, taper off in the thirties and forties, rise rapidly in the fifties and sixties. Divorced men are three times as likely to kill themselves as married men. In general, two out of three suicides are white males. Other factors identified with suicide include alcoholism, homosexuality, and family history. Using these and other factors, a composite emerges, like a police artist's sketch, of an older, white, divorced, unemployed male who lives alone and is in poor health—the "high-risk paradigm" described by clinicians.

Using this information, researchers have devised scales to quantify suicide risk, so that clinicians will have something "objective" to help them in making decisions. Pointing out that test chestnuts like the Rorschach, TAT, and MMPI have been no help in predicting suicide risk, Aaron Beck developed "The Beck Depression Inventory," a "Hopelessness Scale," and a scale to measure the intentions of suicide ideators, with nineteen questions including "Wish to Die," (scored "moderate to strong," "weak," or "none") and "Wish to Live" (which offers the same options). Psychiatrist Robert Litman, reviewing 26,000 cases at LASPC, devised the "Suicide Potentiality Rating Scale," now used in both telephone screening and clinical interviews.

While these scales provide useful checklists for clinicians, playing "the numbers game," says Jacobs, can be dangerous. "You have to be careful. On a percentage basis, young people are less likely to kill themselves, but if one does, for that person, it's a hundred percent." The categories of risk that most scales use have become less reliable. Suicide has increased dramatically among the

young; the rate for women is increasing faster than that for men; for young blacks, faster than for whites.

If the scales have not proved their worth, neither has clinical intuition. In a 1974 study, a computer proved more accurate than experienced clinicians in predicting suicide attempters. In addition, more than half the patients preferred computer to doctor as interviewer.

Ed Shneidman looks for other signs; he believes suicidal people communicate their intentions, but most people don't know how to listen. From "psychological autopsies" conducted with family and friends, he claims that four of five suicides give "clues" to their intentions, verbally ("You won't see me around much longer") or by their behavior (giving away prized possessions, putting their affairs in order). A sixteen-year-old Californian asked a teacher, "Do you have to be crazy to kill yourself?", wrote the word "death" on his hand, and one morning told a friend he was "going to heaven very soon." That day in class, he pulled a revolver out of a paper bag and shot himself. He was one of twelve high-school suicides in San Mateo County that year; most had left clues recognizable to "trained suicidologists."

"Education is the single most important item in lowering the suicide rate," says Shneidman. "I don't just mean suicide prevention classes. I mean a general heightening of awareness, so that if I give you my watch, you won't simply take it and thank me. You ought to say 'Ed, sit down, tell me what is happening.'" Shneidman advocates mass-media campaigns like that which helped 30 million Americans give up smoking in the past decade. "It's like V.D. and cancer," he says. "Education is more important than crisis intervention. I don't like putting out fires; I think it's more important to build a hotel where fires won't occur."

Yet even to trained suicidologists, clues are often recognizable only in retrospect—and in hindsight, almost anything can look like a clue. "I'm sure if you or I went out the window right now, somebody might say, 'I knew that was going to happen someday,'" says Harvard psychologist Douglas Powell. Powell did counseling work in 1975 when a Harvard junior ran through a dormitory window to his death during reading period. For weeks, the boy's friends wondered why, agreeing there had been no apparent reason, no clues. Then his roommate recalled one detail: Eric* had always set ashtrays, mugs, and postcards on the window sill. And, for several weeks before his death, each time Eric sat in front of that window, another object had been removed.

Clues are especially well camouflaged in adolescents. In *Vivienne*, the story of a talented, insecure fourteen-year-old who hung herself in her mother's silversmithing studio, psychiatrist John Mack lists warning signs that families and teachers should recognize: suicidal talk, moodiness, loss of self-worth, turning excessively to a diary instead of friends, failure in school, increased drug and alcohol consumption, hypochondria, and philosophical preoccupation with death and dying. But what looks like "presuicidal syndrome" to a psychiatrist may look to parents like "a phase."

People often recognize clues but fail to respond, through fear or ignorance. Nor is it known what proportion of people who leave clues go on to kill themselves.

"All the students come in at some point and talk about suicide," says one high-school social worker. "I can't put them all in the hospital." Her biggest difficulty is persuading parents to let her help: "I get screaming parents who say 'You're nuts—my child's fine, not depressed.'"

"It's often difficult to get parents to acknowledge the problem, because they *are* the problem," says Peter Saltzman, child psychiatrist at McLean Hospital. Even suicide attempts may have no effect. One of Saltzman's patients, following an attempt, was told by his father, "Next time, jump off the Bourne Bridge." Failure to get attention may lead to a "face-saving" success. Leon Eisenberg describes a college student having a turbulent affair with a classmate. "He said, 'If you don't go steady with me, I'll jump off this building.' She said, 'You don't have the guts.'" He did. "He ran right up the steps to the eighth floor, out on the roof, and jumped off," says Eisenberg. "And I might add the young lady showed no remorse at all."

TREATING THE SUICIDAL PATIENT

Extending treatment to all potential suicides is clearly impossible. "We've reached the point of no return in defining vulnerable populations. Finding that vulnerable person amounts to looking for the proverbial needle in a haystack," says Herbert Hendin, who knocks Shneidman's proposed educational campaign. "I don't follow the logic of putting millions into educating the lay public in something that psychiatrists haven't proven *they* can identify. It makes more sense to do something for the people you *do* find. A lot of seriously suicidal people present themselves in ways nobody can miss—they jump from five-story buildings—and nobody does anything for them."

Previous attempt is the highest predictor of suicide risk; between 30 and 40 percent of suicides have tried before, and about 10 percent of those who attempt will succeed within ten years. Yet most attempters, says Hendin, are returned to the community after brief hospitalization, without provision for follow-up treatment. "If you could even identify 20 percent of the seriously suicidal from those who make attempts, and cure 10 percent, you could literally change the suicide rate," he declares.

Can clinicians actually "cure" suicidal people? Patients who have made attempts and entered into treatment have the highest suicide rate of any patient group. Yet many in the prevention movement assume that once we tag people as suicidal, we can help them.

Suicide is a symptom, not a diagnosis. Because a clinician can't treat it, he must treat the "underlying illness"—the patient's closest diagnosable ailment. Some clinicians believe that if they successfully do so, they've treated the suicidal patient, as if suicidality were a nasty side effect of the underlying illness. Yet many suicidal patients have no diagnosable underlying illness, and patients often kill themselves shortly after coming out of depression—or long after depression has lifted. "Suicide proneness is primarily a psychodynamic matter; the formal elements of mental illness only secondarily intensify it, release it, or immobilize it," writes psychi-

atrist John Maltsberger. "The urge to suicide is largely independent of the observable mental state and it can be intense despite the clearing of symptoms of mental illness."

There's little agreement on how to treat any mental illness. A person suffering from some form of depression (an estimated 25 to 30 percent of suicides) may be treated with antidepressant drugs, shock treatment (making a comeback after decades of disrepute), cognitive therapy, Yoga, or any of more than 200 brands of psychotherapy practiced today. Of the 200, only psychoanalysis is agreed to be inappropriate for suicidal patients: "Most are either too anxious, too depressed, or just not well enough put together to stand it," says Hendin. Although suicidal patients come in different diagnoses with different needs, they're likely to get whatever the therapist stocks. "One would hope that clinicians had a number of strings to their therapeutic bow and would change depending on the nature of the problem," says Leon Eisenberg. "Unfortunately, this field is characterized by people who do a type of treatment for every customer

> *H*ERBERT HENDIN: "I don't follow the logic of putting millions into educating the lay public in something that psychiatrists haven't proven they can identify."

that comes along." A therapy that works with suicidal patients will be ignored by most clinicians if it's not their *modus operandi*. Group therapy with suicide attempters appears to generate a caring bond which, like Alcoholics Anonymous, can extend beyond formal sessions to times of crisis. Family therapy has been effective, and some therapists won't see a suicidal patient without seeing the family. But many more worry about overstepping patient-therapist confidentiality.

There is no pill for suicide, but there are scores for depression. Most therapists supplement drug therapy with psychotherapy, or vice versa, but drugs are the *only* treatment method for some. "In state hospitals, community mental-health centers, and as practiced by welfare doctors, it is often the sole therapy," writes Jonas Robitscher in *The Powers of Psychiatry*.

Antidepressants have been a red herring in the treatment of suicidal patients. "They do nothing to alter the underlying vulnerability to suicide," says Maltsberger. The same drug that relieves depression may give a patient sufficient energy to act on his impulses. "It's a complicated issue," says Alan Pollack, director of McLean Hospital's outpatient clinic. "You have to judge whether patients can manage medication, how reliable they are in taking it, and how big a supply you can give them without their OD'ing on it." A 1970 study showed that of 200 people who committed suicide with barbiturates prescribed by physicians, over two-thirds had a history of previous attempts. Notes the study, "One might indeed wonder what kind of a nonverbal communication the suicidal person must feel he is receiving when he is handed a prescription for a potentially lethal quantity of drugs." . . .

SOCIAL STUDIES

A suicidologist visiting China asked her guide about suicide. The guide frowned and said, "Oh no, the state would not approve." In some ways, the United States does. Suicidologists point to a number of measures that might lower the suicide rate: gun control (only in the U.S. are guns the primary method of suicide); tighter controls on prescription drugs; availability of abortion and birth control ("An unwanted child is an unhappy child is a depressed child is a suicidal child," reasons Sol Blumenthal of the New York Department of Public Health); safety features—bridges without walkways, nets for observation decks, windows that don't open wide. Opponents of these proposals argue that denying access to one method will merely force the suicidal person to another, but this overlooks the impulsive nature of many suicides who adopt whatever means is at hand. When a plexiglass barrier was erected atop the campanile at the University of California, Berkeley, the campus suicide rate declined. Moreover, would-be suicides are often particular; many women, for instance, choose not to spoil their looks by jumping or shooting. Suicidologists feel that limiting access to lethal methods would lower the rate: from 1971 to 1976 barbiturate prescriptions declined from 40 to 20 million; the number of suicides by barbiturates was halved. Although suicides by other drugs increased during that time, careful prescription practices clearly had a measurable effect. When the suicide rate in England and Wales declined between 1960 and 1975, the drop was attributed to the changeover from coal gas to less toxic natural gas.

San Francisco has one of the highest suicide rates in America, and one of the most picturesque and lethal sites: the Golden Gate Bridge. Since it was built in 1937, 757 have jumped (21 jumped and lived). In 1973, when the 500th suicide was expected, a circus atmosphere of souped-up surveillance and media publicity prevailed. Fourteen attempts were foiled between numbers 499 and 500. Ten years and almost 300 suicides later, the plan for a million-dollar "suicide fence" has been scratched.*

While we cannot make the world "suicide-proof," nor our lives a 24-hour suicide watch, "we should not make it physically easy to commit suicide," argues Sol Blumenthal. Others aren't persuaded. "Ninety-nine percent of us don't need it," says one San Franciscan. "Is it fair to ruin the view for the sake of a few? If they want to die so much, why not let them?"

---

* A study showed that of 1,440 persons prevented from jumping, only 4 percent went on to kill themselves by other means.

THE RIGHT TO SUICIDE

In 46 B.C., determined to kill himself before an advancing army could, Cato the Younger threw himself on his sword. He revived to find physicians dressing his wound. Shoving them away, he ripped off the bandages and died.

The right to suicide has been debated ever since; the recent appearance of at least four "how-to" manuals has added urgency to the discussion. "A Guide to Self-Deliverance," published by EXIT, a London-based organization, is a 10,000-word booklet offering four bloodless methods of suicide, and including a table of lethal doses for prescription and nonprescription drugs. Formerly the Voluntary Euthanasia Society, EXIT has expanded its credo to include not only the incurably ill but the right to suicide for everyone, and is now the Society for the Right to Die with Dignity. The book is for members only; membership is a matter of $23. *Suicide, Mode D'Emploi* ("suicide operating instructions") offers recipes for fifty lethal cocktails in its 276 pages. It sold 50,000 copies in the first five months after publication in 1982, made the French best-seller list, and has been linked to ten suicides. Efforts to ban it have failed; American publishers are jockeying for the rights. "I feel no remorse," said the publisher. "This is a book that pleads for life. But it also recognizes that the right to suicide is an inalienable right, like the right to work, the right to like certain things, the right to publish. What use is a right without the means to execute it?"

Courts are struggling with that question. In Ledyard, Connecticut, two young men sawed a shotgun barrel so their best friend, partially paralyzed in an accident, could more easily maneuver the gun to shoot himself. They now face trial. A 22-year-old Oklahoman distributed flyers offering to help people kill themselves if they would split their insurance benefits. Arrested before getting any takers, he faces five years in prison.

With increasing attention to death and dying, and the idea that a "good" death is as important as a "good" life, "suicide will gradually become the culturally sanctioned mode of death," writes Robert Kastenbaum, a former president of AAS. He suggests that suicide "facilitation" be linked to prevention, in one center.

Incidence of suicide in fifteen countries

	(per 100,000)	
	Males	Females
Austria	36.7	14.7
Switzerland	34.5	15.4
Denmark	31.8	19.8
West Germany	30.1	15.1
Sweden	28.3	12.9
France	23.3	9.9
Poland	22.8	4.2
Japan	22.6	13.6
Canada	21.2	7.3
UNITED STATES	19.0	6.3
Norway	17.2	7.1
Australia	16.6	6.7
U.K. (England and Wales only)	10.7	6.5
Israel	6.6	4.7
Spain	6.1	2.1

Source: World Health Organization, *World Health Statistics.*

GUIDELINES

Like many psychiatrists, John (Terry) Maltsberger took on suicide in self-defense. When he was a resident at Massachusetts Mental Health Center, a woman patient chloroformed herself. Several of Maltsberger's patients were also threatening suicide. He began meeting weekly with Dan Buie, a young colleague who had lost a patient to suicide and felt equally concerned. They worked their way onto a committee whose charges included the "suicide review"; gradually, their combined perspective emerged in professional journals.

Ask about suicide care in Boston today, and Maltsberger's name is almost sure to come up. The man himself is Napoleon-sized, dapper, with a gracious accent an American might import after a week in London. Bright blue eyes peer from a full-moon face under silver hair. The consulting room on the first floor of his Beacon Hill town house is dimly lit but warm, with a working fireplace at one end, and at the other a long drawerless desk on which fifty books—including *The Savage God,* a staple of every therapist's library—are lined. French doors open onto a jungly garden that seems about to entrap Maltsberger in a three-dimensional Rousseau painting. He is explaining the "psychodynamic formulation" of suicide, which he and Buie evolved.

"It boils down to finding what a person has to live for," says Maltsberger. "Most people live for all sorts of things—friends, a special person, work—and if they lose something on one front, they pick it up on another. But suicidal people are quite deficient in any capacity to keep themselves afloat, on the basis of inner resources. Once somebody threatens suicide, you start looking at what resources the person has."

Maltsberger and Buie specify three areas people may live for: other people, work, their own bodies. "Obviously, when someone who is dependent and depressive loses a girlfriend or a husband, it can precipitate a suicidal crisis. There are people who never have relationships, who lock themselves in the library and devote themselves to scholarship. But when they retire, or can't work anymore, they may kill themselves. A surgeon may live only to operate; if he loses the use of his hands, he may do away with himself. And there are people who may be very dependent and depressive, but as long as they can jog and look in the mirror and say, 'Gee, I'm in great shape,' they can go on.

"So if someone has relied all his life on some capacity to work at Sanskrit, and he goes blind and can't do it anymore, the task becomes to find what this person can substitute as a life-saving activity. It isn't always possible. Many people are quite indifferent to the love of others, for instance. Others may be indifferent to success at work. Suicidal people are very specialized in what they will accept as a reason for living." At first, says Maltsberger, the therapist may have to constitute that reason, "until the patient can regain his balance and stand up again." Maltsberger smiles. "It sounds simpleminded, but it really is like that."

The psychodynamic formulation offers therapists a practical way to decide *when* someone is suicidal, what to do for treatment, and whether hospital admittance or discharge is indicated. In their suicide reviews, Buie and Maltsberger found that such decisions were often based entirely on mental-state examinations, assessments of appearance, speech, and behavior—i.e., experienced gut reactions. "When this is the practice," they write, "it is inevitable that a certain number of otherwise preventable suicides will take place." Like one fifty-year-old widow who was released from a hospital after a course of treatment:

The psychiatrist and staff responsible for her care did not appreciate that she had moved into an attitude of quiet resignation and despair. . . . At no time in the course of her treatment had the patient seemed disturbingly depressed to anyone and there had been no suicidal threatening or preoccupation that the patient had made known. She destroyed herself after discharge when her son announced his plans for marriage. Only in retrospect was it noted that this dependent woman was without support in her social and family context.

The approach developed by Maltsberger and Buie requires a therapist to know patient histories well, and to spot events in a patient's daily calendar that might heighten suicide risk. "Treating suicidal people means being available—intensively—from time to time while they're between supporting figures or research projects," says Maltsberger. "I might call them on the telephone every day, perhaps go to their house . . . you have to be there waiting like a net, hoping that as time goes on the person can widen his repertoire and make room for other sustaining influences.

"People who grow up suicide-vulnerable have failed to get the love they ought to have had from their mothers," says Maltsberger. "My approach looks at suicide in terms of developmental failures that make it impossible to maintain a sense of self-worth." In normal development, he explains, capacity for autonomy increases with age, enabling one to endure degrees of loneliness, depression, and anxiety. Those who don't receive enough mother-love fail to develop sustaining inner resources; they must depend on external supports. When external supports fail, suicide is a danger.

Even the psychodynamic formulation offers only temporary relief. Can vulnerability to suicide be altered? Maltsberger sighs, like the Wizard of Oz after giving out heart, brains, and courage, only to find that Dorothy's still in search of a way to get back to Kansas. "That's most ambitious," he says slowly. "That means helping the patient restructure his mind. There are very few patients where that's possible." He pauses. "Some of us believe you can. Often, psychiatrists don't want to try."

"Suicidology would have two distinct aspects—a continuation of the present effort to understand, predict, and prevent self-destructive behavior, and a sensitive new approach designed to help people in certain circumstances attain the particular death recommended to them by cultural idealizations and their own promptings."

"Certain circumstances," for most advocates, means life-support systems and terminal cancer. (In fact, the suicide rate among terminally ill cancer patients is low; they "tend to cling to what life they have left," says Calvin Frederick.) Model cultures include those of the Greeks, who honored suicide, and the Romans, who promoted it to heroic proportions.

The elderly often figure in discussions about the right to suicide. Doris Portwood, in *Common-Sense Suicide,* writes, "When an older woman leaves a social gathering—perhaps an hour after dinner when younger guests are settling down to a game or a fresh drink—no one urges her to call a cab or offers her a lift. She will receive thoughtful words during the process of departure, but no insistence on her staying. There is the assumption that she has, in fact, some good reason for going." Portwood suggests that to decide when to leave, a person should draw up a list of pros and cons. Though "balance-sheet suicide" is an old concept and Portwood insists she's pushing "only for a right of choice—not trying to eliminate the old," some of her pros are cons. She writes of the high cost of subsidizing the elderly, and points out that by the year 2000 there will be more than 30 million American senior citizens. "Those who have the will to opt out may not (yet) get a public vote of thanks. But who can dare say that they will be missed?" she writes.

Concern with adolescent suicide obscures the fact that older people still make up the majority of suicides; white men over fifty, 10 percent of the population, account for 30 percent of all suicides. "What we tend to forget when we talk about rates is that some of the disparities reflect populations we do not like or do not respond to," says psychiatrist Seymour Perlin, citing skid-row alcoholics and the elderly, who are considered undesirable patients. "Too often the old are written off as treatable with pharmacology while younger patients get psychotherapy," said Herbert Hendin, addressing a sparse crowd on "Suicide over Sixty" at an AAS meeting (across the hall, "Adolescent Suicide" was SRO). "The heart of suicide prevention has always been, and remains, suicide among older people."

Underlying is the assumption that older people have less to live for. But the right to live with dignity may be as neglected as the right to die with dignity; if the first were looked into, the second might not seem so pressing. And what is overlooked in Portwood's own fantasy suicide of a peaceful plunge into a lake with a congenial group of friends (besides the consensus that drowning is one of the most difficult and painful deaths) is that many elderly suicides may not be so "rational." Suicidologists worry that easy access to methods and public approval of "rational suicide" may encourage self-destruction by some for whom it may not be a rational choice. Where do you draw the line? Who draws it?

"Most of us don't think about it very much," says psychiatrist Leon Eisenberg. "If a patient says, 'I'm going to kill myself,' we try to stop him." Yet there is the feeling that if someone *really* wants to die, he will.

"Killing yourself is not a great problem," allows Eisenberg. "There are buildings you can jump off, you can throw yourself on subway tracks . . . there's just no way to stop someone unless you lock them off in a corner, and even then people find a way.

"The reason most clinicians don't take the right to kill yourself seriously," continues Eisenberg, "is that all of us have seen people who have failed, been treated for depression, and are now grateful to be alive." (Research shows that 10 to 12 percent of attempters try again within a year; 1 percent succeed; and 10 percent will kill themselves within ten years.) Over the past thirty years, Hendin has interviewed four people who survived six-story jumps. Two changed their minds in midair, two did not, and only one tried again. "It's clear that suicide must not be a mature, thoughtful, highly motivated decision," says Eisenberg, "because once a person made that decision, there's not a damn thing we could do to stop it, because they could keep going after it, after it, after it. And they don't."

To clinicians, there may be no such thing as a "rational suicide," which to them precludes any "right." Countless studies purport to show what percentage of suicides are "mentally ill": Eli Robins in *The Final Months* goes as high as 94 percent. "The argument connecting suicide and mental illness is tautologically based upon our cultural bias against suicide," Zigrids Stelmachers, director of a Minneapolis prevention center, has said. "We say, in essence, 'all who attempt suicide are mentally ill.' If someone asks, how do you know?, the implied answer is, because only mentally ill persons would try to commit suicide." A study at Harvard found that the highest estimate of mental illness when a sample had been diagnosed *before* suicide was 22 percent. Afterward, the highest estimate was 90 percent.

Finding suicidal people mentally ill has practical implications. In recent years, efforts by civil libertarians to abolish involuntary commitment has led to a maze of different standards in different states. Most statutes specify that the individual must be considered dangerous to himself and also mentally ill—criteria determined by the admitting psychiatrist. This gives psychiatrists another reason to staple suicide to a traditional diagnosis. But there is a less tangible, perhaps more dangerous consequence. Writes psychiatrist Thomas Szasz:

In regarding the desire to live as a legitimate aspiration but not the desire to die, the suicidologist stands Patrick Henry's famous exclamation "Give me liberty or give me death!" on its head. In effect, he says, "*Give him* commitment, *give him* electroshock, *give him* lobotomy, *give him* lifelong slavery, *but do not let him choose death!*" By so radically invalidating another person's (not his own!) wish to die, the suicide-preventer redefines the aspiration of the Other as not an aspiration at all. The wish to die thus becomes something an irrational, mentally diseased being displays, or something that happens to a lower form of life. The result is a far-reaching infantilization and dehumanization of the suicidal person. . . .In short, the suicidologist's job is to try to convince people that wanting to die is a disease.

"One of the underlying assumptions in a lot of this work is that there are 'sickies' and 'wellies'—that I must be a 'wellie' but you look sort of like a 'sickie,'" says

Merton Kahne, psychiatrist in chief at M.I.T. "A lot of people in my trade spend a lot of time trying to demonstrate how sick you are, as if you must be sick or you would never kill yourself . . . which I think is nonsense." He adds, "I don't know what a suicidal person is—I thought virtually everyone who's gotten beyond the seventh grade had thought about killing themselves." Calling suicidal behavior "sick" may help distance us from an act that strikes a disturbing chord.

"I remember dealing with my first suicidal patient. I found it very difficult to understand that a person could really choose this," says Nancy Kehoe. "I had to go out for a long walk and try to take in how much pain that person must feel to want to take his own life." Now, dealing with suicidal patients, "I let myself get in touch with the times I've felt pretty desperate, the fleeting moments of driving down the Mass Pike and wishing a truck would hit you. We've all had those moments where we say 'Enough—I can't take it anymore.'"

Perhaps in an ideal world, people would not want to die, but as Stelmachers says, "Some of the things that happen to these people give them pretty rational reasons for ending their lives." If the cry for help can be translated "help me live," it can also be translated "help me die." "A totally open existential therapeutic relationship must make room for everything, including suicide," writes philosopher Peter Koestenbaum. "Only in such a way can the freedom of the patient be recognized and nurtured." Writes Robert Neale, "Suicide can be prevented only by permitting it." Making room for suicide does not mean a clinician must set up facilitation services in a prevention center, or refuse treatment to a ten-year-old who's tried to hang himself, but that he must respect the possibility of suicide at least as much as he fears it. "If the person says, 'I'm going to kill myself,'" says Stelmachers, "one way to respond is to say, 'Well, maybe suicide is the best way out for you, but let's talk about it first.' This says many things—first, 'I am really interested in you and your problems.' It also negates a sneaking suspicion he might have had about himself, that he must be crazy to even consider such an act."

At a recent Harvard conference on suicide, the ethics of prevention came up: "Do we have the right to say no?" wondered an audience member. There were appreciative chuckles; it's the oldest and, by clinicians, least seriously discussed topic in suicide. "Tough question," said John Mack. "Shall we refer that one to God?" More chuckles. "We have a right to take a different position," continued Mack. "Our responsibility as clinicians is to choose life." Someone else spoke up: "I think the philosophical answer is different from the clinical one." Until they are part of the same answer, the study of suicide and its prevention may never be complete.

THE UNANSWERED QUESTION

It was the last day of the fifteenth annual meeting of the American Association of Suicidology. More than 500 suicidologists from dozens of states and countries had gathered at the Vista International Hotel in New York City for a four-day smorgasbord of workshops on "Suicide and the Big City," "Women and Suicide,"

"The Question of the Right to Die When Pain is Very Intense," and 55 other topics.

A who's who of suicide had assembled. Norman Farberow and Ron Maris were there. So were Herbert Hendin, whose *Suicide in America* was about to appear in bookstores, Ari Kiev, and Nancy Allen, the public-health worker who was instrumental in organizing the first "National Suicide Prevention Week" in 1974. And everywhere you looked was Ed Shneidman, peripatetic ringmaster of the suicide prevention movement in America. Heady company; at one point, twelve past presidents of the AAS sat at the dais. Their combined efforts represented over a hundred books, a thousand articles, and more than 200 years of experience in the study and prevention of suicide.

Now, while volunteers took down posters in the lobby (a blank brick wall—"suicide is a dead end") and the silver-haired proprietor of the Thanatology Book Club ("Save time, save money, receive a free book just for joining") closed up shop, the day's first meeting was getting underway downstairs in the Nieuw Amsterdam Ballroom. It was nine o'clock. Less than a third of the registrants were in attendance. Some were recovering from a "Backstage on Broadway" tour arranged by the entertainment committee, while others opted for last-minute sightseeing or confirming flights home, rather than this session on "Borderline Personality Disorders and Suicidal Behavior."

Grisly fare for a Sunday morning. Several people in back slept through presentations by two mildly eminent psychiatrists. (My notes are hieroglyphs: "central organizing fantasy of narcissistic union" and "objective scrutiny of object relations.") As Otto Kernberg, who pioneered the study of the borderline patient, read a dense, theoretical paper, a group of psychiatrists in front gazed up with adoration and a prevention-center volunteer joked about marketing the speech as a sedative.

When Kernberg finished, the moderator, a young psychiatrist who'd been alternating pensive nods with glances at his watch—it was his job to herd everyone upstairs in time for "Is There Room for Self-Help in Suicide Prevention?"—invited questions. Hands shot up in front, and their owners raised progressively complex issues. But a hand in back, belonging to a shabby fellow with a ponytail, persisted. And the moderator finally gave in.

The man stood. His jeans and flannel shirt were worn but not dirty. His ruddy face couldn't decide on a beard or a shave, and his eyes were as cloudy as his question, a stammering ramble proposing meditation as a panacea for suicide. Eyes started to roll in the audience, and there were tolerant chuckles. The moderator flashed the panel an embarrassed collegial smile. When the ponytailed man slowed for a moment, the moderator broke in: "That's an interesting question, but let's move on. We've got time for one more." He looked for another hand; the man remained standing. The moderator began his thank-you-very-much-I'm-sure-we-all-learned-a-lot speech and the man was beginning to sit, bewildered, when Kernberg reached for the microphone, said "I'd like to answer that question," and in his textbook Viennese accent, began responding with care and respect.

The Emotional Brain

Laurence Miller

Laurence Miller teaches neuropsychology at Seton Hall University and is writing a book about personality and the brain.

The idea that the brain's left and right hemispheres are specialized for different kinds of thinking is a familiar one, backed by lots of research. Now it's beginning to look as if the hemispheres are also specialized for different emotional experiences. It's not as simple as, say, a sad side and a happy side, but there's growing evidence that parts of the two hemispheres play different roles in our emotions and moods.

Support for this idea comes from research on some people with brain disorders and on some with apparently perfectly functioning brains. The research is still controversial and sketchy. But the picture that's emerging is intriguing and may lead to new ways of helping people whose emotional lives are out of control.

One line of evidence comes from research on people with neurological problems such as epilepsy, in which storms of abnormal electrical activity occur in certain parts of the brain. If the seizures happen in areas involved in movement, twitches and convulsions may result; if sensory areas are involved, patients may feel tinglings, see flashes or hear strange noises. But if the seizures affect the brain structures involved in emotions and memory, patients may feel rage or fear, have déjà vu sensations or dreamy states or even have mystical experiences.

In most people who suffer seizures, the abnormal electrical activity only occurs within one hemisphere. But this was not true of a woman we'll call Mary, who had a trouble spot in each hemisphere. At 33 she suffered her first seizure and thereafter began having what she described as "mild spells" in which she felt "fullness" in her head. For 20 years anticonvulsive medica-

tion kept her seizures under control, but she then appeared at a Canadian neuropsychiatric clinic complaining that her head again felt full. She also had bizarre sensations of movement and pressure that made her fear she was going crazy. The clinic staff found her to be depressed and suicidal, although she lacked some of the classic symptoms of major depression such as disturbed appetite. She also had periods of almost manic elation.

A team of clinicians headed by the University of British Columbia's Trevor A. Hurwitz, a psychiatrist, recorded Mary's brain waves with an electroencephalograph (EEG) and her behavior with videotapes. They found that her depression was part of the seizure syndrome itself, not just a reaction to her upsetting symptoms. Moreover, the seizures in each hemisphere produced different psychological and emotional symptoms.

Following left-hemisphere attacks, Mary became depressed and agitated, suffered from insomnia and considered suicide. Voices were telling her she was bad, and her radio kept saying "googli googli." After right-sided attacks, her mood became manic and her behavior flighty; she danced on her bed and behaved seductively toward the hospital staff. It was as if the left side of her brain had a different mood, even a different personality, from the right.

Studies of people who have had strokes or head injuries suggest a similar emotional division of labor between the hemispheres. Unlike epileptics, who suffer from abnormal bursts of activity in intact parts of the brain, those with strokes or brain injury have lost actual brain tissue.

From *Psychology Today*, February 1988, pp. 34-39, 42. Copyright © 1988 by PT Partners L. P. Reprinted by permission.

Early in this century, neurologists noted that when people had left-hemisphere damage severe enough to impair their speech and movement, their reactions were often depressive, even aggressive. Unable to talk coherently to express their frustration, such patients would frequently cry, yell, curse and sullenly refuse to cooperate with those who tried to help them.

By contrast, patients with comparable right-hemisphere injuries (which rarely affect speech but can devastate sensation and movement) were often strangely indifferent to their disability. In extreme cases, patients would even deny that their paralyzed limbs belonged to them and might try to walk on nonfunctioning legs or make their way through the hospital wards even though they were neurologically blind.

These clinical findings have been confirmed by well-controlled studies of patients with one-sided (left or right) brain damage. But such research leaves many questions unanswered: How much brain tissue has to be destroyed before emotional changes are seen? And where, exactly, does the injury have to occur? Since the front and back parts of each hemisphere are also specialized for handling different functions (front mostly for movement and action, and back mostly for awareness and evaluation), does it also matter emotionally if the injury occurs toward the front or the back of a given hemisphere?

Neuropsychiatrist Robert Robinson of the Johns Hopkins School of Medicine has concentrated on pinpointing the actual brain mechanisms that produce mood changes after a stroke injures the brain. In one study, his research team put patients whose stroke sites were identified by CAT scan through a comprehensive series of physical, neuropsychological, psychiatric and mood tests. "We found that not only the hemisphere but the specific site of the injury was important," Robinson says. The most severe depressive reactions were seen in patients with lesions in the front parts of the left hemisphere. "The closer the injury was to the frontal areas, the more likely was the emotional effect." Damage toward the front of the right hemisphere also produced the most intense reactions: undue cheerfulness and an apathetic attitude toward the disability.

These studies of people with neurological problems raise a further question: Are the hemispheres specialized for different emotions in brains free of seizures or injury? Researchers Geoffrey Ahern and Gary Schwartz of Yale University believe the answer is yes, at least in the front parts. After hooking up 33 healthy college students to an EEG apparatus that measured electrical activity in various parts of the brain, the researchers asked them 60 questions designed to evoke a variety of moods and tap into verbal or spatial thinking as well. Ahern and Schwartz found that when people experienced positive emotions, the left hemisphere was more active than the right but only in the frontal lobes. Conversely, the right-hemisphere frontal lobes were particularly active during negative emotions.

The emotions studied in these experiments were brief and artificially induced. How do the hemispheres behave during emotions that happen naturally, such as feeling depressed? Yale graduate student Carrie Schaffer, while working with colleagues at the State University of New York at Purchase, gave the Beck Depression Inventory to a group of college students. When the researchers then compared the EEG's of students who scored as most characteristically depressed with those of nondepressed students, the depressed students had more electrical activity in the brain's right frontal region. In this case, too, the right hemisphere seemed to play a special role in depressed feelings.

To some researchers, these and other studies are beginning to fit together into a coherent picture but not necessarily the

same one (see "The Emotional Brain in Three Dimensions," this article). Neuropsychologist Harold Sackeim of Columbia University's College of Physicians and Surgeons believes that the two sides of the brain control emotion through what he calls "reciprocal inhibitory control"—a system of hemispheric checks and balances. In his view, each hemisphere's activity helps keep the other's in check. Disturbances in mood occur when one hemisphere's functioning gets too strong or too weak and overrides or gives way to the other hemisphere's control.

An observation by Canadian psychiatrist Hurwitz, who stud-

THE EMOTIONAL BRAIN IN THREE DIMENSIONS

ROBERT ROBINSON'S GROUP at the Johns Hopkins School of Medicine has recently been studying patients with injury to both cerebral hemispheres, with results that conflict with Harold Sackeim's mutual-watchdog theory about how the brain regulates emotions. "If reciprocal inhibitory control is the rule, the effects of injury to both hemispheres should cancel each other out," Robinson says. In other words, a patient with such damage shouldn't seem appreciably happier or sadder than before the injury. But Robinson's team found that such patients became depressed, as if only the injury on the left side counted.

Neuropsychologist Don Tucker of the University of Oregon has another model of how the brain exerts its control over the emotions—one more complex than Sackeim's. As Tucker sees it, "The main division of the brain may be a front-back one," rather than the right-left one. He reasons that since the frontal lobes are the chief brain system involved in regulating behavior as a whole, they probably regulate emotional behavior as well.

But there's more complexity still. Tucker points out that the brain "is a complex of control mechanisms in tight balance." Emotional control may depend not only on the balance of left-right and front-back components but also of bottom-top components: how the top of the brain works with many structures lower down that are also involved in emotions.

It's tempting to think about the hemispheres as opposites: happy versus sad, positive emotions versus negative ones. But Tucker believes their differences may have more to do with activation (being moved to act on the world) versus arousal (being passively aware of the world). These differences stem from the ways each hemisphere is hooked up preferentially to other parts of the brain. The left hemisphere, he speculates, may activate what we do through its stronger links to the frontal lobes and structures near the base of the brain. The right hemisphere may keep us aroused and aware through its stronger links to posterior and topmost brain regions.

As Tucker sees it, either hemisphere can be involved in pleasant or unpleasant emotions, but each has its own characteristic emotional repertoire due to its special linkages. Thus, the left hemisphere is involved in emotional states tinged with alert expectation—positive ones such as happy anticipation or negative ones such as anxious trepidation. The right hemisphere is involved in more reflective emotional states—positive ones such as relaxed awareness and negative ones such as depression.

ied Mary, the epileptic patient described earlier, supports this view. He suggests that seizures in Mary's left hemisphere allowed the right hemisphere's naturally gloomy style to dominate her emotional life. When seizures zapped the right hemisphere, the left's manic-like emotional tendencies took over.

Sackeim speculates that under normal circumstances, internal and environmental factors combine to keep mood fluctuations fairly consistent with what is happening in our lives. But anything that jars the neural system—such as injury to the brain, the biochemical disruptions of a depressive illness or losing a loved one or a job—is likely to release negative emotions.

People who develop depressive illness, he suggests, don't necessarily feel sadder about life's barbs than the rest of us. They just have a harder time inhibiting the depressive reaction once it starts. "Depressives," he says, "have faulty brakes."

Sackeim thinks that in the near future, a new breed of clinician-researchers will be able to pinpoint a variety of mood-disorder syndromes and link them to different patterns of cerebral disorganization. This, in turn, will lead to more effective therapies, custom-tailored for each specific case.

Research on how the brain hemispheres are involved in emotion is in its infancy, says Sackeim. But as theories and techniques improve, such studies promise to help us understand both normal and disordered emotional states. Studies of the link between brain function and emotions—whether in healthy states, mild and severe mood disorders, the neuronal storms of epilepsy or in the traumatic aftermath of strokes and head injuries—may aid in discovering ways to restore balance to the hemispheres and the emotions they regulate. Such research may reveal not just how we speak and reason but also how we love, hate and yearn.

APHASIAS AND APROSODIAS

BEYOND THEIR DIFFERENT ROLES in the emotions we feel, the two hemispheres are unequally involved in the emotions we reveal to and interpret in other people through facial expression and tone of voice. After left-hemisphere strokes or injuries, people often have severe problems in conveying or understanding the verbal content of speech. Their communication problems are called aphasias.

People with right-hemisphere damage are not usually aphasic, but they often have a different cluster of communication problems known as aprosodias. They have difficulty in dealing with prosody, that aspect of speech that enables a listener to tell if a particular statement (for example, "You're the boss") is meant to be a statement of fact or a question, serious or sarcastic, happy or sad, bland or passionate.

Neurologist Oliver Sacks, in his book *The Man Who Mistook His Wife for a Hat*, gives a vivid account of how aphasics and aprosodics differed as they listened to a Presidential speech on TV. The aphasics were, in a sense, like people trying to follow someone talking in a foreign language they didn't know. But since they couldn't grasp the speech's verbal content, they were especially sensitive to its emotional tone and its studied sincerities; they laughed derisively at the Great Communicator's prosodic posings. Sacks speculates that in this sense, "one cannot lie to an aphasic."

The aprosodic patients brought just the opposite skills to the speech. Emily D., a former English teacher whose usual passion for correct grammar, semantics and organization had been heightened by her inability to pick up emotional tone, was visibly disturbed by the speech. Did the President's words, if not his manner, win over such a prosody-deaf but linguistically astute person? No chance. The President "is not cogent," she remarked. "He does not speak good prose. His word use is improper. Either he is brain-damaged or he has something to conceal." Cases such as these remind us that the normal organization of our faculties into intact wholes may sometimes be a disadvantage, hindering our comprehension of the "hidden agenda" behind others' messages.

Patients with injuries to the left or right hemisphere have different communication problems, but both groups often have their feelings and behavior misunderstood—even by health professionals. Robinson and his colleagues have found that people with left-hemisphere damage may continue to feel depressed for up to two years after a stroke, a fact that can be doubly troubling. The depression derails their motivation to participate in rehabilitation programs that might help them recover. And since many rehabilitation professionals focus on physical disabilities and neglect emotional ones, Robinson says, they may view such patients as "difficult" or "resistant" and not give them all the help they need.

Surprisingly, such depressed patients are also likely to get short shrift when it comes to getting mental-health care. According to a study by physicians Richard Rosse and Charles Ciolino of the Georgetown University Medical Center, they are less likely to be seen in psychiatric and counseling ser-

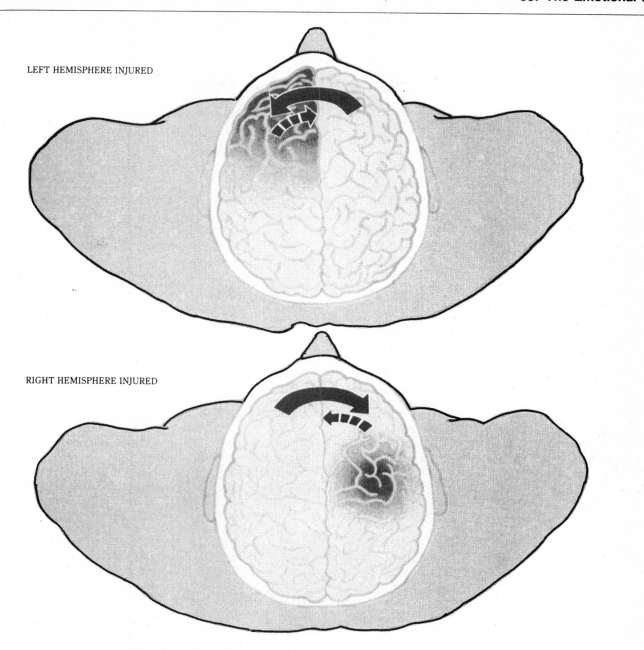

LEFT HEMISPHERE INJURED

RIGHT HEMISPHERE INJURED

When the left hemisphere is injured, depression often results. Injury to the right hemisphere often produces indifference or euphoria. Why? Some researchers propose that in normal states, the left and right hemispheres mutually control one another's emotional tone; disturbance of one hemisphere allows the opposite hemisphere's emotional bias to predominate, thus coloring the person's whole mood.

vices than are people with right-side damage. Rosse and Ciolino point out that because people with left-sided damage have difficulty expressing their feelings verbally, treatment-team members may underestimate the severity of their depression.

The depressive feelings of some aprosodic patients may also be misinterpreted, but for different reasons. Their facial expressions and voice inflection are often strangely blank—almost robotic. Elliott Ross, a neuropsychiatrist at the University of Texas Southwestern Medical Center at Dallas, notes that when aprosodic patients say such things as, "I feel like hell, I wish I were dead," in a flat, deadpan manner, they may not be taken seriously because the comment "lacks feeling."

Development During Middle and Late Adulthood

Developmentalists hold two extreme points of view about the latter part of the life span. One point of view, "disengagement," argues that the physical and intellectual deficits associated with aging are inevitable and should be accepted at face value by the aged. The other point of view, "activity," acknowledges the decline in abilities associated with aging, but also notes that the aged can maintain satisfying and productive lives.

Extreme views in any guise suffer from the problem of homogeneity, which involves stereotyping all individuals within a category or class as having the same needs and capabilities. Whether one's reference group is racial, ethnic, cultural, or age-related, stereotyping usually leads to counterproductive, discriminatory social policy which alienates the reference group from mainstream society.

Evidence obtained during the past decade clearly illustrates the fallacy of extremist views of middle and late adulthood. Development during adulthood and aging is not a unitary phenomenon. Although there are common physical changes associated with aging, there are also wide individual differences in the rates of change as well as the degree to which changes are expressed. It is common to think of the changes associated with aging as solely physical and generally negative. The popular press devotes considerable space to discussions of the causes and treatment of such debilitating disorders as Alzheimer's disease. However, there are also psychological changes associated with aging, and, as is the case at all age levels, some individuals cope well with change and others do not. New research on the aging process suggests that aging is not a unitary phenomenon; physical health and mental health changes do not correlate well. Although a variety of abuses can hasten physical and mental deterioration, proper diet and modest exercise can also slow the aging process. In addition, one cannot understate the importance of love, social interaction, and a sense of self-worth for combating the loneliness, despair, and futility often associated with aging.

Erik Erikson was among the first to draw serious attention to the conflicts associated with each of the age periods in the life cycle. In "Erikson, In His Own Old Age, Expands His View of Life," the man who broke so much ground with his research reassesses the life cycle from the perspective of old age. Other investigators have drawn attention to the specific problems of middle age—that period of development which marks the transition to maturity and, subsequently, to old age. Popular accounts of the pervasiveness of the "midlife crisis" have been tempered by more empirical studies which suggest that the midlife crisis may be a real phenomenon indeed, but real only for a minority of individuals.

The first article in this section presents information on love, a topic that has received surprisingly little attention from developmentalists although it is a favorite subject of poets, novelists, and songwriters. Despite its importance for human behavior, psychologists know remarkably little about love, especially among the elderly.

Nearly everyone is aware of the terrible condition known as Alzheimer's disease. This dysfunction involves progressive loss of memory, often with tragic consequences. Although Alzheimer's disease affects only a small percentage of the elderly, many older adults have memory loss to one degree or another. However, in "The Fear of Forgetting," Jeanne Toal reveals that such memory loss is not as pervasive as is commonly thought.

As Richard Booth points out in "Toward an Understanding of Loneliness," loneliness is an emotional state that nearly every individual experiences at one time or another during the life span. However, this is another area of research that has been underserved by developmentalists.

Developmentalists have not, however, ignored the study of sex differences. While it is clear that men are more aggressive than women, this fact may depend on how one defines aggression. In "Biology, Destiny, and All That," Paul Chance reviews evidence for sex differences and concludes that many differences between the sexes are differences in how characteristics are expressed, rather than absolute differences. What difference do sex differences make? Chance suggests that an analysis of business hiring practices could help to answer this question.

Behavioral gerontology remains a specialization within human development that is absent from most graduate programs in human development. Perhaps this is partly because of the natural tendency of people to avoid confronting the negative aspects of aging, such as loneliness and despair over the loss of one's spouse or over one's own impending death. Nevertheless, contemporary studies do provide fascinating information about the quality of life ("The Vintage Years") and physical changes ("Aging: What Happens to the Body as We Grow Older" and "A Vital Long Life") during aging, and challenge many traditional views about interpersonal relationships and memory processes.

Because the proportion of the population represented by the aged is increasing rapidly, it is imperative that significant advances be made in our knowledge of the

later years of development. Prolonging life and controlling mental and physical illness are only small aspects of the promotion and improvement of the quality of life for the elderly.

Looking Ahead: Challenge Questions

Why has it taken so long for psychologists to attempt a scientific study of love? Do you think they will have as much success as poets and songwriters in detailing the components and dynamics of this pervasive and powerful human emotion?

Many age-related crises seem to be associated with a variety of factors that produce stress. Some stress is situational while other stress is chronic. Loneliness is a stress that can occur at any age and can be either situational or chronic; when chronic, it can easily lead to depression and perhaps to suicide. How would you plan programs to deal with the loneliness of old age? Does any stage theory of adult development provide guidance?

It seems that the nature-nurture (genetic vs. acquired) debate will never be resolved. Just when evidence seems to shift in one direction, new findings emerge which tip the scales the other way. What is your assessment of the evidence presented with respect to sex differences?

Gerontologists suggest that significantly greater life expectancies are possible even without medical break-throughs. Control of childhood diseases, better education, better physical fitness, and proper diet are factors that increase the life span. How would your life differ now if your expected life span was 150 years? How do you think your elderly years will differ from those of your parents?

THE MEASURE OF LOVE

LOVE MAY STILL BE A MYSTERY TO POETS, BUT THE SECRETS OF BOTH ITS STRUCTURE AND ITS WORKINGS ARE NOW YIELDING TO A YALE PSYCHOLOGIST.

ROBERT J. STERNBERG

Yale psychologist **Robert Sternberg** created his first intelligence test for a science project in the seventh grade; it ended badly, when he was disciplined for using it to test his fellow students' IQs. A later project, in tenth grade, showed him that music—specifically, the Beatles' rendition of "She's Got the Devil in Her Heart"—increased test scores.

Since then, Sternberg has continued to hunt for the bases of intelligence and the mental processes that underlie thinking. "Many of them are what are sometimes called executive processes," he says, "such as recognizing problems in life, figuring out what steps to take and monitoring how well the solution is going." He is a recipient of many awards for his work and the author of dozens of articles on professional journals.

Love is one of the most important things in life. People have been known to lie, cheat, steal and kill for it. Even in the most materialistic of societies, it remains one of the few things that cannot be bought. And it has puzzled poets, philosophers, writers, psychologists and practically everyone else who has tried to understand it. Love has been called a disease, a neurosis, a projection of competitiveness with a parent and the enshrinement of suffering and death. For Freud, it arose from sublimated sexuality; for Harlow, it had its roots in the need for attachment; for Fromm, it was the expression of care, responsibility and respect for another. But despite its elusiveness, love can be measured!

My colleagues and I were interested both in the structure of love and in discovering what leads to success or failure in romantic relationships. We found that love has a basic, stable core; despite the fact that people experience differences in their feelings for the various people they love, from parents to lovers to friends, in each case their love actually has the same components. And in terms of what makes love work, we found that how a man thinks his lover feels about him is much more important than how she actually feels. The same applies to women.

When we investigated the structure of love, the first question Susan Grajek, a former Yale graduate student, and I looked at was the most basic one: What is love, and is it the same thing from one kind of relationship to another? We used two scales: One, called a love scale, was constructed by Zick Rubin, a Brandeis University psychologist; the other was devised by George Levinger and his colleagues at the University of Massachusetts. (We used two scales to make sure that our results would not be peculiar to a single scale; the Rubin and Levinger scales turned out to be highly correlated.) Levinger's measures the extent to which particular feelings and actions characterize a relationship (see box). Rubin designed his 13-item scale to measure what he believes to be three critical aspects of love: affiliative and dependent need for another, predisposition to help another, and exclusiveness and absorption in the relationship with another.

Consider three examples of statements Rubin used, substituting for the blanks the name of a person you presently love or have loved in the past. For each statement, rate on a one (low) to nine (high) scale the extent to which the statement characterizes your feelings for your present or previous love.

"If I could never be with ____, I would feel miserable." "If ____ were feeling badly, my first duty would be to cheer him (her) up." "I feel very possessive toward ____."

The first statement measures affiliative and dependent need, the second, predisposition to help, and the third, exclusiveness and absorption.

Validating the Score

Although there is no guarantee that the scale truly does measure love, it seems, intuitively, to be on the right track. What's more important, scores on the Rubin love scale are predictive of the amount of mutual eye gazing in which a couple engages, of the couple's ratings of the probability that they will eventually get married and of the chances that a couple in a close relationship will stay in that relationship. There is thus a scientific basis as well as intuitive support for the scale's validity.

We asked participants to fill out the Rubin and Levinger scales as they applied to their mother, father, sibling closest in age, best friend of the same sex and lover. Thirty-five men and 50 women from the greater New Haven area took part. They ranged in age from 18 to 70 years, with an average

From *Science Digest*, April 1985, pp. 60, 78-79. Reprinted by permission of SCIENCE DIGEST © 1985 by The Hearst Corporation.

age of 32. Although most were Caucasian, they were of a variety of religions, had diverse family incomes and were variously single, married, separated and divorced.

To discover what love is, we applied advanced statistical techniques to our data and used the results to compare two kinds of conceptions, based on past research on human intelligence. Back in 1927, the British psychologist Charles Spearman suggested that underlying all of the intelligent things we do in our everyday lives is a single mental factor, which Spearman called *G*, or general ability. Spearman was never certain just what this general ability was, but he suggested it might be what he referred to as "mental energy." Opposing Spearman was another British psychologist, Godfrey Thomson, who argued that intelligence is not any one thing, such as mental energy, but, rather, many things, including habits, knowledge, processes and the like. Our current knowledge about intelligence suggests that Thomson, and not Spearman, was on the right track.

We thought these two basic kinds of models might apply to love as well as to intelligence. According to the first, Spearmanian kind of conception, love is a single, undifferentiated and indivisible entity. One cannot decompose love into its aspects, because it has none. Rather, it is a global emotion, or emotional energy, that resists analysis. According to the second, Thomsonian kind of conception, love may feel like a single, undifferentiated emotion, but it is in fact one best understood in terms of a set of separate aspects.

Our data left us with no doubt about which conception was correct: Love may feel, subjectively, like a single emotion, but it is in fact composed of a number of different components. The Thomsonian model is thus the better one for understanding love as well as intelligence.

Although no one questionnaire or even combination of questionnaires is likely to reveal all the components of love, we got a good sense of what some of them are: (1) Promoting the welfare of the loved one. (2) Experiencing happiness with the loved one. (3) High regard for the loved one. (4) Being able to count on the loved one in times of need. (5) Mutual understanding of the loved one. (6) Sharing oneself and one's things with the loved one. (7) Receiving emotional support from the loved one. (8) Giving emotional support to the loved one. (9) Intimate communication with the loved one. (10) Valuing the loved one in one's own life.

These items are not necessarily mutually exclusive, but they do show the variety and depth of the various components of love. Based on this list, we may characterize love as a set of feelings, cognitions and motivations that contribute to communication, sharing and support.

To our surprise, the nature of love

proved to be pretty much the same from one close relationship to another. Many things that matter in people's relationship with their father, for example, also matter in their relationship with a lover. Thus, it is not quite correct to say, as people often do, that our love for our parents is completely different from our love for our lover. There is a basic core of love that is constant over different close relationships.

But there are three important qualifications: First, when we asked whom people love and how much, we found that the amounts of love people feel in different close relationships may vary widely. Furthermore, our results differed slightly for men and women. Men loved their lover the most and their sibling closest in age the least. Their best friend of the same sex followed the lover, and their mother and

---

Love for our parents is not so different from love for a lover. A basic core of love is constant.

---

father were in the middle. Women loved their lover and best friend of the same sex about equally. They, too, loved their sibling closest in age the least, with their mother and father in the middle. But whereas men did not show a clear tendency to prefer either their mother or their father, women showed more of a tendency to prefer their mother. These results are good news for lovers and same-sex best friends but bad news for siblings close in age. (Remember, however, that all these results are averages. They do not necessarily apply in any individual case.)

Second, the weights or importances of the various aspects of love may differ from one relationship to another. Receiving emotional support or intimate communication may play more of a role in love for a lover than in love for a sibling.

And third, the concomitants of love—what goes along with it—may differ from one relationship to another. Thus, the sexual attraction that accompanies love for a lover is not likely to accompany love for a sibling. (Although sexual attraction feels like a central component of love, most of us learn, often the hard way, that it is possible to have sexual attraction in the absence of love, and vice versa. As researchers, we decided to keep sexual attraction distinct from our list of central compo-

nents because it enters into some love relationships but not others.)

We did not obtain clear evidence for sex differences in the structure of love for men versus women. However, other evidence suggests that there are at least some. George Levinger and his colleagues, for example, investigated what men and women found to be most rewarding in romantic relationships. They discovered that women found disclosure, nurturance, togetherness and commitment, and self-compromise to be more rewarding than men did. Men, in contrast, found personal separateness and autonomy to be more rewarding than women did. There is also evidence from other investigators to suggest that women find love to be a more integral, less separable part of sexual intercourse than men do.

Some people seem to be very loving and caring people, and others don't. This observation led us to question whether some people are just "all-around" lovers. The results were clear: There is a significant "love cluster" within the nuclear family in which one grows up. Loving one member of this family a lot is associated with a tendency to love other members of the family a lot, too. Not loving a member of this family much is associated with a tendency not to love others in the family much, either. These are only tendencies, of course, and there are wide individual differences. But the general finding is that love seems to run in nuclear families.

Romantic Prediction

These results do not generalize at all outside the nuclear family. How much one loves one's mother predicts how much one loves one's father, but not how much one loves one's lover. So people who haven't come from a loving family may still form loving relationships outside the family—though coming from a loving family doesn't guarantee that you will be successful in love.

Having learned something about the nature of love, we were interested in determining whether we could use love-scale scores to predict and even understand what leads to success or failure in romantic relationships. Because our first study was not directly addressed to this question, Michael Barnes, a Yale graduate student, and I conducted a second study that specifically addressed the role of love in the success of romantic relationships.

In our study, each of the members of 24 couples involved in romantic relationships filled out the love scales of Rubin and Levinger. But they filled them out in four different ways, expressing: (1) Their feelings toward the other member of the couple. (2) Their perceptions of the feelings of the other member of the couple toward them. (3) Their feelings toward an ideal other member of such a couple. (4)

A COST-BENEFIT LOVE SCALE

EXCERPTED AND ADAPTED FROM THE INTERPERSONAL INVOLVEMENT SCALE BY GEORGE LEVINGER, UNIVERSITY OF MASSACHUSETTS, AMHERST

All relationships are both rewarding and costly; caring for someone, while emotionally satisfying, takes time and can be very taxing. The items that follow are excerpted from a scale George Levinger, of the University of Massachusetts, Amherst, devised to rate negative and positive aspects in any given relationship (with lover, spouse, parent, friend, etc.). In his research, author Robert J. Sternberg adapted the scale to a different purpose: He asked people to rate the importance of each statement in terms of both an actual and an ideal relationship. His findings are discussed in the accompanying story.

____ **1.** Doing things together

____ **2.** Having no secrets from each other

____ **3.** Needing the other person

____ **4.** Feeling needed by the other person

____ **5.** Accepting the other's limitations

____ **6.** Growing personally through the relationship

____ **7.** Helping the other to grow

____ **8.** Having career goals that do not conflict

____ **9.** Understanding the other well

____ **10.** Giving up some of one's own freedom

____ **11.** Feeling possessive of the other's time

____ **12.** Taking vacations together

____ **13.** Offering emotional support to the other

____ **14.** Receiving affection from the other

____ **15.** Giving affection to the other

____ **16.** Having interests the other shares

Their perceptions of the feelings of an ideal other member of such a couple toward them. These questions dealt with two basic distinctions: the self versus the actual other and the actual other versus an ideal.

The participants, all of whom were students in college or graduate school and none of whom were married, were asked to think in terms of a realistic ideal that would be possible in their lives, rather than in terms of some fantasy or Hollywood ideal that could exist only in movies or other forms of fiction. In addition to filling out the love-scale questionnaires, participants also answered questions regarding their satisfaction and happiness with their present romantic relationship.

Between Real and Ideal

We compared two different conceptions of how love might affect satisfaction in a romantic relationship. According to the first conception, level of satisfaction is directly related to the amount of love the couple feel: The more they love each other, the more satisfied they will be with their relationship. According to the second and more complex conception, the relation between love and satisfaction is mediated by one's ideal other. In particular, it is the congruence between the real and the ideal other that leads to satisfaction. As the discrepancy between the real and ideal other increases, so does one's dissatisfaction with the relationship.

Consider two couples: Bob and Carol and Ted and Alice. Suppose that Bob loves Carol just as much as Ted loves Alice (at least, according to their scores on the love scales). According to the first conception, this evidence would contribute toward the prediction that, other things being equal, Bob and Ted are equally satisfied with their relationships. But now suppose that Bob, unlike Ted, has an extremely high ideal: He expects much more from Carol than Ted expects from Alice. According to the second conception, Ted will be happier than Bob, because Bob will feel less satisfied.

The need for a loving relationship is paramount to an individual's psychological well-being, and establishing a lasting relationship is one of life's most important goals.

These two conceptions of what counts in a relationship are not mutually exclusive, and our data show that both matter about equally for the success of the relationship. Thus, it is important to remember that although love contributes to a successful relationship, any relationship can be damaged by unrealistically high ideals. At the other extreme, a relationship that perhaps does not deserve to last may go on indefinitely because of low ideals.

In addition to love and the ideal, we found that both how a person feels about his lover and how he thinks she feels about him matter roughly equally to satisfaction in a relationship. But there are three important qualifications.

First, the correlation between the love-scale scores of the two members of a couple is not, on the average, particularly high. In many relationships, the two members do not love each other equally.

Second, it is a person's perception of the way his lover feels about him, rather than the way the lover actually feels, that matters most for one's happiness in a relationship. In other words, relationships may succeed better than one might expect, given their asymmetry, because people sometimes systematically delude themselves about the way their partners feel. And it is this perception of the other's feelings rather than the other's actual feelings that keeps the relationship going.

Third, probably the single most important variable in the success of the relationships we studied was the difference between the way a person would ideally *like* his partner to feel about him and the way he thinks she really does feel about him. We found that this difference is actually more important to the success of a relationship than a person's own feelings: No variable we studied was more damaging to the success of a romantic relationship than perceived under- or overinvolvement by the partner.

Why might this be so? We believe it is because of the ultimate fate of relationships in which one partner is unhappy with the other's level of involvement. If a person perceives his partner to be underinvolved, he may try to bring her closer. If she does not want to come closer, she may react by pulling away. This leads to redoubled efforts to bring her closer, which in turn lead her to move away. Eventually, the relationship dies.

On the other hand, if a person perceives her partner to be overinvolved, she may react by pulling away—to cool things down. This results in exactly the opposite of what is intended: Her partner tries to push even closer. The relationship becomes too asymmetrical to survive.

All our results suggest that the Rubin and Levinger scales could be useful tools to diagnose whether relationships are succeeding or not. There is one important and sobering fact to keep in mind, though: Scores from a liking scale devised by Zick Rubin were even better predictors of satisfaction in romantic relationships than were scores from the love scales, especially for women. Thus, no matter how much a person loves his partner, the relationship is not likely to work out unless he likes her as well.

An End to a Mystery?

Despite its complexity, love can be measured and studied by scientific means. Those who believe that love is, and should remain, one of life's great mysteries will view this fact as a threat; I view it in exactly the opposite way. With a national divorce rate approaching 50 percent and actually exceeding this figure in some locales, it is more important than ever that we understand what love is, what leads to its maintenance and what leads to its demise. Scientists studying love have the opportunity not only to make a contribution to pure science but to make a contribution to our society. At the very least, the study of love can suggest the cause, if not the cure, for certain kinds of failed relationships. I believe that our ability to measure love is contributing to progress both in understanding the nature of love and in suggesting some of the causes of success and failure in close relationships.

Mental Shape-Ups

THE FEAR OF FORGETTING

To improve your memory, don't write yourself a million notes.
Just relax.

Jeanne Toal

Jeanne Toal *thinks she's a science writer who specializes in biomedical advances—but she just can't seem to remember.*

Forty years old and in the best physical shape of his life, Floyd Whiting felt like a man slipping into senility. "It started with small memory lapses," he says. "I'd be in the middle of a sentence and forget what I wanted to say, or I'd make an appointment and never show up." Then the problem worsened. "One day I found myself driving around town and couldn't find a road I knew well. Another time, two of my colleagues came into the lab where I work and I didn't recognize either of them."

An avid athlete who competes regularly in ultramarathons, Whiting was terrified he had Alzheimer's disease. He didn't. In fact, a battery of neuro-psychological tests showed he had no biological problem. His trouble: just a naturally aging memory. Since then, Whiting has coped by protecting him-

self with back-up systems. Conversations are recorded, appointments written down. His car is full of maps. "What else can you do?" he muses. "A failing memory is just something you have to live with."

No, it isn't, says Stanford University psychiatrist Jerome Yesavage: "We've found that most of the memory deficits associated with normal aging are reversible." Eight years ago, Yesavage and colleague Danielle Lapp started a memory program for people over 55. Since then, hundreds of men and women have regained their recall through these classes.

Yesavage and Lapp believe the terror triggered by forgetfulness can itself hurt memory, among both young and old. By learning relaxation techniques, they've found, many people can get an anxious memory back on track.

These researchers belong to a worldwide network exploring just how our brains store information, and how to keep the system working well, at age eight or age 80. The research suggests that, although memory does tend to weaken with age, much of the forgetfulness can be controlled. Emotions, physical condition and anxiety all can affect it. And even when memory

problems are based only on biology, there are strategies (and in the future, perhaps, medications) that can help you compensate for the slowdown.

"When I go to a conference or meeting, I'm often stopped by people worried about their memories," says Lapp. "People are getting frantic over what used to be considered normal memory lapses—like losing their glasses." Fear of Alzheimer's may well underly this panic. Five years ago the disease was relatively unknown; today much of the public is aware that its inexorable decline begins with memory failure.

Some facts about memory loss can alleviate much of the fear. Only about 10% of people over 65 ever experience a biological form of senility such as Alzheimer's.

True, research does show that even normal people tend to forget more as they age. One study, for example, found that more than 65% of those over 75 suffered some sort of memory loss.

What causes this natural forgetfulness? Some scientists still suspect the culprits are gradual hardening of arteries—which lowers blood flow to the brain—and lower overall energy lev-

From *American Health Magazine*, October 1986, pp. 77-78, 80, 82-84, 86. ©, Jeanne Toal.

els. But current research implicates an age-related drop in brain chemicals associated with memory.

Whatever the cause, there's now evidence that memory loss with age may not be as great as once believed—and can be minimized. In the past, gloomy statistics documenting great declines in the intellectual abilities of older people came from studies comparing them with the young. The old-timers always lagged far behind. Now many researchers believe that background—not age—was partially responsible for the poorer performance.

People born during the same era share a common biocultural history, explains Penn State psychologist Sherry L. Willis. When you compare people from different generations—

Fear of Alzheimer's is causing a panic over normal memory loss.

70-year-olds with 30-year-olds, for example—you're comparing people from different environments. "Many of the studies showing a general decline in intellectual abilities with age don't take that into account," she says.

The good news: Recent research suggests that people who turn 65 today are mentally sharper than previous generations, thanks to improvements in education and nutrition. And when researchers track the abilities of an *individual* over a lifetime, says Willis, the results are very

heartening. In work funded by the National Institute on Aging, Willis and psychologist K. Warner Schaie have charted some people for up to 28 years, and, she says, "many people show no intellectual declines into their 70s." Even at that late date, it may be possible to minimize the loss.

For years, West German psychologist Paul B. Baltes has been researching whether and how older people can improve intellectual ability and, more recently, memory. At the Max Planck Institute for Human Development and Education in West Berlin, Baltes and his colleagues have done preliminary work with 20 older people (ages 68 to 82) and 10 young adults (ages 18 to 25), with encouraging results. Baltes was able to train all the people in the

MEMORY SHAPE-UPS

Step One: Relaxation. The best tools for relaxing before flexing your memory include progressive muscle relaxation and focused breathing. Both have long been hailed as effective ways to calm mind and body.

Start with muscle relaxation. Sit comfortably or lie on your back with your palms open and your head in a straight but natural position. Then alternately tense and relax every major muscle group in your body. You can either start at your feet and work your way up, or work down from the top of your head. Either way, by the time you finish, your entire body should be thoroughly relaxed.

Now take a few moments to focus your mind with breathing and visualization exercises. The ability to form mental pictures is essential to most memory techniques, says Stanford researcher Danielle Lapp. In her classes, she suggests students picture a sandy beach with slow waves rolling in and out. "Then begin to breathe with the waves," she coaches. "Inhale gently through the nose as the wave rolls in, and exhale slowly as the wave

rolls away. This creates a nice, soothing rhythm that clears the way for learning anything you want."

(You'll find more relaxation techniques in *American Health's Relaxed Body Video*. See p. 103.)

Step Two: Flexing Your Memory Muscle. The slogan "Use it or lose it" applies to memory as well as biceps. "Memory responds to our needs and interests," says Lapp. "If you get old and slide into sedentary mental habits, you will develop a bad memory."

In her forthcoming book, *Don't Forget* (McGraw-Hill, $6.95), Lapp emphasizes that conscious, voluntary association is the foundation of a good memory. By linking new material with something familiar, you plant memory cues that are easy to retrieve. You can use image association to remember everything from where you parked the car to the name of virtually everyone you meet.

To memorize names and faces, for example, pick out a prominent facial feature and associate it with an image that connects in your mind with the name. Associations also work for memorizing numbers. Example: The height of Mount Fujiyama—12,365 feet—is easy to retain if you associate it with a calendar: 12 months and 365 days. The telephone number 747-1225 becomes a type of jet airplane

and the date of Christmas.

One very effective association trick to remember a list in order is called the Method of Loci. You simply link the items you want to remember with familiar locations, following a predetermined order (clockwise, for instance.) You could, for example, use your living room to remember a grocery list, by connecting each food item to an object in the room. Picture a can of coffee spilling on the fireplace, the first thing you see when entering the room. Imagine ripe tomatoes splattered on the second thing, the television screen, bananas hanging from the lamp (third) and a roll of paper towels wrapped around the couch (fourth). To recall the list, take a quick mental trot around the room, retrieving items as you go.

And if you really want to imprint something on your brain—such as a flower—try processing it with all of your senses: Notice its size, color, shape, texture and scent.

All these methods require more interest and patience than effort, Lapp insists. Use them enough and eventually you'll develop memory reflexes. She's her own star student: "By using these methods I've made my memory 10 times better than it was when I first got into memory research eight years ago."

study to remember strings of names or numbers presented to them in rapid succession. He used a common technique called the Method of Loci. Simply put, the people were taught to link each name or number with an easily remembered historical site in Berlin. (To learn a version of this tactic, see ''Memory Shape-Ups,'' box.)

When the older people pushed memory to the limit, says Baltes, they performed at extraordinarily high levels. They could remember an average of 60 numbers and 35 names that were presented at 20-second intervals. "There is nothing magical about such expert feats of memory," says Baltes. "With practice, most healthy, motivated older adults can reach high levels of performance."

An older person's maximum memory ability, however, will inevitably fall short of a similarly trained young adult's, Baltes cautions. At much faster presentation rates, such as four

Aging may not hurt memory as much as we once believed.

seconds, "you *will* see a definite decline in what older people can do compared with younger people," he says. Researchers are also studying the effects of aging on intellectual abilities related to memory, such as logical skills and spatial orientation. Here, too, results show older people can greatly improve their performance through practice.

"Memory is a complex, multidimensional phenomenon," says psychologist Thomas Crook, formerly of the National Institute of Mental Health (NIMH), and now head of Memory Assessment Clinics, Inc., in Bethesda, MD. "You can be good at remembering one thing and bad at another. Our tests show, for example, that memory

for names declines regularly over a lifetime, while vocabulary memory remains stable." The bottom line: Memory training can compensate for the majority of age-related memory losses. "Practice," says Crook, "is the key."

How Relaxation Helps

Most memory training programs, including Lapp's classes at Stanford and Baltes's in West Berlin, are centered on teaching people memory tips. But Yesavage and Lapp have added a unique approach. Before each memory workout, Lapp leads her classes in a series of relaxation and imagery exercises. "People with memory problems are often anxious about their memories," she says. "And anxiety, we've found, clutters the channels of memory. Relaxation opens those channels." (For samples of Lapp's re-

HOW THE BRAIN REMEMBERS

Where is memory? How is it formed in the changing chemical environment of the brain? These questions continue to baffle scientists despite 60 years of experimentation and effort.

It's no wonder. Mapping the neural circuitry of memory is a task of unsurpassed complexity. The human brain has more neurons, or nerve cells, than the Milky Way has stars, and each cell can communicate with a thousand others. Recently, however, scientists have uncovered clues to the location of memory and to the chemical processes that underlie it.

Memory is not isolated in one area of the brain. The latest memory maps differentiate between memory for *motor*, or "procedural," skills learned through practice—such as holding a fork or flinging a frisbee—and *cognitive*, or intellectual, abilities, such as remembering facts and figures.

Researchers at Stanford have discovered that the cerebellum—the "little brain" attached to the brain stem

that controls motor coordination and posture—also stores our memory for learned motor skills. Psychologist Richard Thompson noted that the cerebellum of rabbits became more active while they were being conditioned to blink after hearing a specific sound.

To confirm that the cerebellum was indeed home to their new memories, Thompson removed a particular cubic millimeter of cells from that part of the brain. Result: The rabbits immediately forgot the eye-blink response.

Higher types of memory and learning (intellectual or cognitive) have long been linked to the hippocampus—a finger-sized bundle of neural tissue that's part of the limbic system, the brain's center for emotions. Mortimer Mishkin of the NIMH and Larry Squire of the VA Medical Center in San Diego, CA, found, for example, that damage to both the hippocampus and the nearby amygdala causes lab monkeys to forget the location of a favorite food.

While these scientists map out *where* the brain stores which type of memory, others are busy deciphering the cell-by-cell details of *how* memory

is stored. Among them: neurobiologist Eric Kandel of the Howard Hughes Medical Institute at Columbia University. He's devoted 20 years to studying the Aplysia, a sea snail with only 20,000 nerve cells. His research suggests memory is retained by a chain of chemical reactions boosting levels of particular neurotransmitters—chemicals brain cells use to communicate with one another—in the junctions, or synapses, between nerve cells.

At the Woods Hole (MA) Marine Biological Laboratory, neurobiologist Daniel Alkon has observed rabbits and another kind of snail, the Hermissenda. He believes learning occurs as a result of changes in cell membranes, but not necessarily from changes in the synapses.

Despite their different approaches, both Kandel's and Alkon's work indicates that for memory and learning to occur, change must happen in the input/output properties of neurons.

Thanks to this collective body of new knowledge, the study of memory is in what Thompson calls the "critical breakthrough" stage, "the most exciting in the history of the field."

Anxiety clutters the channels of memory— relaxing opens them.

laxation exercises, see "Memory Shape-Ups," box.)

In one Stanford experiment, half the subjects, ages 62 to 83, studied relaxation skills before learning a standard mnemonic device, while the rest learned only the memory technique. Those who received relaxation training did significantly better (experimental group: 69%, control group: 44%) on a recall test requiring them to match faces with names.

Yesavage and Lapp base their relaxation theory on a scientific phenomenon called the Yerkes-Dodson law. It holds that "just the right" degree of arousal can actually improve your performance—in this case, the ability to remember. After all, it's tough to learn when you're feeling blah or drowsy. But if you start off *too* highly aroused— inevitable if you're anxious—your performance will drop.

Evidence for this phenomenon comes from studies of memory in animals and people. Separate research by psychologists Paul Gold of the University of Virginia and James McGaugh of the University of California, Irvine has shown that hormones released during stress—such as adrenaline—can either enhance or impair memory in rats. Drugs that give too high a boost in adrenaline (and arousal) cause amnesia, while low-dose injections actually improve recall.

Currently, Gold and colleague Linda Donder-Fredericks are doing similar tests with people—this time using glucose, which, like adrenaline, is released during stress. Their preliminary results are striking. "Elderly people who drink a fruit punch with glucose added to it show en-

hanced memories," says Gold. But sugaring your drinks in order to boost your memory is ill-advised, he warns. Tolerance levels vary from person to person, moment to moment, and too much glucose may hinder memory. Stress, and foods eaten just before taking glucose, can also influence its effect on remembering.

Chemical effects of stress may interfere with remembering.

The Mood Connection

The more scientists explore memory, the more keys they find to how we store experience and later replay it. There's evidence, for example, that *emotions* act as memory preservatives—they may determine what and how much you recall. Some researchers even believe there may be no learning at all without emotion.

"Adding emotional content to a mental image increases the likelihood that it will be processed into memory," says Lapp. "I teach my students to associate things they want to remember with strong emotions." To remember that you put your glasses next to the vase in the hallway, for example, make a mental note that it's your favorite vase.

One reason scientists give for this connection: Brain structures believed to store higher forms of memory are found in the limbic system—the brain's center for emotions. (See "How the Brain Remembers," box.) "We don't think it's coincidental that the limbic system is important in both memory and emotion," says NIMH brain researcher Mortimer Mishkin. "One may well determine the other.

Emotion, environment and fitness may all be gateways to memory.

Emotion could be one of the gateways to memory."

Emotion may also play a role in *retrieving* information from memory. In a series of laboratory studies, Stanford psychologist Gordon Bower demonstrated that, in his words, "it's easier to remember happy events when you're happy and sad events when you're sad." Lately, however, some researchers have challenged Bower's findings because they've been difficult to replicate. The problem, he admits, is that in a lab it's difficult to consistently reproduce specific human emotions. Until the procedures are better refined, the question remains unresolved.

Scientists are also studying other potential gateways to memory. Among them:

Drugs: Studies suggest that memories people acquire while under the influence of drugs, alcohol or tobacco can be triggered more readily when they take the same substance again.

Environment: Context helps organize memories. In one study, scuba divers who learned a list of words under water were able to recall more of those words on later dives than on dry land.

Fit Body: Researchers at the University of Utah found a four-month program of aerobic exercise improved the short-term memory of a group of previously sedentary older adults.

"People don't realize how many factors affect memory," Lapp says. "You can often modify the conditions where it doesn't work well, and create conditions where it does. If there's one thing I emphasize, it's that memory is under your control."

Toward an Understanding of Loneliness

Richard Booth

Richard Booth, MSW, is Associate Professor, Department of Psychology, Sociology, and Social Work, Quad-Cities Campus, Black Hawk College, Moline, Illinois, and a doctoral candidate at the University of Iowa.

WESTERN LITERATURE is filled with themes of people in the grips of emotional turmoil. Sartre carved the image of the alienated individual who must struggle with the essential meaninglessness of existence—alone and with deliberation.[1] One reads of the "marginal man" displaced, suspended from the fringes of the system, struggling not merely for survival but for fundamental personal and social meaning.[2] Romeo and Juliet eagerly offered their lives rather than be separated from each other, and Oedipus punished himself severely for a mistake he could hardly have avoided making. The emptiness of Willy Loman, the curiosity of Huck Finn, the spontaneity of Marie Antoinette, the vanity of Dorian Gray, the misery of Oliver Twist, the persistence of Dorothy in Oz are but a few of the literary examples of the human element coming to terms with life, whether the characters are fictional or real. No matter how it is described, there is a common theme of people attempting to bring emotion under rational control and to eke from their lives some good feeling about themselves and those with whom they interact.

Loneliness is an emotional condition that affects virtually all people from time to time and which is accompanied by such intensity that people sometimes do regrettable things to offset its influence. Loneliness can be found in every therapist's office, every classroom, every workplace, and every clinic and hospital. Every family system experiences it, somehow deals with it, and proceeds with the rest of its life. Yet, how many people have the skills necessary to confront the condition straightforwardly in their own lives, not to speak of the lives of family members, students, clients, and colleagues? Indeed, when this vague although pervasive phenomenon pushes itself to the fore, who can identify it so readily and correctly that it is not confused with moodiness, fleeting bad temper, depression, and even anger or global negativity? What is loneliness, and is it a condition apart from the other conditions just mentioned?

Although literary and psychoanalytic traditions have contributed substantially to knowledge about loneliness, little experimental work has been done in this area until recently. This article suggests a cognitive-emotional approach to understanding loneliness, explores some empirical work on the subject, and offers workers some possible intervention strategies.

CHARACTER OF LONELINESS

Despite its probable high frequency in the population and its compelling intensity, there has been little research on this affective state. Although not much empirical work has been done, profiles of loneliness and of the lonely person have emerged. The bulk of the work has been largely clinical and has come from the psychoanalytic tradition. Horney, Sullivan, and Fromm all speak to the basic need to seek and engage in meaningful relationships.[3] These theorists perceive people as fundamentally interactional and interpersonal and outline the difficulties that may arise if human relationships become disturbed. Fromm goes so far as to suggest that if an individual's basic needs (two of which are rootedness and relatedness) remain unfulfilled, either the person will die or become insane. Even if the individual does not fall prey to those extremes, he or she will become powerless. The need for relatedness about which these theorists speak has long been a central concern of the social work profession. When a person is out of step with his or her context and begins to march to a different drummer, as Slater suggests, problems inevitably will follow. A possible one is loneliness.[4]

From a sociological perspective, Durkheim studied the concept of anomie—that state of "normlessness" in which people are essentially lost in the system and confused about their "fit."[5] Peplau, Russell, and Heim came closer to a definition of loneliness by outlining the conditions within which loneliness is likely to emerge. They stated: "Loneliness occurs when a person's network of social relationships is small or less satisfying than the person desires."[6] From this perspective, it is conceivable that a person with twice the "normal" size friendship group could, nonetheless, be lonely, while the person with a small but satisfying friendship group may not be lonely at all. Peplau, Russell, and Heim seem to be suggesting that loneliness is, to some extent at least, a matter of cognitive expectations rather than merely an affective experience. Weiss agreed with them, saying, ". . . an individual when lonely maintains an organization of emotions, self-definitions, and definitions of his or her relations to others, which is quite different from the ones he maintains when not lonely."[7] Thus, loneliness appears to be a subjective notion—a cognitive-emotional phenomenon that often is of such intense proportions that it may influence overt behavior. Drawing still closer to a definition but remaining general, Weiss added that loneliness is "a gnawing . . . chronic distress without redeeming features."[8] In essence, the lonely individual's interpersonal reality does not fit interpersonal expectations. Why this occurs is unclear.

The Lonely Character

Viewing the situation from a more clinical perspective, one may ask

From *Social Work*, March/April 1983, pp. 116-119. Reprinted by permission.

whether there are any clues by which to identify the condition—any really reliable ways to find the lonely student in the classroom, the lonely family member at home, or the lonely colleague in the agency. In other words, are there any recognized correlates of the condition that may help social workers to intervene early and perhaps counteract potential disabling chronicity? Jones, Freeman, and Goswick suggested that it is difficult to identify a lonely person and found that naive judges tend to rate lonely and nonlonely persons similarly. However, the literature has provided a broad descriptive analysis of the lonely individual. In certain groups of lonely people, high positive correlations have been found between alcoholism and suicide and loneliness. Moreover, lonely people have been found to have significantly lower self-esteem than nonlonely persons and to be significantly more sensitive to rejection by others.[9]

Cutrona found that lonely persons who have passed through situational loneliness into chronicity tend to view themselves as personally responsible for their plight. They appear to have become involved in a cyclic situation in which they reinforce their loneliness through its chronicity—a situation from which they are unable to extricate themselves. Cutrona also found that measures on the Beck Depression Inventory were significantly and positively correlated with loneliness in this group.[10] It is interesting to note that both situationally and chronically lonely people were found to feel negative about human relationships in general. They also tended to reject others with greater frequency than others rejected them. Russell, Peplau, and Cutrona also found that the lonely subjects they studied showed signs of general restlessness, were more self-enclosed, and felt more "empty" than nonlonely persons. These findings were statistically significant.[11]

Jones, focusing specifically on how lonely people report they feel, identified an array of negative emotions. Generally, his subjects indicated they felt helpless, misunderstood, and unloved.[12] They also tended to feel unacceptable to others, separated from meaningful interaction with others, and generally worthless. They described themselves as being bored with life and as feeling unhappy. Although a number of self-concept factors were involved, the strength of the individual factors and the exact nature of their interaction remain far from clear.

Impact on Others

As was noted, the judges used by Jones, Freeman, and Goswick had some difficulty in consistently differentiating the lonely from the nonlonely in their ratings of subjects.[13] However, the judges used by Solano and those used by Sansone, Jones, and Helm had different types of problems in identifying lonely people.[14] Thus, the literature clearly is inconsistent on this matter. Nevertheless, it seems safe to hypothesize that because they are unable to differentiate, completely and consistently, lonely from nonlonely persons through casual interaction with them, observers would tend to deal similarly with both types of persons.

What effect, if any, would knowing that another person is lonely have on the behavior of observers? Fromm-Reichman argued that nonlonely people tend to construct walls of defense against lonely people in an attempt to ward off their own fears.[15] In other words, lonely people seem to frighten nonlonely people and perhaps even threaten the condition of nonloneliness; thus, nonlonely people feel the need to defend against lonely people for their own well-being. Wimer and Peplau found that whether nonlonely people enjoy dealing with lonely people largely depends on what they perceive is causing and maintaining the loneliness. Specifically, when nonlonely people think that lonely people are trying to help themselves get away from their loneliness, they are disposed to feel much more positively toward them than when they perceive that lonely people are not even attempting to rid themselves of their condition. Further, if nonlonely people think that others are lonely through no fault of their own, they are more likely to have positive feelings than if they think that the lonely people are the source of their own problems.[16] Thus, although people may generally find it difficult to identify a lonely person unless the individual tells them, there is some evidence to suggest that once they are aware of the condition in another person, they begin to look for the cause of the loneliness. It follows that whether they are unaware of the condition or whether they perceive the lonely person as perpetuating his or her own dilemma, they are unlikely to intervene spontaneously in the condition of loneliness.

IMPLICATIONS FOR PRACTICE

In linking the descriptive and correlational findings of the literature to practice, the author will consider five of the most central themes. They are described in the following material.

Helplessness

The first theme is the fundamental feeling of helplessness that lonely people report about themselves. As was already stated, some lonely persons do not feel in control of their lives. As with the construct of learned helplessness discussed by Seligman, people with an external locus of control tend to blame the environment for their difficulties and failures.[17] In the extreme, they could believe that no matter what they might do to bring about a change, they would have no influence. Peplau, Russell, and Heim found that if external locus of control and loneliness exist in the same person, chances are good that the loneliness is situational rather than chronic. In cases like these, they suggest, clarification of the perceived causes of loneliness is paramount in the helping process. That is, the client and the therapist should identify what most probably is responsible for the client's condition, focusing on the potential alterability of the "causal" situation. It is possible that lonely clients, because of their preoccupation with their own condition and the pain that accompanies it, pay too little attention to those aspects of loneliness over which they could learn to exert control. They may be overstating the actual degree of their helplessness to the detriment of ameliorating their own condition. They can be taught and helped to explore the possibility that part of their loneliness may be due to their own resistance to taking control of it—a control which they can find within themselves with therapeutic assistance.[18]

Unrealistic Expectations

The second theme is that the expectations of some lonely clients are not realistic. Therefore, it may be helpful in certain cases of loneliness for the therapist to help the client scale down his or her expectations. In reassessing their expectations of interpersonal relationships, clients can be helped to avoid frustration and loneliness by coming to understand that all is not lost if there really is no ideal mate, no infallible spouse, no perfect lover, or no all-fulfilling friendship group. Sometimes clients—and therapists as well—may do well to learn that Santa Claus (the personification and deliverer of one's dreams) is but a limited reality.[19] That is, no one can expect to receive all that he or she may wish to receive.

Overcoming the Past

The third theme is that some lonely

people focus their energy and attention on the events that precipitated their loneliness rather than on the "maintenance factors" of the condition.[20] Such clients need help in overcoming a painful event in the past. For example, lonely single parents may have been deeply affected by the divorce from or the death of a spouse. Even after a considerable time, they may be preoccupied with the possibility of a reunion (in the case of divorce) or may continue to deny the death of the beloved. In either event, these persons may have shut themselves off from possible ameliorative efforts such as dating or other types of social interaction. Ignorant of the effects of this counterproductive process, they may blame their children or other external factors for occupying all their time.

These persons may continue to blame their spouses for divorcing them or for dying and leaving them alone. Thus, they maintain their loneliness, desiring meaningful interaction but disallowing it by focusing on "what could have been if" or "what could be if" or "if only" kinds of statements. With help, such clients can be redirected to recognize the difference between the initial conditions out of which their loneliness emerged and the factors that perpetuate the condition. It is those maintenance factors of the condition that are frequently amenable to change.

Loneliness vs. Depression

The fourth theme is the confusion of loneliness with depression. Because of this confusion, more research is needed to identify the differences and similarities between the two states. As was mentioned earlier, those who pass from situational to chronic loneliness tend to blame themselves for their condition and score higher on the Beck Depression Inventory. It may be that the transitory loneliness—the type of loneliness that comes and goes in the lives of all people—is largely situational, while depression is related more to internal-blame factors. In this regard, Peplau, Russell, and Heim found that depression scores were highest for subjects who felt they were to blame for their loneliness and that the loneliness was not likely to go away.[21] There still remains, however, enough confusion between these two concepts to justify further work at clarification.

Avoidance of Clients

The fifth theme is that in their ef-forts to defend against loneliness, some therapists may shut out significant cues from clients which may indicate that the clients are lonely even though the clients are not telling them so directly. Therefore, it behooves clinicians in all fields of practice to be alert to possible tendencies to avoid clients who present themselves for help. It seems safe to suggest that all clinicians are on the alert for potential psychological as well as physical dangers and hence may avoid certain aspects of a client's problem because they "sense" that he or she is lonely. Clearly, this idea requires further research. Nevertheless, clinicians should bear in mind that clients, students, friends, and family members who manifest general negativity, poor self-esteem, pessimism, social isolation—conditions that make spending time with them less desirable—may be, in fact, lonely. The perceptive clinician will pick up on these correlates and examine the possibilities.

CONCLUSION

The common social and personal problem of loneliness is multifaceted and requires much more empirical investigation. Researchers need to examine more thoroughly the relationship between loneliness and depression in normal persons to separate out the clinical similarities and differences of these two conditions. Further work is necessary to shift from an analysis of correlations to the realm of cause and effect. However, the literature seems clear about one thing: lonely people are negative about social relationships in ways that nonlonely persons do not share. Thus, another significant question for further study is this: Does negativity lead people to be lonely or does loneliness lead to negativity? This question has serious implications for the socialization of children; it may be reasonable to hypothesize that children who are taught to view life negatively may be prime candidates for loneliness in adulthood as well as for the many ripple effects accompanying that condition.

Notes and References

1. Jean-Paul Sartre, *Existentialism and Human Emotions* (New York: Book Sales, 1957).

2. Charles C. McCaghy, James Skipper, Jr., and Mark Lefton, eds., *In Their Own Behalf: Voices from the Margin* (New York: Appleton-Century-Crofts, 1968).

3. See Karen Horney, "Alienation and the Search for Identity," *American Journal of Psychoanalysis*, 21 (1961); Harry Stack Sullivan, *The Interpersonal Theory of Psychiatry* (New York: W. W. Norton & Co., 1953); and Erich Fromm, *Man for Himself* (New York: Fawcett Books, 1947).

4. Philip Slater, *The Pursuit of Loneliness: American Culture at the Breaking Point* (Boston: Beacon Press, 1970).

5. Emile Durkheim, *Suicide: A Study in Sociology*, John Spaulding and George Simpson, trans. and eds. (New York: Free Press, 1951).

6. Letitia Anne Peplau, Daniel Russell, and Margaret Heim, "An Attributional Analysis of Loneliness," in Irene H. Frieze, Daniel Bar-Tal, and John S. Carroll, eds., *Attribution Theory: Application to Social Problems* (New York: Jossey-Bass, 1979), pp. 53—78.

7. Robert Weiss, ed., *Loneliness: The Experience of Emotional and Social Isolation* (Cambridge, Mass.: M.I.T. Press, 1973), p. 11.

8. Ibid., p. 15.

9. Warren H. Jones, J. E. Freemon, and Ruth Ann Goswick, "The Persistence of Loneliness: Self and Other Determinants," *Journal of Personality*, 49 (March 1981), pp. 27–48. See also W. H. Jones, "Loneliness and Social Behavior." Unpublished manuscript, University of Tulsa, 1980.

10. For further insight into the clinical aspects of loneliness and their relationship to depression, see Carolyn E. Cutrona, "Transition to College: Loneliness and the Process of Social Adjustment," in Letitia Anne Peplau and Daniel Perlman, eds., *Loneliness: A Sourcebook of Current Research, Theory, and Therapy* (New York: Wiley Interscience, 1982).

11. David Russell, Letitia Anne Peplau, and Carolyn Cutrona, "The Revised UCLA Loneliness Scale: Concurrent and Discriminant Validity Evidence," *Journal of Personality and Social Psychology*, 39 (September 1980), pp. 472—480.

12. Jones, op. cit.

13. Jones, Freemon, and Goswick, op. cit.

14. See especially Cecilia Solano, "Two Measures of Loneliness: A Comparison," *Psychological Reports*, 46 (February 1980), 23—28; and C. Sansone, Warren Jones, and Bob Heim, "Interpersonal Perceptions of Loneliness," paper presented at the annual meeting of the Southwestern Psychological Association, San Antonio, Tex., April 1979.

15. Frieda Fromm-Reichman, "Loneliness," *Psychiatry*, 22 (February 1959).

16. Scott Wimer and Letitia Anne Peplau, "Determinants of Reactions to Lonely Others." Paper presented at the annual meeting of the Western Psychological Association, San Francisco, April 1978.

17. Martin Seligman, *Helplessness* (San Francisco: W. H. Freeman 1975).

18. Peplau, Russell, and Heim, op. cit.

19. See Eric Berne, *What Do You Say After You Say Hello?* (New York: Bantam Books, 1975).

20. See Cutrona, op. cit., for a discussion of maintenance factors.

21. Peplau, Russell, and Heim, op. cit.

BIOLOGY, DESTINY, AND ALL THAT

Grabbing hold of a tar baby
of research findings, our writer
tries to pull apart truth from myth
in ideas about the differences
between the sexes.

Paul Chance

Paul Chance is a psychologist, writer, and contributing editor of *Psychology Today*.

In the 1880s, scholars warned against the hazards of educating women. Some experts of the day believed that too much schooling could endanger a woman's health, interfere with her reproductive ability, and cause her brain to deteriorate. In the 1980s we laugh at such absurd ideas, but have we (men and women alike) really given up the ancient idea that a woman is fundamentally an inferior sort of man? It seems not.

It's hard to find evidence these days of gross discrimination against women as a company policy. Successful lawsuits have made that sort of prejudice expensive. Yet evidence of more subtle forms of bias abound. The sociologist Beth Ghiloni conducted a study while she was a student at the University of California, Santa Cruz, that shows that some corporations are meeting the demands of affirmative action by putting women into public relations posts. Public relations is important but distant from the activities that generate revenue, so PR assignments effectively keep women out of jobs that include any real corporate power. Thus, women increasingly complain of facing a "glass ceiling" through which they can see, but cannot reach, high level corporate positions.

It seems likely that such discrimination reflects some very old ideas about what men and women are like. Men, the stereotype has it, are aggressive and self-confident. They think analytically, and are cool under fire. They enjoy jobs that offer responsibility and challenge. They are, in other words, ideally suited for important,

high-level positions. Put them on the track that may one day lead to corporate vice president.

Women, the thinking goes, are passive and filled with self-doubt. They think intuitively, and are inclined to become emotionally distraught under pressure. They therefore enjoy jobs that involve working with people. Give them the lower-rung jobs in the personnel department, or send them to public affairs.

Figuring out how truth and myth intertwine in these stereotypes is difficult. The research literature on sex differences is a tar baby made of numbers, case studies, and anecdotal impressions. To paraphrase one researcher, "If you like ambiguity, you're gonna love sex-difference research." Nevertheless, let us be brave and take the stereotypes apart piece by piece.

Aggression

On this there is no argument: Everyone agrees that men are more aggressive than women. In a classic review of the literature, the psychologists Eleanor Maccoby of Stanford University and Carol Jacklin of the University of Southern California found that boys are more aggressive than girls both physically and verbally, and the difference begins to show up by the age of 3. Boys are more inclined to rough-and-tumble play; girls have tea parties and play with dolls.

The difference in aggressiveness is most clearly seen in criminal activity. There are far more delinquent boys than girls, and prisons are built primarily to contain men. The most aggressive crimes, such as murder and assault, are especially dominated by men.

It seems hardly likely that male aggressiveness, as it is documented by the research, would win favor among personnel directors and corporate headhunters. Yet aggressiveness is considered not only a virtue but an essential trait for many jobs. Vice President George Bush learned the value of aggressiveness when he managed to put aside his wimp image by verbally attacking the CBS news anchorman Dan Rather on national television. The assumption seems to be that the same underlying trait that makes for murderers, rapists, and strong-arm bandits also, in more moderate degree or under proper guidance, makes people more competitive and motivated to achieve great things.

But the research suggests that women may be just as aggressive in this more civilized sense as men. For instance, most studies of competitiveness find no differences between the sexes, according to the psychologists Veronica Nieva and Barbara Gutek, co-authors of *Women and Work*. And many studies of achievement motivation suggest that women are just as eager to get things done as men.

Self-confidence

Men are supposed to be self-confident, women full of self-doubt. Again there is some evidence for the stereotype, but it isn't particularly complimentary to males. Various studies show that males overestimate their abilities, while females underestimate theirs. Researchers have found, for instance, that given the option of choosing tasks varying in difficulty, boys erred by choosing those that were too difficult for them, while girls tended to select tasks that were too easy.

Other research shows that women are not only less confident of their ability to do a job, when they succeed at it they don't give themselves credit. Ask them why they did well and they'll tell you it was an easy task or that they got lucky. Ask men the same thing and they'll tell you they did well because of their ability and hard work.

It is perhaps this difference in self-confidence that makes women better risks for auto insurance, and it may have something to do with the fact that almost from the day they can walk, males are more likely to be involved in pedestrian accidents.

Rational Thinking

A great many studies have found that men do better than women on tests of mathematical reasoning. Julian Stanley, a psychologist at Johns Hopkins University, has been using the mathematics portion of the Scholastic Aptitude Test to identify mathematically gifted youths. He and his colleagues have consistently found that a majority of the high scoring students are boys. Stanley reports that "mathematically gifted boys outnumber gifted girls by a ratio of about 13 to 1." Moreover, the very best scores almost inevitably come from boys.

But while some findings show that men are better at mathematical reasoning, there is no evidence that they are more analytical or logical in general. In fact, the superiority of men at mathematical problem solving seems not to reflect superior analytical thinking but a special talent men have for visualizing objects in space. Women are every bit the match of men at other kinds of problems such as drawing logical conclusions from written text. Indeed, girls have an advantage over boys in verbal skills until at least adolescence.

As for the idea that women are more intuitive, forget it. Numerous studies have shown that women are better at reading body language, and it is probably this skill (born, perhaps, of the need to avoid enraging the more aggressive sex) that gives rise to the myth of women's intuition. In reality, women and men think alike.

Emotionality

In Victorian England, to judge by the novels of the day, a woman was no woman at all if she didn't feel faint or burst into tears at least once a week. The idea that women are more emotional than men, that they feel things more deeply and react accordingly, persists. Is it true?

No. Carol Tavris and Carole Wade, psychologists and co-authors of *The Longest War: Sex Differences in Perspective*, write that the sexes are equally likely to feel anxious in new situations, to get angry when insulted, to be hurt when a loved one leaves them, and to feel embarrassed when they make mistakes in public.

The sexes are equally emotional, but there are important differences in how willing men and women are to express emotions, which emotions they choose to express, and the ways in which they express them. If you ask men and women in an emotional situation what they are feeling, women are likely to admit that they are affected, while men are likely to deny it. Yet studies show that when you look at the psychological correlates of emotion—heartbeat, blood pressure, and the like—you find that those strong silent men are churning inside every bit as much as the women.

Men are particularly eager to conceal feelings such as fear, sorrow, and loneliness, according to Tavris and Wade. Men often bottle these "feminine" feelings even with those they hold most dear.

Another difference comes in how emotions are expressed. Women behave differently depending upon what they feel. They may cry if sad, curse if angry, pout if their pride is hurt. Men tend to respond to such situations in a more or less uniform way. Whether they have been jilted, frightened, or snubbed, they become aggressive. (Men, you will recall, are very good at aggression.) As for the notion that keeping one's head is characteristic of men, well, don't get sore fellows, but it ain't necessarily so.

> "It's unlikely that male aggressiveness, as documented by research, would win favor among personnel directors."

Men's math superiority reflects a special talent for visualizing objects in space.

Job Interests

If you ask men and women what they like about their work, you will get different answers. In 1957, Frederick Herzberg published a study of what made work enjoyable to employees. Men, he concluded, enjoyed work that offered responsibility and challenge. For women, on the other hand, the environment was the thing. They wanted an attractive work area, and some pleasant people to talk to while they did whatever needed to be done. Give your secretary an office with some nice wallpaper, put a flower on her desk once in a while, and she'll be happy.

Experts now agree, however, that such differences probably reflect differences in the jobs held by the people studied. You may get such findings, suggest *Women and Work* co-authors Nieva and Gutek, if you compare female file clerks and male engineers—but not if you compare female and male engineers. In a study of workers in various jobs, Daphne Bugental, a psychologist at the University of California, Santa Barbara, and the late Richard Centers found no consistent differences in the way men and women ranked "intrinsic" job characteristics such as responsibility and challenge and "extrinsic" characteristics such as pleasant surroundings and friendly co-workers. In other words, the dif-

ferences in what men and women find interesting about work reflect differences in the kinds of work men and women characteristically do. People in relatively high level jobs enjoy the responsibility and challenge it offers; people in low level jobs that offer little responsibility and challenge look elsewhere for satisfaction.

Where Do the Differences Come From?

So, what differences separate the boys from the girls? Men are reported to be more aggressive than women in a combative sense, but they are not necessarily more competitive. Men are more confident, perhaps recklessly so, and women may be too cautious. Men are not more likely to be cool under fire, but they are more likely to become aggressive regardless of what upsets them. Men are not more analytic in their thinking, nor women more intuitive, but men are better at solving mathematical problems, probably because they are better at spatial relationships. Finally, women are less interested in the responsibility and challenge of work, but only because they usually have jobs that offer little responsibility and challenge.

While this seems to be the gist of the matter, it leaves open the question of whether the differences are due to biology or to environment. Are

THE DEVELOPMENT OF THE "WEAKER SEX"
(AND THE DEMORALIZATION OF THE DUDE).

VASSAR GRADUATE.—"These are the dumb-bells I used last term in our gymnasium; won't one of you gentleman just put them up? It's awfully easy."

men more aggressive because they are born that way or because of lessons that begin in the cradle? Are women less confident than men because different hormones course through their blood, or because for years people have told them that they can't expect much from themselves?

The research on the nature-nurture question is a candy store in which people of varying biases can quickly find something to their liking. Beryl Lieff Benderly, an anthropologist and journalist, critiqued the physiological research for her new book, *The Myth of Two Minds*. The title tells the story. She even challenges the notion that men are stronger than women. "The plain fact is that we have no idea whether men are 'naturally' stronger than women," she writes. Short of conducting an experiment along the lines of *Lord of the Flies*, we are unlikely to unravel the influences of nature and nurture to everyone's satisfaction. Nevertheless, research does offer hints about the ways that biology and environment affect stereotypical behavior. Take the case of aggression. There are any number of studies linking aggression to biological factors. Testosterone, a hormone found in much higher levels in men than in women, has been found in even larger quantities in criminally aggressive males. Castration, which decreases the level of testosterone, has been used for centuries to produce docile men and animals. And girls who have had prenatal exposure to high levels of testosterone are more tomboyish than other girls.

Yet biology is not quite destiny. In her famous *Six Cultures* study, the anthropologist Beatrice Whiting and her colleagues at Harvard University found that in each of the societies studied, boys were more aggressive than girls. But the researchers also found such wide cultural differences that the girls in a highly aggressive society were often more aggressive than the boys in another, less aggressive society.

The same mix of forces applies wherever we look. Biology may bend the twig in one direction, but the environment may bend it in another. But whether biology or environment ultimately wins the hearts and minds of researchers is less important than how the differences, wherever they come from, affect behavior in the workplace. If someone explodes in anger (or breaks down in tears) in the midst of delicate contract negotiations, it matters little to the stockholders whether the lost business can, in the end, be blamed on testosterone or bad toilet training. The more important question is, what are the implications of sex difference research for business?

What Difference Do the Differences Make?

The research on sex differences suggest three points that people in business can usefully con-

sider. First, the differences between the sexes are small. Researchers look for "statistically significant" differences. But statistically significant differences are not necessarily practically significant. There is a great deal of overlap between the sexes on most characteristics and especially on the characteristics we have been considering. Men are, on average, better at solving mathematical problems, but there are many women who are far above the average man in this area. Women are, on average, less confident than men. But there are many men who doubt themselves far more than the average woman. A study of aggressiveness is illustrative. The psychologist D. Anthony Butterfield and the management expert Gary N. Powell had college students rate the ideal U.S. President on various characteristics, including aggressiveness. Then they had them rate people who were running for President and Vice President at the time: Ronald Reagan, Walter Mondale, George Bush, and Geraldine Ferraro. Researchers found that the ideal president was, among other things, aggressive. They also found that Geraldine Ferraro was judged more aggressive than the male candidates. The point is that it is impossible to predict individual qualities from group differences.

Indeed, Carol Jacklin suggests that differences in averages may give a quite distorted view of both sexes. She notes that while, on average, boys play more aggressively than girls, her research shows that the difference is due to a small number of very aggressive boys. Most of the boys are, in fact, very much like the girls. "I'd be willing to bet," Jacklin says, "that much of the difference in aggressiveness between men and women is due to a small number of extremely aggressive men—many of whom are in prison—and that the remaining men are no more aggressive than most women."

Second, different doesn't necessarily mean inferior. It is quite possible that feminine traits (in men or women) are assets in certain situations, while masculine traits may be advantageous in other situations. The psychologist Carol Gilligan, author of *In a Different Voice*, says that women are more comfortable with human relationships than men are, and this may sometimes give them an edge. For instance, Roderick Gilkey and Leonard Greenhalgh, psychologists at Dartmouth University, had business students simulate negotiations over the purchase of a used car and television advertising time. The women appeared better suited to the task than the men. They were more flexible, more willing to compromise, and less deceptive. "Women can usually come to an agreement on friendly terms," says Greenhalgh. "They're better at avoiding impasses."

In another study, the psychologist Wendy Wood of Texas A&M University asked college students to work on problems in groups of three.

"Biology may bend the twig in one direction, but the environment may bend it in another."

Some groups consisted only of men, others of women. The groups tried to solve problems such as identifying the features to consider in buying a house. The men, it turns out, came up with more ideas, while the women zeroed in on one good idea and developed it. Wood concluded that all-male groups might be better for brainstorming, while all-female groups might be better for finding the best solution to a problem.

Third, sometimes people lack the characteristics needed for a job until they are in the job. Jobs that offer responsibility and challenge, for example, tend to create a desire for more responsibility and challenge. While there are research studies to support this statement, an anecdote from the sociologist Rosabeth Moss Kanter is more telling. Linda, a secretary for 17 years in a large corporation, had no interest in being anything but a secretary, and when she was offered a promotion through an affirmative action program, she hesitated. Her boss persuaded her to take the job and she became a successful manager and loved the additional responsibility and challenge. She even set her sights on a vice president position. As a secretary, Linda would no doubt have scored near the female stereotype. But when she became a manager, she became more like the stereotypical male.

Kanter told that story a dozen years ago, but we are still struggling to learn its lesson. A discrimination case against Sears, Roebuck and Company recently made news. The Equal Employment Opportunity Commission (EEOC) argued that Sears discriminated against women because nearly all of the employees in the company who sell on commission are men. Sears presented evidence that women expressed little interest in commission-sales work, preferring the less risky jobs in salaried sales. In the original decision favoring Sears, the U.S. District Court had found that "noncommission saleswomen were generally happier with their present jobs at Sears, and were much less likely than their male counterparts to be interested in other positions, such as commission sales...." But in the U.S. Court of Appeals, Appellate Judge Cudahy, who dissented in part from the majority, wrote that this reasoning is "of a piece with the proposition that women are by nature happier cooking, doing the laundry, and chauffeuring the children to softball games than arguing appeals or selling stocks."

The point is not that employees must be made to accept more responsible positions for their own good, even if it is against their will. The point is that business should abandon the stereotypes that lock men and women into different, and often unequal, kinds of work. If it finally does, it will discover the necessity—and value—of finding ways of enticing men and women into jobs for which, according to the stereotypes, they are not suited. If business doesn't do that, it may discover that differences between men and women really do separate the sexes.

The Vintage Years

THE GROWING NUMBER OF HEALTHY, VIGOROUS OLDER PEOPLE HAS HELPED OVERCOME SOME STEREOTYPES ABOUT AGING. FOR MANY, THE BEST IS YET TO COME.

Jack C. Horn and Jeff Meer

Jack C. Horn is a senior editor and Jeff Meer is an assistant editor at the magazine.

Our society is getting older, but the old are getting younger. As Sylvia Herz told an American Psychological Association (APA) symposium on aging last year, the activities and attitudes of a 70-year-old today "are equivalent to those of a 50-year-old's a decade or two ago."

Our notions of what it means to be old are beginning to catch up with this reality. During the past several decades, three major changes have altered the way we view the years after 65:

• The financial, physical and mental health of older people has improved, making the prospect of a long life something to treasure, not fear.

• The population of older people has grown dramatically, rising from 18 million in 1965 to 28 million today. People older than 65 compose 12 percent of the population, a percentage that is expected to rise to more than 20 percent by the year 2030.

• Researchers have gained a much better understanding of aging and the lives of older people, helping to sort out the inevitable results of biological aging from the effects of illness or social and environmental problems. No one has yet found the fountain of youth, or of immortality. But research has revealed that aging itself is not the thief we once thought it was; healthy older people can maintain and enjoy most of their physical and mental abilities, and even improve in some areas.

Because of better medical care, improved diet and increasing interest in physical fitness, more people are reaching the ages of 65, 75 and older in excellent health. Their functional age—a combination of physical, psychological and social factors that affect their attitudes toward life and the roles they play in the world—is much younger than their chronological age.

Their economic health is better, too, by almost every measure. Over the last three decades, for example, the number of men and women 65 and older who live below the poverty line has dropped steadily from 35 percent in 1959 to 12 percent in 1984, the last year for which figures are available.

On the upper end of the economic scale, many of our biggest companies are headed by what once would have been called senior citizens, and many more of them serve as directors of leading companies. Even on a more modest economic level, a good portion of the United States' retired older people form a new leisure class, one with money to spend and the time to enjoy it. Obviously not all of America's older people share this prosperity. Economic hardship is particularly prevalent among minorities. But as a group, our older people are doing better than ever.

In two other areas of power, politics and the law, people in their 60s and 70s have always played important roles. A higher percentage of people from 65 to 74 register and vote than in any other group. With today's increasing vigor and numbers, their power is likely to increase still further. It is perhaps no coincidence that our current President is the oldest ever.

Changing attitudes, personal and social, are a major reason for the increasing importance of older people in our society. As psychologist

Bernice Neugarten points out, there is no longer a particular age at which someone starts to work or attends school, marries and has children, retires or starts a business. Increasing numbers of older men and women are enrolled in colleges, universities and other institutions of learning. According to the Center for Education Statistics, for example, the number of people 65 and older enrolled in adult education of all kinds increased from 765,000 to 866,000 from 1981 to 1984. Gerontologist Barbara Ober says that this growing interest in education is much more than a way to pass the time. "Older people make excellent students, maybe even better students than the majority of 19- and 20-year-olds. One advantage is that they have settled a lot of the social and sexual issues that preoccupy their younger classmates."

Older people today are not only healthier and more active; they are also increasingly more numerous. "Squaring the pyramid" is how some demographers describe this change in our population structure. It has always been thought of as a pyramid, a broad base of newborns supporting successively smaller tiers of older people as they died from disease, accidents, poor nutrition, war and other causes.

Today, the population structure is becoming more rectangular, as fewer people die during the earlier stages of life. The Census Bureau predicts that by 2030 the structure will be an almost perfect rectangle up to the age of 70.

The aging of America has been going on at least since 1800, when half the people in the country were younger than 16 years old, but two factors have accelerated the trend tremendously. First, the number of old people has increased rapidly. Since 1950 the number of Americans 65 and older has more than doubled to some 28 million—more than the entire current population of Canada. Within the same period, the number of individuals older than 85 has quadrupled to about 2.6 million (see "The Oldest Old," this article).

Second, the boom in old people has been paired with a bust in the proportion of youngsters due to a declining birth rate. Today, fewer than one American in four is younger than 16. This drop-off has been steady, with the single exception of the post-World War II baby boom, which added 76 million children to the country between 1945 and 1964. As these baby boomers reach the age of 65, starting in 2010, they are expected to increase the proportion of the population 65 and older from its current 12 percent to 21 percent by 2030.

The growing presence of healthy, vigorous older people has helped overcome some of the stereotypes about aging and the elderly. Research has also played a major part by replacing myths with facts. While there were some studies of aging before World War II, scientific

> *A man over 90 is a great comfort to all his elderly neighbours: he is a picket-guard at the extreme outpost; and the young folks of 60 and 70 feel that the enemy must get by him before he can come near their camp.*
> —Oliver Wendell Holmes,
> *The Guardian Angel.*

BY THE YEAR 2030 MORE THAN 20 PERCENT OF THE POPULATION IS EXPECTED TO BE 65 OR OLDER.

interest increased dramatically during the 1950s and kept growing.

Important early studies of aging included three started in the mid or late 1950s: the Human Aging Study, conducted by the National Institute of Mental Health (NIMH); the Duke Longitudinal Studies, done by the Center for the Study of Aging and Human Development at Duke University; and the Baltimore Longitudinal Study of Aging, conducted by the Gerontological Institute in Baltimore, now part of the National Institute on Aging (NIA). All three took a multidisciplinary approach to the study of normal aging: what changes take place, how people adapt to them, how biological, genetic, social, psychological and environmental characteristics relate to longevity and what can be done to promote successful aging.

These pioneering studies and hundreds of later ones have benefited from growing federal support. White House Conferences on Aging in 1961 and 1971 helped focus attention on the subject. By 1965 Congress had enacted Medicare and the Older Americans Act. During the 1970s Congress authorized the establishment of the NIA as part of the National Institutes of Health and NIMH created a special center to support research on the mental health of older people.

All these efforts have produced a tremendous growth in our knowledge of aging. In the first (1971) edition of the *Handbook of the Psychology of Aging,* it was estimated that as much had been published on the subject in the previous 15 years as in all the years before then. In the second edition, published in 1985, psychologists James Birren and Walter Cunningham wrote that the "period for this rate of doubling has now decreased to 10 years...the volume of published research has increased to the almost unmanageable total of over a thousand articles a year."

Psychologist Clifford Swenson of Purdue

University explained some of the powerful incentives for this tremendous increase: "I study the topic partly to discover more effective ways of helping old people cope with their problems, but also to load my own armamentarium against that inevitable day. For that is one aspect of aging and its problems that makes it different from the other problems psychologists study: We may not all be schizophrenic or neurotic or overweight, but there is only one alternative to old age and most of us try to avoid that alternative."

One popular misconception disputed by recent research is the idea that aging means inevitable physical and sexual failure. Some changes occur, of course. Reflexes slow, hearing and eyesight dim, stamina decreases. This *primary aging* is a gradual process that begins early in life and affects all body systems.

But many of the problems we associate with old age are *secondary aging*—the results not of age but of disease, abuse and disuse—factors often under our own control. More and more older people are healthy, vigorous men and women who lead enjoyable, active lives. National surveys by the Institute for Social Research and others show that life generally seems less troublesome and freer to older people than it does to younger adults.

In a review of what researchers have learned about subjective well-being—happiness, life satisfaction, positive emotions—University of Illinois psychologist Ed Diener reported that "Most results show a slow rise in satisfaction with age. . .young persons appear to experience higher levels of joy but older persons tend to judge their lives in more positive ways."

Money is often mentioned as the key to a happy retirement, but psychologist Daniel Ogilvie of Rutgers University has found another, much more important, factor. Once we have a certain minimum amount of money, his research shows, life satisfaction depends mainly on how much time we spend doing things we find meaningful. Ogilvie believes retirement-planning workshops and seminars should spend more time helping people decide how to use their skills and interests after they retire.

A thought that comes through clearly when researchers talk about physical and mental fitness is "use it or lose it." People rust out faster from disuse than they wear out from overuse. This advice applies equally to sexual activity. While every study from the time of Kinsey to the present shows that sexual interest and activity diminish with age, the drop varies greatly among individuals. Psychologist Marion Perlmutter and writer Elizabeth Hall have reported that one of the best predictors of continued sexual intercourse "is early sexual activity and past sexual enjoyment and frequency. People who have never had much pleasure from sexu-

WHILE THE OLD AND THE YOUNG MAY BE EQUALLY COMPETENT, THEY ARE DIFFERENTLY COMPETENT.

ality may regard their age as a good excuse for giving up sex."

They also point out that changing times affect sexual activity. As today's younger adults bring their more liberal sexual attitudes with them into old age, the level of sexual activity among older men and women may rise.

The idea that mental abilities decline steadily with age has also been challenged by many recent and not-so-recent findings (see "The Reason of Age," *Psychology Today,* June 1986). In brief, age doesn't damage abilities as much as was once believed, and in some areas we actually gain; we learn to compensate through experience for much of what we do lose; and we can restore some losses through training.

For years, older people didn't do as well as younger people on most tests used to measure mental ability. But psychologist Leonard Poon of the University of Georgia believes that researchers are now taking a new, more appropriate approach to measurement. "Instead of looking at older people's ability to do abstract tasks that have little or no relationship to what they do every day, today's researchers are examining real-life issues."

Psychologist Gisela Labouvie-Vief of Wayne State University has been measuring how people approach everyday problems in logic. She notes that older adults have usually done poorly on such tests, mostly because they fail to think logically all the time. But Labouvie-Vief argues that this is not because they have forgotten how to think logically but because they use a more complex approach unknown to younger thinkers. "The [older] thinker operates within a kind of double reality which is both formal and informal, both logical and psychological," she says.

In other studies, Labouvie-Vief has found that when older people were asked to give concise summaries of fables they read, they did so. But when they were simply asked to recall as much of the fable as possible, they concentrat-

The pleasures that once were heaven Look silly at sixty-seven.
—Noel Coward, *"What's Going to Happen to the Tots?"*

Old age consoles itself by giving good precepts for being unable to give bad examples.
—La Rochefoucauld, *The Maxims.*

THE OLDEST OLD: THE YEARS AFTER 85

"Every man desires to live long, but no man would be old," or so Jonathan Swift believed. Some people get their wish to live long and become what are termed the "oldest old," those 85 and older. During the past 22 years, this group has increased by 165 percent to 2.5 million and now represents more than 1 percent of the population.

Who are these people and what are their lives like? One of the first to study them intensively is gerontologist Charles Longino of the University of Miami, who uses 1980 census data to examine their lives for the American Association of Retired People.

He found, not surprisingly, that nearly 70 percent are women. Of these, 82 percent are widowed, compared with 44 percent of the men. Because of the conditions that existed when they were growing up, the oldest old are poorly educated compared with young people today, most of whom finish high school. The average person now 85 years and older only completed the eighth grade.

Only one-quarter of these older citizens are in hospitals or institutions such as nursing homes, and more than half live in their own homes. Just 30 percent live by themselves. More than a third live with a spouse or with their children. There are certainly those who aren't doing well—one in six have incomes below the poverty level—but many more are relatively well-off. The mean household income for the group, Longino says, was more than $20,000 in 1985.

What of the quality of life? "In studying this group, we have to be aware of youth creep," he says. "The old are getting younger all the time." This feeling is confirmed by a report released late last year by the National Institute on Aging. The NIA report included three studies of people older than 65 conducted in two counties in Iowa, in East Boston, Massachusetts, and in New Haven, Connecticut. There are large regional differences between the groups, of course, and

they aren't a cross-section of older people in the nation as a whole. But in all three places, most of those older than 85 seem to be leading fulfilling lives.

Most socialize in a variety of ways. In Iowa, more than half say they go to religious services at least once a week and the same percentage say they belong to some type of professional, social, church-related or recreational group. More than three-quarters see at least one or two children once a month and almost that many see other close relatives that often.

As you would expect, many of the oldest old suffer from disabilities and serious health problems. At least a quarter of those who responded have been in a hospital overnight in the past year and at least 8 percent have had heart attacks or have diabetes. In Iowa and New Haven, more than 13 percent of the oldest old had cancer, while in East Boston the rate was lower (between 7 percent and 8 percent). Significant numbers of the oldest old have suffered serious injury from falls. Other common health problems for this group are high blood pressure and urinary incontinence. However, epidemiologist Adrian Ostfeld, who directed the survey in New Haven, notes that "most of the disability was temporary."

Longino has found that almost 10 percent of the oldest old live alone with a disability that prevents them from using public transportation. This means that they are "isolated from the daily hands-on care of others," he says. "Even so, there are a surprising number of the oldest old who don't need much in the way of medical care. They're the survivors.

"I think we have to agree that the oldest old is, as a group, remarkably diverse," Longino says. "Just as it is unfair to say that those older than 85 are all miserable, it's not fair to say that they all lead wonderful lives, either."
—*Jeff Meer*

ed on the metaphorical, moral or social meaning of the text. They didn't try to duplicate the fable's exact words, the way younger people did. As psychologists Nancy Datan, Dean Rodeheaver and Fergus Hughes of the University of Wisconsin have described their findings, "while [some people assume] that old and young are equally competent, we might better assume that they are differently competent."

John Horn, director of the Adult Development and Aging program at the University of

Southern California, suggests that studies of Alzheimer's disease, a devastating progressive mental deterioration experienced by an estimated 5 percent to 15 percent of those older than 65, may eventually help explain some of the differences in thinking abilities of older people. "Alzheimer's, in some ways, may represent the normal process of aging, only speeded up," he says. (To see how your ideas about Alzheimer's square with the facts, see "Alzheimer's Quiz" and "Alzheimer's Answers," this article.)

Generalities are always suspect, but one generalization about old age seems solid: It is a different experience for men and women. Longevity is one important reason. Women in the United States live seven to eight years longer, on the average, than do men. This simple fact has many ramifications, as sociologist Gunhild Hagestad explained in *Our Aging Society.*

For one thing, since the world of the very old is disproportionately a world of women, men and women spend their later years differently. "Most older women are widows living alone; most older men live with their wives. . .among individuals over the age of 75, two-thirds of the men are living with a spouse, while less than one-fifth of the women are."

The difference in longevity also means that among older people, remarriage is a male prerogative. After 65, for example, men remarry at a rate eight times that of women. This is partly a matter of the scarcity of men and partly a matter of culture—even late in life, men tend to marry younger women. It is also a matter of education and finances, which, Hagestad explains, "operate quite differently in shaping remarriage probabilities among men and women. The more resources the woman has available (measured in education and income), the less likely she is to remarry. For men, the trend is reversed."

The economic situations of elderly men and women also differ considerably. Lou Glasse, president of the Older Women's League in Washington, D.C., points out that most of these women were housewives who worked at paid jobs sporadically, if at all. "That means their Social Security benefits are lower than men's, they are not likely to have pensions and they are less likely to have been able to save the kind of money that would protect them from poverty during their older years."

Although we often think of elderly men and women as living in nursing homes or retirement communities, the facts are quite different. Only about 5 percent are in nursing homes and perhaps an equal number live in some kind of age-segregated housing. Most people older than 65 live in their own houses or apartments.

We also think of older people as living alone. According to the Census Bureau, this is true of 15 percent of the men and 41 percent of the women. Earlier this year, a survey done by Louis Harris & Associates revealed that 28 percent of elderly people living alone have annual incomes below $5,100, the federal poverty line. Despite this, they were four times as likely to give financial help to their children as to receive it from them.

In addition, fewer than 1 percent of the old people said they would prefer living with their children. Psychiatrist Robert N. Butler, chairman of the Commonwealth Fund's Commission

AMONG OLDER PEOPLE TODAY, REMARRIAGE IS STILL LARGELY A MALE PREROGATIVE, DUE TO THE SEX DIFFERENCE IN LONGEVITY.

on Elderly People Living Alone, which sponsored the report, noted that these findings dispute the "popular portrait of an elderly, dependent parent financially draining their middle-aged children."

There is often another kind of drain, however, one of time and effort. The Travelers Insurance Company recently surveyed more than 700 of its employees on this issue. Of those at least 30 years old, 28 percent said they directly care for an older relative in some way—taking that person to the doctor, making telephone calls, handling finances or running errands—for an average of 10 hours a week. Women, who are more often caregivers, spent an average of 16 hours, and men five hours, per week. One group, 8 percent of the sample, spent a heroic 35 hours per week, the equivalent of a second job, providing such care. "That adds up to an awful lot of time away from other things," psychologist Beal Lowe says, "and the stresses these people face are enormous."

Lowe, working with Sherman-Lank Communications in Kensington, Maryland, has formed "Caring for Caregivers," a group of professionals devoted to providing services, information and support to those who care for older relatives. "It can be a great shock to some people who have planned the perfect retirement," he says, "only to realize that your chronically ill mother suddenly needs daily attention."

Researchers who have studied the housing needs of older people predictably disagree on many things, but most agree on two points: We need a variety of individual and group living arrangements to meet the varying interests, income and abilities of people older than 65; and the arrangements should be flexible enough that the elderly can stay in the same locale as their needs and abilities change. Many studies have documented the fact that moving itself can be stressful and even fatal to old people, particularly if they have little or no influence over when and where they move.

This matter of control is important, but more complicated than it seemed at first. Psychologist Judith Rodin and others have demonstrated that people in nursing homes are happier, more alert and live longer if they are allowed to take responsibility for their lives in some way, even in something as simple as choosing a plant for their room, taking care of a bird feeder, selecting the night to attend a movie.

Rodin warns that while control is generally beneficial, the effect depends on the individuals involved. For some, personal control brings with it demands in the form of time, effort and the risk of failure. They may blame themselves if they get sick or something else goes wrong. The challenge, Rodin wrote, is to "provide but not impose opportunities. . . . The need for self-determination, it must be remembered, also calls for the opportunity to choose not to exercise control. . . ."

An ancient Greek myth tells how the Goddess of Dawn fell in love with a mortal and convinced Jupiter to grant him immortality. Unfortunately, she forgot to have youth included in the deal, so he gradually grew older and older. "At length," the story concludes, "he lost the power of using his limbs, and then she shut him up in his chamber, whence his feeble voice might at times be heard. Finally she turned him into a grasshopper."

The fears and misunderstandings of age expressed in this 3,000-year-old myth persist today, despite all the positive things we have learned in recent years about life after 65. We don't turn older people into grasshoppers or shut them out of sight, but too often we move them firmly out of the mainstream of life.

In a speech at the celebration of Harvard

> *If I had known when I was 21 that I should be as happy as I am now, I should have been sincerely shocked. They promised me wormwood and the funeral raven.*
>
> —Christopher Isherwood, letter at age 70.

University's 350th anniversary last September, political scientist Robert Binstock decried what he called The Spectre of the Aging Society: "the economic burdens of population aging; moral dilemmas posed by the allocation of health resources on the basis of age; labor market competition between older and younger workers within the contexts of age discrimination laws; seniority practices, rapid technologi-

ALZHEIMER'S QUIZ

Alzheimer's disease, named for German neurologist Alois Alzheimer, is much in the news these days. But how much do you really know about the disorder? Political scientist Neal B. Cutler of the Andrus Gerontology Center gave the following questions to a 1,500-person cross section of people older than 45 in the United States in November 1985. To compare your answers with theirs and with the correct answers, turn to the next page.

	True	False	Don't know
1. Alzheimer's disease can be contagious.	___	___	___
2. A person will almost certainly get Alzheimer's if they just live long enough.	___	___	___
3. Alzheimer's disease is a form of insanity.	___	___	___
4. Alzheimer's disease is a normal part of getting older, like gray hair or wrinkles.	___	___	___
5. There is no cure for Alzheimer's disease at present.	___	___	___
6. A person who has Alzheimer's disease will experience both mental and physical decline.	___	___	___
7. The primary symptom of Alzheimer's disease is memory loss.	___	___	___
8. Among persons older than age 75, forgetfulness most likely indicates the beginning of Alzheimer's disease.	___	___	___
9. When the husband or wife of an older person dies, the surviving spouse may suffer from a kind of depression that looks like Alzheimer's disease.	___	___	___
10. Stuttering is an inevitable part of Alzheimer's disease.	___	___	___
11. An older man is more likely to develop Alzheimer's disease than an older woman.	___	___	___
12. Alzheimer's disease is usually fatal.	___	___	___
13. The vast majority of persons suffering from Alzheimer's disease live in nursing homes.	___	___	___
14. Aluminum has been identified as a significant cause of Alzheimer's disease.	___	___	___
15. Alzheimer's disease can be diagnosed by a blood test.	___	___	___
16. Nursing-home expenses for Alzheimer's disease patients are covered by Medicare.	___	___	___
17. Medicine taken for high blood pressure can cause symptoms that look like Alzheimer's disease.	___	___	___

Alzheimer's Answers National Sample

	True	False	Don't know
1. False. There is no evidence that Alzheimer's is contagious, but given the concern and confusion about AIDS, it is encouraging that nearly everyone knows this fact about Alzheimer's.	3%	83%	14%
2. False. Alzheimer's is associated with old age, but it is a disease and not the inevitable consequence of aging.	9	80	11
3. False. Alzheimer's is a disease of the brain, but it is not a form of insanity. The fact that most people understand the distinction contrasts with the results of public-opinion studies concerning epilepsy that were done 35 years ago. At that time, almost half of the public thought that epilepsy, another disease of the brain, was a form of insanity.	7	78	15
4. False. Again, most of the public knows that Alzheimer's is not an inevitable part of aging.	10	77	13
5. True. Despite announcements of "breakthroughs," biomedical research is in the early laboratory and experimental stages and there is no known cure for the disease.	75	8	17
6. True. Memory and cognitive decline are characteristic of the earlier stages of Alzheimer's disease, but physical decline follows in the later stages.	74	10	16
7. True. Most people know that this is the earliest sign of Alzheimer's disease.	62	19	19
8. False. Most people also know that while Alzheimer's produces memory loss, memory loss may have some other cause.	16	61	23
9. True. This question, like number 8, measures how well people recognize that other problems can mirror Alzheimer's symptoms. This is crucial because many of these other problems are treatable. In particular, depression can cause disorientation that looks like Alzheimer's.	49	20	30
10. False. Stuttering has never been linked to Alzheimer's. The question was designed to measure how willing people were to attribute virtually anything to a devastating disease.	12	46	42

	True	False	Don't know
11. False. Apart from age, research has not uncovered any reliable demographic or ethnic patterns. While there are more older women than men, both sexes are equally likely to get Alzheimer's.	15	45	40
12. True. Alzheimer's produces mental and physical decline that is eventually fatal, although the progression varies greatly among individuals.	40	33	27
13. False. The early and middle stages of the disease usually do not require institutional care. Only a small percentage of those with the disease live in nursing homes.	37	40	23
14. False. There is no evidence that using aluminum cooking utensils, pots or foil causes Alzheimer's, although aluminum compounds have been found in the brain tissue of many Alzheimer's patients. They may simply be side effects of the disease.	8	25	66
15. False. At present there is no definitive blood test that can determine with certainty that a patient has Alzheimer's disease. Accurate diagnosis is possible only upon autopsy. Recent studies suggest that genetic or blood testing may be able to identify Alzheimer's, but more research with humans is needed.	12	24	64
16. False. Medicare generally pays only for short-term nursing-home care subsequent to hospitalization and not for long-term care. Medicaid can pay for long-term nursing-home care, but since it is a state-directed program for the medically indigent, coverage for Alzheimer's patients depends upon state regulations and on the income of the patient and family.	16	23	61
17. True. As mentioned earlier, many medical problems have Alzheimer's-like symptoms and most of these other causes are treatable. Considering how much medicine older people take, it is unfortunate that so few people know that medications such as those used to treat high blood pressure can cause these symptoms.	20	19	61

cal change; and a politics of conflict between age groups."

Binstock, a professor at Case Western Reserve School of Medicine, pointed out that these inaccurate perceptions express an underlying ageism, "the attribution of these same characteristics and status to an artificially homogenized group labeled 'the aged.'"

Ironically, much ageism is based on compassion rather than ill will. To protect older workers from layoffs, for example, unions fought hard for job security based on seniority. To win it, they accepted mandatory retirement, a limitation that now penalizes older workers and deprives our society of their experience.

A few companies have taken special steps to utilize this valuable pool of older workers. The Travelers companies, for example, set up a job

GREAT EXPECTATIONS

SOURCE: U.S. NATIONAL CENTER FOR HEALTH STATISTICS

If you were born in 1920 and are a . . .

	. . .white man	*. .white woman*
your life expectancy was . . .		
at birth	*54.4 years*	*55.6 years*
at age 40	*71.7*	*77.1*
at age 62	*78.5*	*83.2*

If you were born in 1940 and are a . . .

	. . .white man	*. . .white woman*
your life expectancy was . . .		
at birth	*62.1 years*	*66.6 years*
at age 20	*70.3*	*76.3*
at age 42	*74.7*	*80.7*

If you were born in 1960 and are a . . .

	. . .white man	*. . .white woman*
your life expectancy was . . .		
at birth	*67.4 years*	*74.1 years*
at age 22	*73.2*	*80.0*

bank that is open to its own retired employees as well as those of other companies. According to Howard E. Johnson, a senior vice president, the company employs about 175 formerly retired men and women a week. He estimates that the program is saving Travelers $1 million a year in temporary-hire fees alone.

While mandatory retirement is only one example of ageism, it is particularly important because we usually think of contributions to society in economic terms. Malcolm H. Morrison, an authority on retirement and age discrimination in employment for the Social Security Administration, points out that once the idea of retirement at a certain fixed age was accepted, "the old became defined as a dependent group in society, a group whose members could not and should not work, and who needed economic and social assistance that the younger working population was obligated to provide."

We need to replace this stereotype with the more realistic understanding that older people are and should be productive members of society, capable of assuming greater responsibility for themselves and others. What researchers have learned about the strengths and abilities of older people should help us turn this ideal of an active, useful life after 65 into a working reality.

Erikson, In His Own Old Age, Expands His View of Life

In partnership with his wife, the psychoanalyst describes how wisdom of the elderly is born.

Daniel Goleman

In his ninth decade of life, Erik H. Erikson has expanded the psychological model of the life cycle that he put forward with his wife, Joan, almost 40 years ago.

Their original work profoundly changed psychology's view of human development. Now, breaking new ground, they have spelled out the way the lessons of each major stage of life can ripen into wisdom in old age. They depict an old age in which one has enough conviction in one's own completeness to ward off the despair that gradual physical disintegration can too easily bring.

"You've got to learn to accept the law of life, and face the fact that we disintegrate slowly," Mr. Erikson said.

On a recent afternoon, in a rare interview, they sat in their favorite nook in a bay window of Mrs. Erikson's study on the second floor of their Victorian house near Harvard Square in Cambridge, Mass. "The light is good here and it's cozy at night," Mrs. Erikson told a visitor.

Although Mr. Erikson has a comfortable study downstairs, and Mrs. Erikson, an artist and author in her own right, has a separate workroom, they prefer to spend their time together in this quiet corner, in the spirit of their lifelong collaboration.

Mr. Erikson, who never earned an academic degree (he is usually called Professor Erikson), deeply affected the study of psychology. Many believe that his widely read books made Freud pertinent to the struggles of adult life and shaped the way people today think about their own emotional growth. He gave psychology the term "identity crisis."

When Mr. Erikson came to this country in 1933 from Vienna, he spoke little English. Mrs. Erikson, a Canadian, has always lent her editorial hand to those writings of her husband on which she did not act as co-author.

As Mr. Erikson approaches 87 years of age and Mrs. Erikson 86, old age is one topic very much on their minds.

Their original chart of the life cycle was prepared in 1950 for a White House conference on childhood and youth. In it, each stage of life, from infancy and early childhood on, is associated with a specific psychological struggle that contributes to a major aspect of personality.

In infancy, for instance, the tension is between trust and mistrust; if an infant feels trusting, the result is a sense of hope.

In old age, according to the new addition to the stages, the struggle is between a sense of one's own integrity and a feeling of defeat, of despair about one's life in the phase of normal physical disintegration. The fruit of that struggle is wisdom.

"When we looked at the life cycle in our 40's, we looked to old people for wisdom," Mrs. Erikson said. "At 80, though, we look at other 80-year-olds to see who got wise and who not. Lots of old people don't get wise, but you don't get wise unless you age."

Originally, the Eriksons defined wisdom in the elderly as a more objective concern with life itself in the face of death. Now that they are at that stage of life, they have been developing a more detailed description of just what the lessons of each part of life lend to wisdom in old age. For each earlier stage of development they see a parallel development toward the end of life's journey.

For instance, the sense of trust that begins to develop from the infant's experience of a loving and supportive environment becomes, in old age, an appreciation of human interdependence, according to the Eriksons.

"Life doesn't make any sense without interdependence," Mrs. Erikson said. "We need each other and the sooner we learn that the better for us all."

The second stage of life, which begins in early childhood with learning control over one's own body, builds the sense of will on the one hand, or shame and doubt on the other. In old age, one's experience is almost a mirror image of what it was earlier as the body deteriorates and one needs to learn to accept it.

In "play age" or preschool children, what is being learned is a sense of initiative and purpose in life, as well as a sense of playfulness and creativity, the theory holds.

Two lessons for old age from that stage of life are empathy and resilience, as the Eriksons see it.

"The more you know yourself, the more patience you have for what you see in others," Mrs. Erikson said. "You don't have to accept what people do, but understand what leads them to do it. The stance this leads to is to forgive even though you still oppose."

The child's playfulness becomes, too, a sense of humor about life. "I can't imagine a wise old person who can't laugh," said Mr. Erikson. "The world is full of ridiculous dichotomies."

At school age, the Erikson's next stage, the child strives to become effective and industrious, and so develops a sense of competence; if he or she does not, the outcome is feelings of inferiority.

HUMILITY IN OLD AGE

In old age, as one's physical and sensory abilities wane, a lifelong sense of effectiveness is a critical resource. Reflections in old age on the course one's life has taken—especially comparing one's early hopes and dreams with the life one actually lived—foster humility. Thus, humility in old age is a realistic appreciation of one's limits and competencies.

The adolescent's struggle to overcome confusion and find a lifelong identity results in the capacity for commitment and fidelity, the Eriksons hold. Reflections in old age on the complexity of living go hand in hand with a new way of perceiving, one that merges sensory, logical and esthetic perception, they say. Too often, they say, people overemphasize logic and ignore other modes of knowing.

"If you leave out what your senses tell you, your thinking is not so good," Mrs. Erikson said.

In young adulthood, the conflict is between finding a balance between lasting intimacy and the need for isolation. At the last stage of life, this takes the form of coming to terms with love expressed and unexpressed during one's entire life; the understanding of the complexity of relationships is a facet of wisdom.

"You have to live intimacy out over many years, with all the complications of a long-range relationship, really to

The Completed Life Cycle

In the Eriksons' view, each stage of life is associated with a specific psychological conflict and a specific resolution. In a new amplification, lessons from each of the earlier stages mature into the many facets of wisdom in old age, shown in column at right.

Conflict and resolution	Culmination in old age
Old Age Integrity vs. despair: wisdom	Existential identity; a sense of integrity strong enough to withstand physical disintegration.
Adulthood Generativity vs. stagnation: care	Caritas, caring for others, and agape, empathy and concern.
Early Adulthood Intimacy vs. isolation: love	Sense of complexity of relationships; value of tenderness and loving freely.
Adolescence Identity vs. confusion: fidelity	Sense of complexity of life; merger of sensory, logical and aesthetic perception.
School Age Industry vs. inferiority: competence	Humility; acceptance of the course of one's life and unfulfilled hopes.
Play Age Initiative vs. guilt: purpose	Humor; empathy; resilience.
Early Childhood Autonomy vs. shame: will	Acceptance of the cycle of life, from integration to disintegration.
Infancy Basic trust vs. mistrust: hope	Appreciation of interdependence and relatedness.

understand it," Mrs. Erikson said. "Anyone can flirt around with many relationships, but commitment is crucial to intimacy. Loving better is what comes from understanding the complications of a long-term intimate bond."

She added: "You put such a stress on passion when you're young. You learn about the value of tenderness when you grow old. You also learn in late life not to hold, to give without hanging on; to love freely, in the sense of wanting nothing in return."

In the adult years, the psychological tension is between what the Eriksons call generativity and caring on the one hand and self-absorption and stagnation on the other. Generativity expresses itself, as Mrs. Erikson put it, in "taking care to pass on to the next generation what you've contributed to life."

Mr. Erikson sees a widespread failing in modern life.

"The only thing that can save us as a species is seeing how we're not thinking about future generations in the way we live," he said. "What's lacking is generativity, a generativity that will promote positive values in the lives of the next generation. Unfortunately, we set the example of greed, wanting a bigger and better everything, with no thought of what will make it a better world for our great-grandchildren. That's why we go on depleting the earth: we're not thinking of the next generations."

UNDERSTANDING GENERATIVITY

As an attribute of wisdom in old age, generativity has two faces. One is "caritas," a Latin word for charity, which the Eriksons take in the broad sense of caring for others. The other is "agape," a Greek word for love, which they define as a kind of empathy.

The final phase of life, in which integrity battles despair, culminates in a full wisdom to the degree each earlier phase of life has had a positive resolution, the Eriksons believe. If everything has gone well, one achieves a sense of integrity, a sense of completeness, of personal wholeness that is strong enough to offset the downward psychological pull of the inevitable physical disintegration.

Despair seems quite far from the Eriksons in their own lives. Both continue to exemplify what they described in the title of a 1986 book, "Vital Involvement in Old Age." Mr. Erikson is writing about, among other things, the sayings of Jesus. Mrs. Erikson's most recent book, "Wisdom and the Senses," sets out evidence that the liveliness of the senses throughout life, and the creativity and playfulness that this brings, is the keystone of wisdom in old age.

"The importance of the senses came to us in old age," said Mr. Erikson, who now wears a hearing aid and walks with a slow, measured dignity.

In her book, Mrs. Erikson argues that modern life allows too little time for the pleasures of the senses. She says: "We start to lose touch with the senses in school: we call play, which stimulates the senses and makes them acute, a waste of time or laziness. The schools relegate play to sports. We call that play, but it isn't; it's competitive, not in the spirit of a game."

The Eriksons contend that wisdom has little to do with formal learning. "What is real wisdom?" Mrs. Erikson asked. "It comes from life experience, well digested. It's not what comes from reading great books. When it comes to

understanding life, experiential learning is the only worthwhile kind; everything else is hearsay."

Mr. Erikson has been continuing a line of thought he set out in a Yale Review article in 1981 on the sayings of Jesus and their implications for the sense of "I," an argument that takes on the concept of the "ego" in Freudian thought.

"The trouble with the word 'ego' is its

The Eriksons contend that wisdom has little to do with formal learning.

---

technical connotations," Mr. Erikson said. "It has bothered me that 'ego' was used as the translation of the German word 'Ich.' That's wrong. Freud was referring to the simple sense of "I."

Another continuing concern for the Eriksons has been the ethics of survival, and what they see as the urgent need to overcome the human tendency to define other groups as an enemy, an outgrowth of the line of thinking Mr. Erikson began in his biography of Gandhi.

Mr. Erikson was trained in psychoanalysis in Vienna while Freud was still there, and worked closely with Freud's daughter Anna in exploring ways to apply psychoanalytic methods to children. That expertise made him welcome at

Harvard, where he had his first academic post.

There he began the expansion of Freud's thinking that was to make him world famous. By describing in his books "Childhood and Society" and "Identity and the Life Cycle" how psychological growth is shaped throughout life, not just during the formative early years that Freud focused on, Mr. Erikson made a quantium leap in Freudian thought.

Over the years since first coming to Harvard, Mr. Erikson has spent time at other universities and hospitals, including Yale in the late 1930's, the University of California at Berkeley in the 40's, the Austen Riggs Center in Stockbridge, Mass., in the 50's, and again at Harvard through the 60's. Until last year, the Eriksons lived in Marin County near San Francisco, but it is to Cambridge that they returned.

One lure was grandchildren nearby. Their son Kai, with two children, is a professor of sociology at Yale, and their daughter Sue, with one child, also lives nearby.

Informally, Mr. Erikson still continues to supervise therapists. "The students tell me it's the most powerful clinical supervision they've ever had," said Margaret Brenman-Gibson, a professor of psychology in the psychiatry department at Cambridge City Hospital, a part of Harvard Medical School.

In Cambridge, the Eriksons share a rambling three-story Victorian with three other people: a graduate student, a professor of comparative religion and a psychologist. The housemates often take meals together.

"Living communally," said Mrs. Erikson, "is an adventure at our age."

A VITAL LONG LIFE

NEW TREATMENTS FOR COMMON AGING AILMENTS

Medical science enables us to live to a healthy ripe old age —and enjoy it.

Evelyn B. Kelly

Evelyn B. Kelly is vice president of the Florida chapter of the American Medical Writers' Associaiton and a consultant on psychological and gerontological concerns.

Bill, seventy-seven, is a new model of older adult. He manages a vast network of athletic camps, flies to the World Series, and spends his spare time working with his church and the Gideons. He walks two miles a day and is very careful about his diet. Bill lives his life with zest and vigor and still contributes to society.

Bill is prototypical of the new elder culture predicted for the next cohort of older adults. According to Ken Dychtwald, a gerontological consultant, the next generation of older adults will be healthier, more mobile, better educated, and more politically astute than today's seniors. They will be part of a more powerful and energetic elder culture.

Like the old gray mare, aging is definitely not what it used to be. Time was when people believed they should eat, drink, and be merry—for tommorow, they'd retire to their rockers. Invariably, it was held, the passing years meant steady mental and physical decline until one died from "old age." While many still cling to these myths, researchers marvel at the human potential to extend physical and intellectual capacities in later life.

Many factors have contributed to this greatly improved forecast for the aging. Medical advances of the last decade have made it possible for present and future generations to live longer lives. Demonstrably, healthier life-styles have played a major role in extending life span. The effects of disease, abuse, or disuse should never be called "normal aging." Put simply, people do not die of old age, but of specific conditions, over which they may have some control.

The sheer numbers of older adults are overwhelming. Although the maximum life span has not increased, the number of persons sixty-five and older has increased from 4 percent of the population in 1900 to 12 percent in 1985. Even more important, the fastest-growing segment of the population is the

Physical therapy, including exercise and weight reduction, is an important way to manage arthritis.

group eighty-five and older.

In 1900, only 25 percent of deaths occurred in the group over sixty-five. Advances in preventing childhood diseases had pushed the number to almost 70 percent by 1980. Impressive gains have been made in treating and preventing the four leading causes of death in the sixty-five plus population: heart disease, strokes, cancer, and pneumonia.

But this bounty of years has brought mixed blessings. As our aged population continues to grow, survivors may encounter nonfatal disorders, such as arthritis, diabetes, and dementia. Specified in the blueprint for a model healthy old age are education and healthy life-style choices.

Heart disease and strokes: silent stalkers

A diagnosis of high blood pressure surprises most people, who find the knowledge that they are potential candidates for stroke or heart failure traumatic. About 75 percent of hypertensives are over forty, but their condition is seldom diagnosed until they are fifty or sixty. Associated with excess weight, and found more commonly in black people than white and men than women, hypertension has been called the silent stalker.

According to Dr. Harvey Simon of Harvard Medical School, hearts may wear down a little, but they probably do not wear out. He adds that disease and disuse, rather than age, account for most of the deterioration. The heart's ability to pump blood declines about 8 percent each decade after adulthood, and blood pressure increases as fatty deposits clog the arteries. By middle age, the opening of the coronary arteries is 29 percent narrower than in the twenties.

Life-style changes, drugs, or a combination can reduce blood pressure. With more than fifty blood-pressure drugs on the market, if one produces side effects, another can be prescribed. A number of drugs are still being tested. In August 1987, the U.S. Food and Drug Administration (FDA) approved the use of lovastatin, a cholesterol-reducing drug, but recommended it only for patients who do not respond to diet and exercise.

Heart attacks are caused by a blood clot in one of the coronary arteries. Most deaths caused by heart attacks occur within minutes or a few hours. Until recently, doctors could do little to treat an attack and prevent damage to the heart muscle. Two recent developments, however, mean good news for older adults. A new class of drugs called "clotbusters" may be injected into the veins to dissolve clots, and the FDA has recently approved streptokinase, a blood thinner, and another "clotbuster," TPA—tissue plasminogen activator.

Other weapons against heart attacks are balloon angioplasty—a procedure that opens clogged arteries with small balloonlike structures—and laser angioplasty, an experimental method where light vaporizes plaque deposits on the arteries.

Diseases of the cerebral blood vessels causing stroke rank high among the disabilities of the elderly. Reducing blood pressure can cut the risk of stroke in half. In stroke, the blood flow to part of the brain stops. The effect is similar to that of a heart attack; without blood, the tissue has no oxygen and dies.

Although about half of the elderly are hypertensive, physicians do not agree on their treatment. In a recent American Medical Association report, Richard Davidson and George Caranosos of the University of Florida announced that treating elderly hypertensives appears to be effective in stroke prevention. Older adults are not more susceptible to the side effects of antihypertensive drugs than other groups.

Risk factors leading to heart disease include heredity, diabetes, high-fat diets, high blood pressure, and smoking. Changes in life-style dramatically reduce the risk of these silent stalkers.

Cancer and older adults

There's no question about it: Cancer occurs more frequently with increasing age. The passing of years may lengthen exposure to carcinogens, alter host immunity or chro-

The next generation of older adults will be healthier. Aging may occur from disuse, not from disease. This farmer and his wife stay active. He is 100 and she is 95 years old.

The Fountains of Youth

In 1513, Ponce de Leon combed Florida in search of mythical waters that would prevent aging. Today, Dr. J. Michael McGinnis, who launched the Healthy Older People program, has found that up to 50 percent of all premature mortality can be related to life-style habits, and that older adults are willing and able to make life-style changes to improve their health. Even in late age, behavioral changes, especially in the realms of exercise and nutrition, can produce health benefits.

Every cell in our bodies is dependent on oxygen. We can go weeks without food and days without water, but only a few minutes without oxygen. The heart pumps the blood that carries oxygen throughout the body, so anything that improves circulation is beneficial. Exercise fulfills this important function—and more!

Exercise

The symptoms of "old age"—no energy, stiff joints, poor circulation—may really be the signs of inactivity. You don't stop exercising because you are becoming old; you become old because you stop exercising. The same conditions attributed to aging can be induced in young people who sit around doing nothing. Aging may occur not from dis*ease* but from dis*use*.

● The relationship between inactivity and circulatory problems is immutable. The less active you are,

the more chances you run of developing heart problems and high blood pressure.

● Exercise and the lowered risk of heart disease are connected with chemicals in blood cholesterol called lipoproteins. The so-called good lipoproteins—the HDLs (high-density lipoproteins)—are found in the blood of those on solid exercise programs. The bad lipoproteins—the LDLs (low-density lipoproteins)—outnumber the good in those who are inactive. Letter carriers—who walk all day—have fewer heart attacks than sedentary post office clerks.

● It is never too late to start exercising. Studies at the University of Toronto have shown that men and women over the age of sixty who began to exercise regularly were found to have achieved the fitness level of persons ten to twenty years younger in only seven weeks.

With exercise alone, cholesterol and triglyceride levels drop markedly. Appetite is suppressed and digestion improves. When people diet to lose weight, their metabolism slows down, and they reach a plateau of weight loss. With exercise, however, the rate of body metabolism increases, and body fat is absorbed. Because carbohydrate and complex-sugar metabolism is improved, diabetics who exercise can maintain lower insulin levels.

Nutrition

It sounds obvious, but a balanced diet is still the most important over-

all eating strategy. That means including something from each of the four food groups daily—grains; fruits and vegetables; meat, fish, and poultry; and dairy products. The emphasis today is on low-fat, high-fiber foods. By eating more fiber, you lower your cholesterol level and lessen the chances of getting heart disease.

The relationship between diet and many chronic conditions such as heart disease, stroke, hypertension, and cancer has been proven. However, Healthy Older People revealed that older adults are confused about what constitutes good nutrition. They seemed very knowledgeable about what *not* to eat, but were unable to describe the elements of a balanced diet.

A focal point of controversy in nutrition is the official U.S. recommended dietary allowances, or RDAs. One problem is that RDAs are established for only two adult groups—those under fifty and those over fifty. Certainly, the physiological differences between fifty-five- and eighty-five-year-olds reflect different dietary needs that must be addressed in education programs.

Also, nutrition and exercise act as natural tranquilizers that relieve tension and promote mental well-being. While exercise and nutrition can't reduce health risks to zero, through our behavior and life-style choices we can indeed create an inner fountain of youth.

mosomal linkage, or increase exposure to ontogenic viruses.

But the very group that needs screening is often disregarded because of preconceived notions about cancer in older adults. Some think older adults are too fragile to tolerate proper treatment. The American Cancer Society emphasizes that early detection is important at all ages.

In many parts of the body, such as the skin, the lungs, and digestive tract, the chances of cancer rise steadily with age. Prostate cancer

is even more directly connected to age. Some cancers, theoretically preventable, are caused by habits or environmental factors, such as smoking or exposure to industrial chemicals. The following chart shows the screening recommendations of the American Cancer Society.

Pneumonia: friend of the old?

Friend of the old: That's what Sir William Osler called pneumonia in 1912. He stated that a death from

pneumonia is short, relatively painless, and peaceful.

Today, few would call pnuemonia a friend: We have little admiration for the No. 1 infectious killer of older adults.

And what's more—admissions to nursing homes and hospitals increase the chance of encounter with this "enemy." The yearly incidence of pneumonia, 20 to 40 cases per thousand, sharply rises with institutionalization to about 250 cases per thousand.

Hospitals are associated with an

increased risk for lower respiratory infections. Nosocomial pneumonia (the name given to infections acquired in a hospital) is involved in about 15 percent of pneumonia cases; the attack rate increases with age. In addition, nosocomial pneumonias are stubborn and hard to treat.

Pneumonia in older adults is a diagnostic and therapeutic nightmare. In many cases, pneumonia may not be suspected because the common signs of respiratory disease are absent. The ill person suspected of having pneumonia is often treated with a "broad-spectrum" antibiotic to kill different kinds of bacteria. This shotgun approach may or may not work. Despite the availability of potent antibiotics, mortality due to pneumonia in the aged remains substantial and comes with a high price tag. Over $550 million is spent each year in hospital care alone for elderly adults with pneumonia.

Education for prevention is imperative. Despite a well-recognized association between outbreaks of influenza and deaths from pneumonia, only about 20 percent of high-risk persons are vaccinated against the flu each year. Lederle, a pharmaceutical company, is marketing a vaccine that protects against twenty-three types of pneumococcal bacteria.

Future pharmaceutical development will help in the prevention and treatment of pneumonia in older people. Possible breakthroughs include vaccines against more bacterial strains, more potent oral antibiotics, and methods to keep people from breathing in microbes.

Arthritis: common and still confusing

Bury your body in horse manure up to your neck. . . . Sit in an abandoned uranium mine. . . . Take cobra and krait venom. . . . With such remedies, people have desperately sought relief from the pain of arthritis. Neanderthal cave paintings drawn more than forty thousand years ago show stooped beings with bent knees. Although medical science has conquered many old and exotic diseases, this condition continues to plague over thirty-one million Americans. In older people, its two most common forms are rheumatoid arthritis and osteoarthritis.

Rheumatoid arthritis (RA) is the most serious, painful, and disabling form. Affecting three times as many women as men, RA first affects the linings (synovial membranes) of small joints—such as those in the hands and feet—then moves to tissues and organs, such as the heart and lungs. Osteoarthritis (OA) is often called the "wear and tear" disease and is by far the commonest form of arthritis. The condition begins with the thinning, wearing down, or roughening of cartilage, which may lead to chem-

Cancer Screening Tests

Procedure	Sex	Age	Frequency
Sigmoidoscopy	M&F	over 50	every 3–5 years
Stool guaiac slide test	M&F	over 50	every year
Digital rectal examination	M&F	over 40	every year
Pap test	F	20–65	at least every 3 years
Pelvic examination	F	20–40 over 40	every 3 years every year
Endometrial tissue sample	F	at menopause	at menopause
Breast self-examination	F	over 20	every month
Breast exam by physician	F	20–40 over 40	every 3 years every year
Mammography	F	35–40 40–50 over 50	initial test 1–2 years every year
Health counseling and cancer checkup	M&F M&F	20–40 over 40	every 3 years every year

(Recommended by the American Cancer Society)

ical changes that cause the joint to become inflamed. Usually seen in older people, the condition can develop in anyone whose joints have taken a lot of punishment: the obese, those injured in accidents, or those subject to unusual stress in work or sports.

According to Dr. Paul Nickerson of the Geriatric Diagnosis Clinic in Cleveland, Ohio, "The association between age and OA is striking, but it should not be assumed that OA is caused by normal aging." He quickly adds that despite our ignorance of the etiology of arthritis, there is still much to offer the patient.

Dr. Nickerson outlines three areas of management: (1) physical therapy, including exercise and weight reduction; (2) medication with nonsteroid antiinflammation drugs (NSAID), which include aspirin and ibuprophen; and (3) surgery for those with severe pain. Advanced age is sometimes a positive factor in joint replacement, because highly active young people run a greater risk of loosening new joints.

Dr. Nickerson foresees research leading to more and safer NSAIDs and, ultimately, a drug that will inhibit cartilage-damaging enzymes.

In the next generation, research will bring relief from pain and help people learn to live with intractable discomfort. Pain clinics are using new techniques, such as nerve stimulation and biofeedback. Until arthritis is fully understood and conquered, the key to treatment lies in changing attitudes and developing a satisfying life-style despite arthritis.

Diabetes: dare to discipline

Not since the discovery of insulin in 1921 by Banting and Best has there been so much good news about diabetes for older Americans. Recent breakthroughs have led to better understanding of this maligned disease—and diabetes deserves its reputation. Heart attacks, blindness, limb loss, and kidney failure are only a few of its complications.

Diabetes mellitus is a disorder in which the body fails to convert the food we eat into the energy we need. Type I (insulin-dependent diabetes mellitus (IDDM)) may appear early in life and results from a failure of the pancreas to produce insulin. Type II (noninsulin-dependent diabetes mellitus (NIDDM)), found in 90 percent of diabetics, is more common with age. Some researchers believe diabetes is directly related to excess weight, because reducing body fat appears to make body cells more receptive to insulin. By daring to discipline themselves, type II diabetics can usually keep blood glucose levels normal by controlling weight, exercising, and taking oral medication.

Recent research at the National Institute on Aging revealed that the body's ability to handle glucose decreases with age. The finding has led to a revision of the official guidelines for diagnosing diabetes. As a result, fewer elderly people are at risk of being considered diabetic. The new guidelines mean that many persons once considered borderline have been freed from the emotional burden and consequent problems of a chronic degenerative disease.

Fear of the dementing Alzheimer's disease

Publicity about Alzheimer's disease has bred a widespread, morbid fear of the condition. Believing that memory decline is inevitable, middle-aged and older adults interpret any slip of memory as symptomatic of decline.

Minor forgetfulness is normal at any age. The key difference is that in Alzheimer's, dementia worsens until those affected cannot function normally. Fear of dementia will not prevent the condition: Remember, eighty percent of older people remain alert and active.

Alzheimer's disease is a real heartbreaker for the 2.5 million men and women plagued by it, and their families. Although no cure has been discovered, much can be done to make life more bearable for both patients and families.

Alzheimer's disease was discovered by German physician Alois Alzheimer in 1906. While symptoms may vary, patients can go from severe forgetfulness to personality changes with loss of all verbal skills and physical control. The course of the disease may run from less than three years to fifteen years or more. Death usually results from malnutrition or infection.

As some conditions produce symptoms similar to Alzheimer's, the disease is usually diagnosed by exclusion. For example, depression is closely correlated with memory loss. If the depression is successfully treated, the person may regain memory and dignity. Other conditions with similar symptoms include: malnutrition, stroke, drug reaction, metabolic change, or head injury.

An autopsy of the brain of an Alzheimer's patient reveals an abnormal disarray of litter. Abnormal neuron masses called "neurofibrillary tangles" appear in the outer cortex of the brain. Plaques of scarlike structures mark deteriorated nerve endings.

Some experts are looking for a genetic connection. Abnormal genes have been identified in most dementias, such as Huntington's chorea and Down's syndrome. Scientists thought they might be close to a cause when a protein amyloid present in the brains of Alzheimer's patients seemed to be leading them to the causative gene. However, the Winter 1987 issue of the Alzheimer's Disease and Related Disorders Association newsletters reported that Alzheimer's is probably not due to the replication of the gene that produces the amyloid. Another recent disappointment was the suspension of a test on tetrahydroaminoacridine (THA)—a drug that had alleviated some of the symptoms of memory loss. THA was found to cause liver damage in a significant number of patients. Scientists still look with hope for similar drugs to treat this disease associated with aging.

Adults of this and future generations can look forward to greater medical miracles and increased life expectancy. More knowledge and improved life-style choices will result in a new norm: living to a healthy old age. The nineteenth-century poet Robert Browning could have been addressing us today, when he wrote, "Grow old along with me! The best is yet to be."

Aging

What Happens to the Body as We Grow Older?

Given our medical ignorance and the fact that the body does not age all at once—we can have a young kidney and an old heart—the whole concept of aging needs careful re-examination.

Is it possible to describe "typical" aging? Not really. We talk about someone who has begun to be forgetful, whose skin shows signs of losing elasticity, whose lung capacity is diminished because of emphysema, whose cardiac reserve is diminished because of atrophy of the heart muscle, whose organ functions (for instance, kidney or liver function) are at a fraction of what they once were, whose skeletal structure is softened, whose hair is grey, whose eyes are clouded by cataracts and whose hearing is diminished—that is a caricature of a typical old person.

This stereotypical image of aging, however, does not hold true among all individuals. The debate continues over how much of elderly appearance is the result of natural aging and how much is the result of abuse of the body. According to William Kannel, one of the principal investigators of the Framingham Heart Study, "The issue of what constitutes aging and normal aging is an enigma that has never been satisfactorily solved." In epidemiology, there are several approaches that try to answer the question.

One approach is employed by the Veterans Administration Normative Aging Study, which seeks to find people who might eventually develop certain diseases but, at the beginning of the study are free of any ailment. By studying what happens to people over time, project director Pantel Vokonas and co-workers are trying to identify the effects of age. The researchers are looking at the signs of growing older that can be attributed to age rather than illness or disease.

The normative aging approach is, in a sense, a quest for immortality. The assumption is that if we could remove all these diseases, people would live, if not forever, at least for much longer than they now do. Unfortunately, it is virtually impossible to find people totally free of the disabilities sooner or later associated with old age. Even if a person seems to be free of heart or kidney damage, for example, there is no way of being sure that these organs are still in their pristine state.

Another approach to aging is that taken by the Framingham Heart Disease Epidemiology Study, underway since 1949, in which a whole population is followed as they age to see what problems they encounter. "Ours is a more pragmatic approach," said Kannel. "We are interested in seeing what kinds of things cause people who reach advanced age to no longer have much joy in living. We don't care whether it is cardiovascular disease, opaqueness of the lens, poor hearing, soft bones, arthritis, strokes, mental deterioration or normal aging. We're studying the ailments that afflict an aging population and take the joy out of reaching a venerable stage in life."

According to Kannel, "The reward for reaching a venerable stage of life is too often a cardiovascular catastrophe." Cardiac function, muscle, skeleton and so on all decline with time, although recent evidence suggests that cardiac function in a non-diseased heart remains amazingly stable well into old age. Some of this decline must be due to wear and tear, but according to Kannel, "it is just too difficult to dissociate from the long-term effects of noxious influences."

From *Bostonia*, February/March 1986, pp. 17-20. Reprinted with the permission of Bostonia Magazine 1986.

With respect to cardiac function, for example, decline is not necessarily unpreventable. It has been shown that 65-year-olds can be trained to improve their levels of performance. It is easy, for example to train somebody to restore his or her exercise capacity and measure cardiac function and oxygen utilization. We have the technology to measure these things. But how do you train a kidney? Moreover, it is still not clear what noxious influences cause the decline in function in most of the organ systems. If we did know, we would be able to remove the noxious influences and watch the recovery. "For many of the organ declines, we really have poor information," said Kannel. "It just so happens that for cardiovascular disease, we have a good body of data on what risk factors there are. And it turns out that many of these are modifiable."

Given our medical ignorance and the fact that the body does not age all at once—we can have a young kidney and an old heart—the whole concept of aging needs careful re-examination. The assumption that all the organs fail in concert is not borne out by experience. There are many people who are alert and showing few signs of diminished intellectual capacity, but have a failing heart or damaged liver. People have different rates of organ decline.

The Framingham Heart Study, however, is showing that many of the risk factors for the young are still operative in the elderly. Even though it may take decades for the disastrous effects of a habit like smoking to show up, there is still good reason to quit. One might think that once a lifetime of smoking has put someone on the track for cancer, eliminating smoking in advanced years will not remove that risk of cancer. The risk does remain, but there is no good reason to multiply the risk by continuing to smoke.

But beyond that, there are other good reasons for elderly people to give up smoking. Smoking contributes to chronic bronchitis, emphysema and may precipitate coronary attacks, peripheral vascular disease and perhaps even stroke. Quitting smoking will not bring the person back to total normal function; but it helps slow deterioration. With coronary disease, in particular, the advantage seems to occur regardless of how long one has smoked. According to Kannel, the data show an immediate 50 percent reduction in the risk of coronary disease whether

> "The issue of what constitutes aging and normal aging is an enigma that has never been satisfactorily solved."
>
> WILLIAM KANNEL

the person has smoked 10 years, 20 years, 30 or 40 years. In terms of coronary attacks, there is trouble showing the benefit of quitting. But for coronary *deaths* and peripheral vascular disease, there is no difficulty at all showing that quitting helps.

"I think that the elderly are becoming increasingly health conscious," said Kannel. "It's curious: One would think that young people, who have so much life ahead of them, would take things more seriously. But the elderly, they are the ones who are driving more carefully, avoiding doing stupid and reckless things because they more acutely feel the approach of the grim reaper."

Some of the other effects commonly attributed to aging may well be preventable. Data from the Framingham studies show that a great deal of the high-frequency hearing loss in the elderly *male* can be laid at the door of noisy industries. We can predict, from the popularity of loud music among the young today, a generation of deaf elderly in 50 or 60 years. Osteoporosis, too, could be reduced if people were kept more active and ate more calcium-rich foods. In other words, the conclusions drawn from the Framingham Heart Study are that prevention is possible, that it must be started early and that it takes sustained effort. The burden of these common and disabling conditions, whether or not we term them aging phenomena, are the sources of a great deal of discontent in the elderly.

There is also a strong genetic component in the process of aging. "People with superb genes are able to withstand a lifetime of abuse because they may be better able to cope with an overload of fat in the diet, too many calories, too much salt, too much trauma, too little exercise, smoking. If they have been blessed with superb metabolic machinery, they somehow survive. Other people, with inferior metabolic machinery, may avoid all these risks and live longer than the great risk taker. There is a lot to be said for genetics."

"Even with bad genes, one can do something effective to reduce the liability," stated Kannel.

There is also a mistaken notion, he continued, that following a healthy lifestyle entails considerable sacrifice. "Diet, exercise and the like need not be so austere that they are painful. We are only recommending a Mediterranean or Asian diet. If you follow the specifics of those, you get the fat content, lower cholesterol, lower calories that you need. That is hardly a gastronomic nightmare. These are good foods. Ham and eggs need not be the epitome of gastronomic experience. One can eat very well following a prudent diet, as recommended by the American Heart Association."

"Exercise is something we need to build back into daily living," added Kannel. "We have taken it out by all the modern conveniences. A better way of living is to exercise naturally without the contriving. If you can walk to work instead of driving and parking right next to the door, you are better off. Try not to use escalators in two-story buildings. We have engineered exercise out of our lives; the time has come to engineer it back."

The Effects of Aging

It is difficult to measure the rate of aging. One study (Hodgson and Buskirk) has shown that the maximal oxygen intake declined after age 25 at the rate of 0.40 to 0.45 ml of oxygen taken in per minute per kilogram of body weight each year. Grip strength went down about 0.20 kg per year. The investigators also found, however, that training at age 60 could improve the maximal oxygen intake by about 12 percent.

Average decline in human male, from age 30 to age 75

Factor	% Decline
Brain weight	44
Number of axons in spinal nerve	37
Velocity of nerve impulse	10
Number of taste buds	64
Blood supply to brain	20
Output of heart at rest	30
Number of glomeruli in kidney	44
Vital capacity of lungs	44
Maximum oxygen uptake	60

*There is controversy about how much of the decline reported is a result of aging as opposed to disease.

Of all the biological changes associated with growing old, young people are probably most acutely aware of the cosmetic changes. Hair greys, wrinkles become pronounced and shoulders tend to narrow with advancing age.

In American society, these changes are greeted with less than enthusiasm because, in a society that seems to cherish youth, these changes make one look old.

Cosmetic changes affect the sexes differently. Women may be outraged or humiliated by physical changes that lead to what Susan Sontag has called the process of "sexual disqualification." Women may be forced into roles of helplessness, passivity, compliance and non-competitiveness. Men, on the other hand, may

enjoy the assertiveness, competency, self-control, independence and power—all signs of "masculinity"—that come with age.

What creates the common cosmetic changes in aging? Wrinkles begin below the skin's outer layer (epidermis) when the dermis, a layer of tissue filled with glands, nerve endings and blood vessels, begins to shrink. At the same time, the dermis begins to atrophy, changes in the fat, muscle and bone create the deep wrinkles. Other factors—exposure to the sun, environmental toxins, heredity and disease—also affect the wrinkling of the skin.

Greying of the hair is the result of progressive loss of pigment in the cells that

give hair its color. Age spots are caused by the accumulation of pigment in the skin. But the shortened stature and flabby muscles are the result of lack of exercise and other behavioral factors. These, and many other so-called effects of aging, may be reversed.

After a while, time does take its toll. Between the ages of 35 and 80, the maximum work a person can do goes down by 60 percent. The strength of the grasp by the dominant hand (right in right-handers) goes down 50 percent and the endurance to maintain the strongest grasp goes down 30 percent. For some reason, not understood, the other (subordinate) hand, which was weaker to begin with, does not lose as much strength and

---

Alzheimer's Disease: The Search for a Cure Continues

It is difficult to paint anything but a bleak picture of Alzheimer's disease.

Named after the German neurologist Alois Alzheimer (1864–1915), it is a relentless and irreversible form of dementia that has been known to strike adults as young as 25, but most often appears in people over 70. For the estimated one to three million afflicted Americans, the early symptoms often involve memory loss, apathy and difficulties with spatial orientation and judgment. As these problems worsen, victims become increasingly depressed, confused, restless and unable to care for themselves.

In the final stages, Alzheimer patients may become so helpless that they are bedridden until they die from secondary problems such as pneumonia caused by accidentally inhaling food. Unfortunately, the victims of Alzheimer's often include the patient's stressed family and caregivers who, despite their devoted efforts, must watch loved ones turn into unmanageable strangers.

At this juncture, the exact cause, diagnosis and treatment of Alzheimer's continues to elude medical researchers. An abundance of new clues and insights into this mysterious disease is being uncovered, however. In addition, growing public awareness has led to the development of support groups and experi-

mental programs to help families and caregivers cope with caring for Alzheimer patients. "I'm amazed that researchers have progressed so far in such a short time," noted Professor Marott Sinex, from the School of Medicine's Department of Biochemistry, who has been researching the disease for the past 12 years.

Sinex pointed to many advances in the past decade, including improvements in drug therapy that ease symptoms. There is an increased understanding of possible genetic causes, of the subtle differences in neurotransmitters in the brains of victims and how they change over time and of anatomical changes in the brain that relate to memory loss and other cognitive problems. In addition, improved medical technology—such as the PET scan, which allows scientists to visualize the living brain—is adding new insights. "We now have a dynamic anatomy of the disease," said Sinex. "We can visualize its progression, which we couldn't do before. In other words we're dealing with a real, three-dimensional problem now, whereas before we only had two-dimensional understanding."

Although the exact cause of Alzheimer's is not known, Sinex noted that researchers now know it is associated with an excess of genetic material on a particular chromosome—an abnor-

mality that, interestingly, has already been linked to Down's syndrome. (Several Boston University researchers are independently investigating biochemical, genetic and physical similarities seen in Down's syndrome—a congenital disease whose victims are born moderately to severely retarded with distinctive physical traits—and Alzheimer's disease.) Scientists also know the disease can be inheritable and that—despite the fact that it can strike relatively young adults—it is "strongly age-dependent," with most cases not appearing until people are in their 70s. In addition to these factors, researchers are looking at other possible causes; for example, Sinex is investigating the possiblity that a virus may be involved.

As for diagnosis, the best physical evidence researchers have is the "plaques and tangles" —filamentous material whose nature and origin scientists are not exactly sure of—found in the brains of autopsied victims. (Although similar plaques and tangles are found in the brains of normally aging people, in Alzheimer victims the structures appear more frequently and in specific areas.) Sinex pointed out, however, that today "a really good clinic" can accurately diagnose the disease about 85 percent of the time by eliminating other problems such as stroke, a tumor,

drug poisoning or unrelated depression. Fortunately, research is advancing in this area as well. For example, Mark Moss, an assistant research professor in the School of Medicine's Department of Anatomy, recently developed a "gamelike" diagnostic test that the National Institute on Aging has recommended for clinical use and that "will help us understand the brain structures responsible for memory impairment."

The best method of treating Alzheimer victims today, explained Sinex, involves prescribing medications that "fall in the general category of anti-depressants." Apart from drug therapy, Sinex pointed out that organizations such as the Eastern Massachusetts Chapter of the Alzheimer's Disease and Related Disorders Association, of which he is president, can be helpful to both victims and their caregivers by providing information and connecting people with support groups.

Given this variety of recent advances, Sinex concluded optimistically, "It's simply a lot less traumatic to have Alzheimer's now than it was 10 or 15 years ago."

JON QUEIJO

endurance. The speed of nerve conduction is slower. The volume of blood pumped throughout the body goes down 50 percent. The maximum volume of air a person can inhale goes down 50 percent and oxygen diffuses from the lungs to the red cells of the blood 30 percent more slowly. Blood flow to the kidneys at age 80 is considerably less than that of age 20 in many, but not all. And, by age 70, the bones of the coccyx (the "tail bone" at the base of the spine) fuse. In short, our bodies slow down and stiffen as we age. This is natural and occurs even in the absence of disease.

Vision, hearing, taste, smell and touch have all been reported to change with age. New research suggests, however, that the effects of aging per se may not be as major a factor as originally believed on declining senses.

Probably the most familiar changes are those in seeing. Presbyopia (presby = old + opia = vision) is a sign of the gradual inability of the lens of the eye to focus on near objects—hence, the growing need for "reading glasses" or bifocals as people age. There are other relatively harmless changes, as well. Almost from birth, the lens of the eye begins to get more rigid. By around age 45, printed pages must be held at arm's length or farther to get them into focus. But, of course, at that distance, the letters are usually too small to read. As people age, their eyes may become more sensitive to glare and bright lights. They

> Smoking contributes to emphysema and may precipitate coronary attacks. Quitting smoking will not bring the person back to absolutely normal function; but it helps slow the process down.

may also be less able to discriminate between gradations of color.

More serious eye conditions increase with age as well. Approximately seven percent of all people between the ages of 65 and 74 have serious visual deficits. After age 75, the proportion more than doubles to 16 percent. Approximately two-thirds of all severe visual impairments occur in people 65 or older.

Macular degeneration is the most serious cause of low vision in the elderly in the United States. The macula is a spot on the retina of the eye needed for very fine focusing. With age, this region can degenerate and become obstructed with fine blood vessels.

Other visual impairments include: glaucoma, which is a dangerous and

painful increase of pressure within the eye; diabetic retinopathy, which is a destruction of the fine blood vessels in the eye, destroying parts of the retina, associated with uncontrolled diabetes; and cataracts, which are cloudy eye lenses.

Just as presbyopia is a vision deficit associated with advancing age, so presbyacusis (presby = old + acousis = sound) is a progressive hearing loss associated with aging. This is especially true for the higher frequencies. Of the estimated 14.2 million Americans with measurable hearing loss, about 60 percent of those with the most severe hearing problems are older than 65.

Hearing loss associated with aging may be of several causes. Genetic factors, infection and a lifetime of noise certainly contribute. Poor personal hygiene, build-up of ear wax, may also reduce hearing. Certain medications (for example, drugs of the streptomycin group) can injure the hair cells in the ear and interfere with both hearing and the sense of balance.

While these physical and sensory changes occur in all people as they age, keep in mind that the degree to which they affect individuals varies greatly. The key is how we take care of ourselves. The pleas from the medical profession to cut down smoking, drinking and to exercise are grounded in heavy evidence. Keeping yourself active and healthy throughout your life can result in an old age that is productive and rewarding.

Osteoporosis
The Stooping Disease

An estimated five million people in the United States are afflicted with osteoporosis—a disease marked by loss of bone mass that weakens the skeleton and may result in spontaneous fractures. Although everyone loses bone mass as they age, this process is accelerated in women. In the first 10 years after menopause, women lose bone at twice the rate men do.

Healthy bone is constantly changing. In childhood and adolescence, bones form at a faster rate than they are reabsorbed. In healthy adults, up until age 40, peak bone mass is maintained through a balance between the processes of bone formation and loss. Peak bone mass, which usu-

ally occurs around age 30, is influenced by hormones, calcium intake, level of physical activity and the stress of weight bearing. Heredity also plays a role. From about age 40, bone absorption is more rapid than bone formation.

There is no way to diagnose the early stages of osteoporosis. The bone loss that characterizes the disease does not show up on x-rays until a substantial portion of the bone is already lost. At its later stages, however, osteoporosis produces extreme visible changes: loss of height, rounding of the upper back ("dowager's hump"), forward thrust of the head, protruding abdomen and expansion of the chest. These symptoms are all due to

the collapsed vertebrae weakened by osteoporosis.

What causes osteoporosis is not yet understood. Because post-menopausal women are at elevated risk, some investigators have suggested that it involves estrogen deficiency and should be treated with hormones. This cannot be the case, however, because not all post-menopausal women develop osteoporosis.

At highest risk are white women with a family history of osteoporosis, of northern European descent, with small bone frames and of normal or less than average weight. Certain dietary and behavioral factors can increase the risk: drinking more than four to six cups of coffee a day, smoking, heavy use of alcohol and lack of calcium in the diet all increase the risk. On the other hand, physical exercise

decreases the risk.

University Hospital is currently planning to open a clinic at the end of February to treat osteoporosis victims. The clinic will be multidisciplinary involving orthopedic surgery, endocrinology, nutrition and internal medicine. All patients referred to the clinic will be prescreened by a special x-ray test which will determine bone mineral content. A specialized blood work-up will be done to rule out metabolic causes for metabolic bone disease and other appropriate studies as indicated by each case. Also a program of functional bracing will be started. Questions concerning the clinic should be directed to the Department of Orthopedic Surgery, University Hospital, (617) 638-8905.

Index

abortion, and genetic testing, 13, 15, 16, 17, 18

adolescents: aggression in, 87; depression in, 83–86; materialism of, 167; memory development in, 165; and puberty, 172–176; social adjustment of, 37, 160, 170; suicide of, 190, 194, 198; as parents, 152, 159

adulthood, ages and stages of, 182–189; see also, aging, men, parents, women

affective disorder, depression as, 83–86

Africa, as source of human evolution, 6, 7, 9, 11

ageism, 228, 229

aging: 222–229; and health, 233–237, 238–241; and intelligence, 39; and memory, 210, 211–212; and suicide, 198; and wisdom, 230–232

aggression: 79–82; in day-care settings, 71–72; levels of, in men and women, 217, 219, 220; in symbolic play, 98–99

allergies, and shyness in children, 61

alphabetic mapping, and dyslexia, 118

alpha-feto-protein, and genetic testing, 15

alternate lifestyles, fathering in, 141–142

altruism, development of, and fathering, 141, 142

Alzheimer's disease, 14, 111, 210, 225, 227, 228, 237, 240

amniocentesis, 12, 15, 16, 18

animism, in children, 101, 102, 103

anorexia, 84, 85, 192

antibiotics, and infectious disease, 127

aphasia, 201

aprosodia, 201, 202

arthritis, and the elderly, 236–237

Asia, as source of human evolution, 10–11

associative learning, and dyslexia, 119

attention-deficit hypothesis, of dyslexia, 119

autonomy vs. shame and doubt, Erikson's theory of, 184, 230, 231

babies, see infants

baby boomers, 188–189

"baby buster" generation, 189

bad-tempered aggression, 81

balance-sheet suicide, 198

BEAM, 122

benign neglect, as cause of alienation, 159

blacks, and hidden obstacles to success, 152–156

bonding, of infants and parents, 44

brain, development of newborn's, 36–39; and emotions, 200–203; and memory, 210–213; new models for working of, 111–114

brain electricity activity mapping, see BEAM

Brazelton, T. Berry, 43, 44, 139

bulimia, 84, 85

cancer: in the elderly, 234–235, 236; and suicide, 198; treatment of, and life expectancy, 128–129

cardiovascular illness, see heart disease

cerebral cortex, development of newborn's, 36, 42–43

CF, see cystic fibrosis

child care, see day care

children: aggression in, 71–72, 82; alienation in, 157–161; changing views toward, 130, 132–137; and compliance,

discipline, and control, 64–70, 71–75; depression in, 83–86; development of, and socioeconomic background, 45; effect on, of day care, 50–53, 148–151; importance of fathers to, 138–143; importance of physical attractiveness of, 54–57; mortality of, 126; and Piaget's theory of cognitive development, 101–104; and play, 92–100; shyness in, 60–63; see also, adolescents, infants, newborns

chorionic villus sampling, 12, 15, 16

chromosomes: disorders of, and inherited genetic diseases, 12, 13, 14, 17; and gene mapping, 6–11

cohorts, 185, 188, 233

college students: changes in, 182; materialism of, 167

compliance, discipline, and control, 64–70, 71–75

componential theory of human intelligence, 106, 107, 108

conception, science of, 19–23

concrete operational stage, Piaget's theory of, 101, 102, 103, 164

contextual theory of human intelligence, 107, 108, 109

control, compliance, and discipline, 64–70, 71–75

coping strategies, for jealousy and envy, 181

counseling, genetic, 15, 16

creativity, and memory development, 165, 166, 170, 231

critical periods, for brain development, 36–39

cross-modal transfer, and dyslexia, 119

crystallized intelligence, 39

cultural differences: and aggression, 79–80; and bias against suicide, 198–199; and children, 133; as obstacle to black achievement, 153–154; and physical attractiveness, 56, 57

CVS, see chorionic villus sampling

cystic fibrosis, and genetic testing, 12, 13, 14, 17, 18

day care: and child development, 45, 55; dilemma of, 50–53, 148–151; peer aggression in, 71–72

deoxyribonucleic acid, see DNA

depression: childhood, 83–86, 145; and loneliness, 216; and suicide, 191, 192, 195, 200, 201

diabetes, and the elderly, 237

disability rights activists, and genetic testing, 17

disarray, and newborn's cortical development, 42–43

discipline: compliance, and control, 64–70, 71–75; after divorce, 146

divorce: effect of, on children, 133, 144–147, 159; and fathering, 141–142

DNA: and brain development, 38; and gene mapping, 6–11; and genetic testing, 12, 13, 14; mitochondrial, 6–11

Down syndrome: and Alzheimer's disease, 240; genetic testing for, 15, 16, 17, 18

Duchenne muscular dystrophy, and genetic testing, 12, 14, 17, 18

Durkheim's theory, of suicide, 190

dyslexia, 115–123

education: of children, 134, 137; fetal, 32–35; of infants, 40, 44; remedial, for dyslexia, 122–123

egg, human: fertilization of, 8; and in vitro fertilization, 20; and twins, 25, 26

eidetic memory, 38

emotion: and adolescence, 168–169, 174, 175; and the brain, 113, 200–203; of jealousy and envy, 177–181; and memory, 212, 213; and sex differences, 218

empathy, development of, 77, 141, 142, 231

employers, as day-care sponsors, 51–52, 53

envy, 177–181

enzymes: and DNA, 14; restriction, 9; testing of, for cystic fibrosis, 13

Erikson, Erik, and theory of adult development, 183, 184, 230–232

ethics: and genetic testing, 12, 17; and in vitro fertilization, 18–23

evolution: human, and gene mapping, 6–11; of jealousy and envy, 177

expectancy communications, 154–156

experience: and brain development, 38–39; and intelligence, 107, 109

experiential theory of human intelligence, 107, 108

family: and day care, 148–151; and longevity, 126–131; and resilient children, 88–89

family day care, 50, 51, 150, 151

fathers: importance of, 44–45, 138–143; newborn's preference for voice of, 33–34

female transmission of mitochondrial DNA, 6–11

feminists, attitudes of: toward day care, 51; toward genetic testing, 12, 13, 17

feminization of poverty, 149

fertilization, in vitro, 19–23

fetuses: genetic testing of, 12; hearing of, 41; prenatal education of, 32–35

forgetting, and memory, 210–213

formal operational stage, Piaget's theory of, 101

fraternal twins, 25, 26

functional age, 222

games, and play, 94, 95, 98, 100

gender: babies' ability to distinguish, 42; and depression, 85; see also, sex

generativity vs. self-absorption, Erikson's theory of, 184, 186, 231–232

genes: for Alzheimer's disease, 237; and brain development, 38, 42; for depression, 84–85, 86; for dyslexia, 121, 122; as explanation for black intellectual inferiority, 154, 155; mapping and testing of, 6–11; for shyness, 60–63; and twins, 24, 26, 27, 28, 29

good and evil, battle between, and play, 99, 100

half-identical twins, 25, 26

handicaps, and genetic testing, 12, 13, 17, 18

hand preference, and dyslexia, 115

Harlow, Harry and Margaret, 37, 61, 62

health care, and life expectancy, 126–131
hearing: fetal development of, 34; in newborns and children, 37, 40, 41
heart disease: and the elderly, 234; and life expectancy, 128, 129
helpfulness: development of, and fathering, 141; required, 88, 90
helplessness, sense of: of adolescent's parents, 175; and life expectancy, 130; and loneliness, 215; and suicide, 192
heredity: and depression, 86; and shyness, 60–63
home-based day care, 52, 150
human sensory ability, development of, 34, 35, 38
humility, Erikson's views on aging and, 231
Huntington's chorea, and genetic testing, 12, 14, 17, 18

identical twins, 24, 25, 26, 27
identity crisis, 230
identity vs. identity confusion, Erikson's stage of, 184, 231
immunization, and life expectancy, 128, 129
industry vs. inferiority, Erikson's stage of, 184, 231
infants: competent, 134–135; in day care, 50–53; importance of fathers to, 138–143; importance of physical attractiveness of, 54–57; intelligence and learning ability of, 40–45; mortality of, 126, 129, 152; shyness in, 60–63; talking to, 46–49; see also, newborns
inferiority, myth of black intellectual, 154
infertility, treatment for, 19–23
initiative vs. guilt, Erikson's stage of, 184, 231
insight, and intelligence, 108, 109
integrity vs. despair and disgust, Erikson's theory of, 184, 230, 231
intellectual work ethic, 156
intelligence: crystallized, 39; development of, in children, 133, 134, 135, 136, 137, 139–140; obstacles to blacks fulfilling potential for, 152–156; triarchic theory of, 105–110
intervention programs: for children, 150; for parents, 142
intimacy vs. isolation, Erikson's theory of, 184, 231
in vitro fertilization, 19–23
IQ tests: and blacks, 154; and intelligence, 105, 107, 109, 135, 136, 150
IVF, see in vitro fertilization

jealousy, 177–181

Kagan, Jerome, and research on shyness, 60–61, 62, 63
kindergarten, 93–94

language: child's development of, 46–49, 103; deficits and dyslexia, 115–123; fetal and infant sensitivity to, 33–34; private, of twins, 26; see also, speech
latchkey children, 149
leisure: children's lack of, and play, 94; use of, by adolescents, 168
Levinger's love scale, 206, 207, 208, 209
Levinson's ladder, 183, 184, 185
life expectancy, and family ties, 126–131
limbic system, 85, 86, 212, 213

lithium treatment: for aggression, 79, 82; for depression, 84
loneliness, 214–216
love, 207–209
love scale: Levinger's 206, 207, 208, 209; Rubin's, 206, 207, 209

macular degeneration, and aging, 241
malleable child, the, 133–134
manic depression, 84, 85, 86
marker, genetic, 13, 14
maternal inheritance, and gene mapping, 6–11
maturational lag, as cause of dyslexia, 115
medicine, modern, and longevity, 126–131
memory: and brain function, 112; development of, in children, 36, 38, 164, 165, 170; and dyslexia, 119; eidetic, 38; and forgetting, 210–213; and prenatal education, 32–35
men: aggression in, 81, 217, 219, 221; and differences between the sexes, 217–221; envy and jealousy in, 181; and love, 207; older, and remarriage, 226; and social change, 182; and stages of adulthood, 183–187; and suicide, 194; see also, fathers, parents
messenger RNA, 8
metacomponents, 107, 108
method of loci, 211, 212
midlife crisis, 183, 184
mirror writing, 115, 116, 118, 120
mitochondria, and gene mapping, 6–11
mneumonic strategies, 165, 213
molecular biology, and gene mapping, 6–11; and genetic testing, 12–18
molecular polymorphisms, 8–9
Moro reflex, 42
mortality rates, and family ties, 126–131; infant, 152
mothers: and day care, 50–53; importance of, to infant, 33, 34, 41, 44–45, 138, 139, 142; mortality of, 126, 129; and sex role development of children, 140; single, 19; surrogate, 22, 23
mtDNA, see mitochondria, and gene mapping
muscular dystrophy, see Duchenne muscular dystrophy
mutation, of mitochondrial DNA, 8, 9, 11

natural selection: brain development of, 38; toward helping behaviors, 77; and intelligence, 108; and resistant bacteria, 126–127
nature vs. nurture: and aggression, 80–81; and sex differences, 219–220; and twins, 28–29
Neonatal Behavioral Assessment Scale, Brazelton's, 43
nerve growth, development of human, 34, 35, 38, 111, 112, 113, 171
neurological development: fetal, 32–35; newborn, 36–39
neurons, 111, 112, 113, 212
newborns: development of brain of, 36–39; effect of prenatal education on, 33; see also, infants
nursing, implications of Piaget's theory of cognitive development for, 101–104
nutrition, and life expectancy, 129–130, 235

object permanence, Piaget's theory of, 43
oldest old, 225
osteoarthritis, 236
osteoporosis, and aging, 239, 240

parents: anxieties of, 43; and child's suicide, 195; and compliance, control, and discipline of children, 64–70, 71–75; and day care, 50–53; and divorce, 144–147; education for, 132, 133, 137, 142; effect of child's puberty on, 172–176; influence of, on genetic makeup of offspring, 6–11; influence of physical attractiveness of children on, 54–57; and play, 93, 96, 97–98; and prenatal education, 32–35; of resilient children, 88–89; single, 131, 138, 139, 141–142; of twins, 28; see also, fathers, mothers, single parents
Passages, by Gail Sheehy, 183, 184
peers, importance of: in adolescence, 173, 174; for alienated children, 159
perception: in children, 103, 164; intermodal, 42; visual, and dyslexia, 115, 120
performance gap, of blacks, 153
phonetic decoding, and dyslexia, 118
phonetic segmentation, and dyslexia, 115
phonological-coding components, and dyslexia, 115
physical attractiveness, importance of, in infancy and childhood, 54–57
Piaget, Jean, and development of intellect: 41–42, 43, 65, 93, 94, 101–104, 137, 164, 166; criticism of theories of, 165, 166, 171
play, importance of, 92–100, 133, 170, 230, 232
pneumonia, and the elderly, 234, 235–236
polymorphisms: molecular, 8–9, 10; restriction fragment-length, 14
poverty: and alienated children, 159, 160; and blacks, 152; and day care, 51, 149; and intellectual development, 136; and life expectancy, 130–131
predatory aggression, 81
pregnancy, tentative, and genetic testing, 12
prenatal education, 32–35
prenatal testing, promise and peril of, 12–18
preoperational thinking, Piaget's theory of, 65, 101–102, 103
presbyopia, and aging, 241
presuicidal syndrome, 194
primary aging, 224
private language, of twins, 26
proactive controls, as form of discipline, 69
prosocial child behaviors: and fathering, 141; and selfishness, 76, 77, 78
psychodynamic formulation of suicide, 197
psychological autopsies, of suicide victims, 194
psychoneuroimmunology, 131
puberty, and parents, 172–176
public schools, use of, for day care, 148–151
punishment, and aggression, 80–82

racism, as obstacle to success of blacks, 152–156
rational suicide, 198

reading, to babies, and language development, 49
reading problems, *see* dyslexia
reasoning: in children, 103; and development of self-regulation, 73, 74
recessive genes: and cystic fibrosis, 13; and inheritance, 16
reciprocal inhibitory control, 203
recombinant DNA, 8–9
reference group expectancies, 155
reflex: development of human, 34; Moro, 42; respiratory occlusion, 42; rooting, 37
rehearsal technique, 165
relatedness, and loneliness, 214
required helpfulness, 89, 90
resiliency, in children, 67, 87–91
respiratory occlusion reflex, 42
restriction fragment-length polymorphisms, 14
reversal errors, and dyslexia, 120
rheumatoid arthritis, 236
ribonucleic acid, *see* RNA
ribosomal RNA, 8
right to suicide, 196, 198–199
RNA, 8
role-playing, for children: to develop empathy, 78; to develop self-control, 74
romantic jealousy, 179
rootedness, and loneliness, 214
rooting reflex, 37
Rubin's love scale, 206, 207, 209
rules, and play, 94, 95

sanitation, and life expectancy, 129–130
schools: and alienated children, 158, 160; and creativity, 166; and memory development, 165; public, and day care, 149; and teenagers, 168
secondary aging, 224
secret language, of twins, 26
selective ignoring, as coping strategy for jealousy and envy, 181
self-bolstering, as coping strategy for jealousy and envy, 181
self-control, development of, 64, 67, 73–74
self-esteem: and jealousy and envy, 178; levels of, and sex differences, 218; and life expectancy, 130; and loneliness, 215; in resilient children, 88
selfishness, and children, 76–78
self-reliance, as coping strategy for jealousy and envy, 181

self-talk, 49
sensorimotor stage, Piaget's theory of, 101
sensory ability: and aging, 240–241; development of human, 34, 35, 37, 41
sensual child, the, 133
serial-deficit theory, of dyslexia, 120
serotonin: and aggression, 81, 82; and depression, 86; and shyness, 61; and suicide, 192
sex: and aging, 224; babies' ability to distinguish, 42; and compliance, 72–73; and dyslexia, 121; prediction of fetal, and genetic testing, 15; and resilient children, 89; role development and fathers, 140–141; *see also,* gender, men, women
shyness, as inherited trait, 60–63
sight: childhood development of, 37, 41, 42; and dyslexia, 115; fetal development of, 34
single parents: 131, 138, 139, 141–142, 159; and day care, 149, 150; and divorce, 144–147; and loneliness, 216; *see also,* parents, mothers, fathers
situational loneliness, 215, 216
smell, sense of, and fetal development, 34
social capacity: of infants, 45; and compliance, 65
social-comparison jealousy, 179
social deprivation, Harlow's experiments of, on rhesus monkeys, 37, 61, 62
social skills, development of, in children, 170
socioeconomic background: and child development, 45, 139–140, 142; and infant mortality, 126; and longevity, 127
speech: child's development of, 46–49; fetal sensitivity to, 33–34; newborn sensitivity to, 36–37
sperm: fertilization of human egg by, 8; and in vitro fertilization, 20; and twins, 25, 26
squaring the pyramid, 223
stereotypes, sex, 217–221
Sternberg, Robert: and study of love, 207–209; and triarchic theory of intelligence, 105–110
stress: and aging, 226; circuits, 60, 62; and depression, 86; and families, 158; in language, and babies, 47; and life expectancy, 130; and memory, 213; and resilient children, 87, 89
strokes, and the elderly, 234, 239
subintentioned death, suicide as, 193

suicide: 190–199, 200; attempts, 84, 85; and loneliness, 215
Suomi, Stephen, 60, 62, 63, 86
surrogate motherhood, 22, 23
symbolic play, aggression in, 98–99

tacit knowledge, and intelligence, 109
talking, art of, to a baby, 46–49
taste, sense of, and fetal development, 34
Tay-Sachs disease, and genetic testing, 15, 18
teachers: and classroom discipline, 71; and day care, 149; influence of child's physical attractiveness on, 56
teenagers, *see* adolescents
testosterone, and aggression, 79, 82, 220
toddler's speech, characteristics of, 47
touch, sense of: and fetal development, 34; in newborns, 42
toys as symbols, 96–98
transfer RNA, 8
transitory loneliness, 216
triarchic theory of intelligence, 105–110
trust vs. mistrust, Erikson's theory of, 184, 230, 231
twins: all about, 24–29; and dyslexia, 121–122
typo, DNA, 14

"use it or lose it" attitude, toward physical and mental fitness, 224

vanishing-twin syndrome, 26–27
vigilantism, 80
vintage years, later life as, 222–229
vision, *see* sight

whole-word learning strategy, and dyslexia, 118
wisdom, Erikson's views on aging and, 230–232
work ethic, intellectual, 156
women: and differences between the sexes, 217–221; and genetic testing, 13, 15, 16, 18; increase in working, 139, 149, 158; and infertility, 21; jealousy and envy in, 181; and love, 207; and social change, 182–183; and stages of adulthood, 184, 186; and suicide, 194; *see also,* mothers, parents

Yerkes-Dodson law, 213
youth creep, 225

Credits/ Acknowledgments

Cover design by Charles Vitelli

1. Psychobiology of Human Development
Facing overview—The Dushkin Publishing Group, Inc. 27—Courtesy of Mr. & Mrs. Langenderfer.

2. Infancy and Early Childhood
Facing overview—United Nations photo by John Isaac.

3. Childhood
Facing overview—United Nations photo by Milton Grant.

4. Family, Social, and Cultural Influences
Facing overview—The Dushkin Publishing Group, Inc.

5. Adolescence and Early Adulthood
Facing overview—United Nations photo by John Isaac.

6. Adulthood and Aging
Facing overview—United Nations photo by F. B. Grunzweig. 208—Cheryl Kinne, The Dushkin Publishing Group, Inc. 233—Colonial Penn Group, Inc. 234—United Nations photo by Shelly Rotner.

ANNUAL EDITIONS: HUMAN DEVELOPMENT 89/90
Article Rating Form

Here is an opportunity for you to have direct input into the next revision of this volume. We would like you to rate each of the 43 articles listed below, using the following scale:

1. **Excellent: should definitely be retained**
2. **Above average: should probably be retained**
3. **Below average: should probably be deleted**
4. **Poor: should definitely be deleted**

Your ratings will play a vital part in the next revision. So please mail this prepaid form to us just as soon as you complete it.
Thanks for your help!

Annual Editions revisions depend on two major opinion sources: one is our Advisory Board, listed in the front of this volume, which works with us in scanning the thousands of articles published in the public press each year; the other is you—the person actually using the book. Please help us and the users of the next edition by completing the prepaid article rating form on this page and returning it to us. Thank you.

Rating	Article	Rating	Article
	1. The Mitochondrial Eve		24. The Child Yesterday, Today, and Tomorrow
	2. Perfect People?		25. The Importance of Fathering
	3. The New Origins of Life		26. Helping Children Cope with Divorce
	4. All About Twins		27. Project Day-Care
	5. The First 9 Months of School		28. Rumors of Inferiority
	6. Making of a Mind		29. Alienation and the Four Worlds of Childhood
	7. What Do Babies Know?		30. The Magic of Childhood
	8. The Art of Talking to a Baby		31. Puberty and Parents: Understanding Your Early Adolescent
	9. The Child-Care Dilemma		32. Jealousy and Envy: The Demons Within Us
	10. "What Is Beautiful Is Good": The Importance of Physical Attractiveness in Infancy and Early Childhood		33. The Prime of Our Lives
	11. Born to Be Shy?		34. Suicide
	12. Compliance, Control, and Discipline, Part 1		35. The Emotional Brain
	13. Compliance, Control, and Discipline, Part 2		36. The Measure of Love
	14. Beyond Selfishness		37. The Fear of Forgetting
	15. Aggression: The Violence Within		38. Toward an Understanding of Loneliness
	16. Depression at an Early Age		39. Biology, Destiny, and All That
	17. Resilient Children		40. The Vintage Years
	18. The Importance of Play		41. Erikson, In His Own Old Age, Expands His View of Life
	19. Practical Piaget: Helping Children Understand		42. A Vital Long Life: New Treatments for Common Aging Ailments
	20. Three Heads Are Better Than One		43. Aging: What Happens to the Body as We Grow Older?
	21. How the Brain Really Works Its Wonders		
	22. Dyslexia		
	23. Family Ties: The Real Reason People Are Living Longer		

(Continued on next page)

ABOUT YOU

Name_____ Date_____

Are you a teacher? ☐   Or student? ☐

Your School Name _____

Department _____

Address _____

City _____ State _____ Zip _____

School Telephone # _____

YOUR COMMENTS ARE IMPORTANT TO US!

Please fill in the following information:

For which course did you use this book? _____

Did you use a text with this Annual Edition?   ☐ yes   ☐ no

The title of the text? _____

What are your general reactions to the Annual Editions concept?

Have you read any particular articles recently that you think should be included in the next edition?

Are there any articles you feel should be replaced in the next edition? Why?

Are there other areas that you feel would utilize an Annual Edition?

May we contact you for editorial input?

May we quote you from above?

ANNUAL EDITIONS: HUMAN DEVELOPMENT 89/90

BUSINESS REPLY MAIL

First Class   Permit No. 84   Guilford, CT

Postage will be paid by addressee

The Dushkin Publishing Group, Inc.
Sluice Dock
DPG **Guilford, Connecticut 06437**

No Postage
Necessary
if Mailed
in the
United States